T0330094

Network Convergence

Network Convergence

Services, Applications, Transport, and Operations Support

Hu Hanrahan

University of the Witwatersrand, Johannesburg, South Africa

John Wiley & Sons, Ltd

Other Wiley Editorial Offices

John Wiley & Sons Inc., 111 River Street, Hoboken, NJ 07030, USA

Jossey-Bass, 989 Market Street, San Francisco, CA 94103-1741, USA

Wiley-VCH Verlag GmbH, Boschstr. 12, D-69469 Weinheim, Germany

John Wiley & Sons Australia Ltd, 42 McDougall Street, Milton, Queensland 4064, Australia

John Wiley & Sons (Asia) Pte Ltd, 2 Clementi Loop #02-01, Jin Xing Distripark, Singapore
129809

John Wiley & Sons Canada Ltd, 6045 Freemont Blvd, Mississauga, ONT, L5R 4J3, Canada

Library of Congress Cataloging-in-Publication Data

Hanrahan, Hu.
   Network convergence : services, applications, transport, and operations
support / Hu Hanrahan.
      p. cm.
   Includes bibliographical references and index.
   ISBN-13: 978-0-470-02441-6 (cloth : alk. paper)
   ISBN-10: 0-470-02441-0 (cloth : alk. paper)
1. Convergence (Telecommunication) 2. Internetworking (Telecommunication)
3. Wireless communication systems I. Title.
   TK5101.H2837 2007
   621.382–dc22

                                                                    2006033562

British Library Cataloguing in Publication Data

A catalogue record for this book is available from the British Library

ISBN 9780470024416

Typeset by the author using LaTeX software.
Printed and bound in Great Britain by Antony Rowe Ltd, Chippenham, Wiltshire.
This book is printed on acid-free paper responsibly manufactured from sustainable forestry
in which at least two trees are planted for each one used for paper production.

Contents

Preface

The present information age is enabled by telecommunications and information technology (IT) and the continued convergence of their services, technologies and business models. Within telecommunications, the historic separations between fixed networks, mobile telephone networks, data communications and enterprise networks are diminishing. The Internet initially supported the academic community but now provides communications for public information services and business IT operations. The Internet depends on telecommunications infrastructure for user access and core networks. Telecommunications increasingly adopts technology developed primarily for the Internet.

Within the telecommunications world, the evolution of technology and the enhanced services that can be offered through synergies with IT have been captured in the concept of *next generation networks* (NGN). This concept does not mean a particular network but rather the process of evolution from present technology and services to new technologies, enabling new services and applications with both telecoms and IT characteristics. The NGN concept also captures the need for structured thinking to deal with the complexity of systems, applications and business models in *information and communications technologies* (ICT).

This book takes as a starting point present day networks, services and operations support and maps their evolution toward specific future forms, collectively termed next generation networks. In the process, we examine the nature, process and results of convergence. Many candidate technologies are offered but few are adopted. The process of assessment, selection and integration of technology is complex, particularly when constraints exist due to the ongoing presence of legacy technology. We therefore examine the process of migration to new technologies.

Many recent texts describe various next generation network technologies. The approach of this book is analytical rather than descriptive. We establish *concepts, principles and architectural frameworks* rather than conventional descriptions of specific technologies. Individual architectures and technologies are depicted within a framework, facilitating comparisons and drawing contrasts and assisting in integrating different technologies to form a system. The framework aids the understanding of next generation networks, their potential for supporting new, enhanced applications and their relationships with legacy networks.

We draw analytical and descriptive methods from several sources, as diverse as the classical Intelligent Network, software engineering and selected telecommunications standards.

Telecommunications systems must be based on standards. In most cases, standards or draft standards exist for emerging networks and services. The book analyses and explains the concepts and principles of a number of standards for networks, services and applications. The analytical approach supports assessing, designing, integrating and operating technologies in next generation networks. This understanding also informs research and development leading to new technologies.

Acknowledgments

A book of this nature, trying as it does to get to the fundamentals of a rapidly moving subject, cannot be a success without the contribution of the masters and doctoral students who have studied the material, whose comments have enriched it and whose research has informed it. The inputs of Rolan Christian and David Vannucci are especially valued. The masters students who have taken my Advanced Telecommunications Service Architectures course have contributed to refinement of this book. Research students over the years, and Dr Russell Achterberg in particular, contributed to the body of learning captured in the book.

The project to write this book was initiated by David Hatter. His interest and advice is gratefully acknowledged. At John Wiley & Sons, Sally Mortimore launched the project and Richard Davies saw it to a conclusion. The advice of Birgit Gruber was always valuable.

Trademarks and registered names referred to in the text are acknowledged, including but not limited to:

ETSI:	3GPP, DECT, TIPHON, UMTS
Sun Microsystems:	J2EE, JAIN, Java
Telecommunications Industry Association:	cdma2000
Telemanagement Forum:	Enhanced Telecom Operations Map (eTOM)

Conventions

Names of standards and entities specifically defined in various standards as well a concepts that are core to the model used in the book are capitalised. Defined functional entities (e.g. Service Control Point, H.323 Gatekeeper) are capitalised while generic types are not. Thus, we have a *media gateway* in general, but there are specific *Media Gateways* defined in the H.323 and IMS standards. Similarly, the names of protocols are capitalised (e.g. Address Resolution Protocol), as are operations within protocols, generally indicated by alternate type faces, (Setup or POST) and formally defined parameters (e.g. CallID, Accept-encoding, Circuit Identification Code).

The names of generic entities, concepts, categories or classes of things are not capitalised even if they emanate from standards (e.g. enterprise viewpoint, platform-independent model).

San Serif font indicates an operation, method call or parameter in an UML definition, an applications programming interface or application layer protocol. Internet protocols and ASN.1 definitions are generally indicated by Typewriter font.

Companion Website

Resources to aid the use of this book, including PowerPoint presentation files, a selection of problems, a guide to useful Web pages and an expanded on-line glossary, are available via the website `http://www.wiley.com/go/hanrahan_network`

Abbreviations

2.5G	GPRS enhanced second generation mobile telephone network
2G	Second generation, generally applied to mobile telephone networks
3G	Third generation, generally applied to mobile telephone networks
3GPP	The 3rd. Generation Partnership Project
3GPP2	The 3rd. Generation Partnership Project Two.
4G	Fourth generation
AAA	Authentication, authorisation and accounting
AAL1/2	ATM Adaption Layer 1, 2, ...
ACD	Automatic call distribution
ACK	Acknowledge control bit in TCP
ACM	Address Complete Message in ISUP
ACSE	Application context service element
ADC	Analogue-to-digital converter
ADM	Add-drop multiplexer
ADPCM	Adaptive differential pulse code modulation
ADSL	Asymmetric Digital Subscriber Loop
AIN	Advanced Intelligent Network
AN	Access network
ANM	Answer Message in ISUP
ANSI	American National Standards Institute
AOR	Address of record
APDU	Application protocol data unit
API	Application programming interface
App	Application layer
ARP	Address Resolution Protocol
AS	Autonomous system
AS	Application server
ASN.1	Abstract Syntax Notation One
ASP	Application service provider
ATM	Asynchronous Transfer Mode
AuC	Authentication Centre
B2BUA	Back-to-back user agent
B3G	Beyond 3G
BC	Bearer Control in TIPHON
BCSM	Basic Call State Model
BER	Basic Encoding Rule
BGCF	Border Gateway Control Function
BGP	Border Gateway Protocol

BICC	Bearer Independent Call Control
B-ISDN	Broadband Integrated Services Digital Network
BORSCHT	Battery, overvoltage, ringing, signalling, hybrid, codec and testing
BSC	Base Station Controller
BSS	Base Station Subsystem
BSS	Business support system
BTS	Base Transceiver Station
C3G	Converged 3G
CA	Call agent
CAMEL	Customized Applications for Mobile network Enhanced Logic
CAP	CAMEL Application Part
CC	Call control
CC	Call Controller in TIPHON
CC	Crossconnect
CC	Connection Coordinator in TINA
CCAF	Call Control Agent Function
CCF	Call Control Function
CCITT	Predecessor to the ITU-T
CD	Compact disk
CDMA	Code division multiple access
CDR	Call data record
CE	Circuit emulation
CIC	Circuit Identification Code
CIM	Computationally independent model
CL	Connectionless
CM	Call manager
CMIP	Common Management Information Protocol
CMISE	Common Management Information Service Element
CN	Core network
CO	Connection-oriented
Conf	Conference
Cont	Content layer
COPS	Common Open Policy Service
CORBA	Common Object Request Broker Architecture
CoS	Class of service
COTS	Commercial off-the-shelf
CPE	Customer premises equipment
CPH	Call party handling
CS	Circuit switched
CS-1/2/3	Capability Set One, Two, ...
CSCF	Call Session Control Function
CSD	Circuit-switched data
CSE	CAMEL Service Environment
CSI	CAMEL Subscriber Information
CSM	Communication Session Manager in TINA
CS-MGW	Media gateway for the circuit switched domain

CSRC	Contributing Source
CSTA	Computer Supported Telephony Applications
CTI	Computer–telephony integration
CV	Computational viewpoint
CVS	Connection View State
DECT	Digital Enhanced Cordless Telephony
DFP	Distributed Functional Plane
DHCP	Dynamic Host Configuration Protocol
Diameter	An extension to the Radius AAA protocol
DiffServ	Differentiated Services
DLL	Data link layer
DNS	Domain Name Service
DP	Detection Point
DPE	Distributed processing environment
DPP	Detection Point Processing
DSLAM	Digital Subscriber Line Access Module
DSS1	Digital Subscriber Signalling System No. 1
DTD	Document Type Definition
DTMF	Dual Tone Multifrequency
DVB	Digital Video Broadcast
DWDM	Dense wavelength division multiplexing
ECMA	European Computer Manufacturers Association
EDGE	Enhanced Data rates for GSM Evolution
EDI	Electronic data interchange
EDP	Event Detection Point
EDP-N	Event Detection Point, Notification Type
EDP-R	Event Detection Point, Response Type
EIR	Equipment Identity Register
EJB	Enterprise Java Beans
EngV	Engineering viewpoint
ENUM	The E.164 to Uniform Resource Identifiers Dynamic Delegation Discovery System Application
ESB	Enterprise service bus
eTOM	Enterprise Telecommunications Operations Map
ETSI	European Telecommunications Standards Institute.
EV	Enterprise viewpoint
FAB	Fulfillment,assurance and billing
FCAPs	Fault, configuration, accounting, performance and security
FCC	Flow Connection Coordinator in TINA
FE	Functional entity
FM	Frequency modulation
FR	Frame Relay
FS	Feature server
FTP	File Transfer Protocol

GCR	General Call Register
GDMO	Generic Description of Managed Objects
GERAN	GSM/EDGE Radio Access Network
GFP	Global Functional Plane
GGSN	Gateway GPRS Support Node
GII	Global Information Infrastructure
GIOP	General Inter-orb Protocol
GMM	GPRS Mobility Management
GMPLS	Generalised Multiprotocol Label Switching
G-MSC	Gateway Mobile Switching Centre
GoS	Grade of service
GPRS	General Packet Radio Service
GSM	Global System for Mobile Communications
gsmSCF	Service Control Function for CAMEL Environment
gsmSRF	Special Resource Function for CAMEL Environment
gsmSSF	Service Switching Function for CAMEL Environment
GTP	GPRS Tunnel Protocol
GTP-U	GPRS Tunnel Protocol for the User Plane
GUI	Graphical user interface

H.323	The H.323 suite of protocols, the H.323 protocol
HDLC	High-level Data-link Control
HLR	Home Location Register
HSCSD	High-speed Circuit-switched Data
HSS	Home Subscriber System
HTML	Hypertext Markup Language
HTTP	Hypertext Transfer Protocol

IA	Initial Agent in TINA
IAD	Integrated access device
IAP	Internet access provider
ICA	Information Communications Architecture
I-CSCF	Interrogating-Call Session Control Function
ICT	Information and communications technologies
IDL	Interface Definition Language
IDRP	Interdomain Routing Protocol
IEC	International Electrotechnical Commission
IETF	The Internet Engineering Task Force
IGP	Interior gateway protocol
IIOP	Internet Inter-orb Protocol
IMS	IP Multimedia Subsystem
IMSI	International Mobile Subscriber Identity
IM-SSF	IMS Service Switching Function
IN	Intelligent Network
INAP	Intelligent Network Application Protocol
INCM	Intelligent Network Conceptual Model
IntServ	Integrated Services

IP	Intelligent Peripheral in IN context
IP	Internet Protocol
IPDR	Internet Protocol Data Record
IPEN	IP edge node
IPv4	Internet Protocol version 4
IPv6	Internet Protocol version 6
ISC	IP Multimedia Subsystem Service Control Interface
ISDN	Integrated Services Digital Network
ISO	International Organisation for Standardisation
ISP	Internet service provider
ISUP	ISDN User Part
IT	Information technology
ITU-T	International Telecommunications Union–Telecommunications Standardization Sector
IUA	ISDN User Adapter for SCTP
IV	Information viewpoint
IVR	Interactive voice response
J2EE	Java 2 [Platform], Enterprise Edition
JAIN	Java API for Integrated Networks
JSLEE	JAIN Service Logic Execution Environment
JTAPI	Java-based Telephony Applications Programming Interface
JVM	Java Virtual Machine
L2	Layer 2
LAN	Local area network
LBS	Location-based service
LCG	Logical Connection Graph in TINA
LDAP	Lightweight Directory Access Protocol
LDP	Label Distribution Protocol
LMDS	Local Multipoint Distribution System
LSA	Link State Advertisements
M2UA	MTP Level 2 User Adapter for SCTP
M3UA	MTP Level 3 User Adapter for SCTP
MAP	Mobile Application Part
MC	Media Control in TIPHON
MCU	Multipoint Control Unit in H.323
MDA	Model Driven Architecture
Megaco	The H.248 Gateway Control Protocol
MG	Media gateway
MGC	Media gateway controller
MGCF	Media Gateway Control Function
MGCP	Media Gateway Control Protocol
MIB	Management information base
MMDS	Multichannel, Multipoint Distribution System
MMS	Multimedia Messaging Service
MOM	Message-oriented middleware

MOS	Mean opinion score
MPLS	Multiprotocol Label Switching
MRFC	Media Resource Function Controller
MRFP	Media Resource Function Processor
MS	Mobile Station
MSC	Message sequence chart
MSC	Mobile-system Switching Centre
MSISDN	Mobile Subscriber ISDN Number
MT	Mobile termination
MTP	Message Transfer Part
NAS	Network access server
NAT	Network Address Translation
NE	Network Element in TMN
NEF	Network Element Function in TMN
NGN	Next generation network
NGOSS	New Generation Operations Support System
N-ISDN	Narrowband Integrated Services Digital Network
NNI	Network-to-Network Interface
NRA	Network Resource Architecture in TINA
O-BSCM	Originating Basic Call State Model
ODP	Open Distributed Processing
OMA	Open Mobile Alliance
OMG	Object Management Group
OMT	Object Modelling Technique
ORB	Object request broker
OS	Operations System in TMN
OS	Operations support
OS/BS	Operations support and business support
OSA	Open Service Access
OSE	OMA Service Environment
OSF	Operations System Function in TMN
OSI	Open Systems Interconnection
OSPF	Open Shortest Path First Protocol
OSS	Operations support system
PA	Provider Agent in TINA
PAM	Presence and Availability Management
PBX	Private [automatic] branch exchange
PCM	Pulse code modulation
P-CSCF	Proxy Call Session Control Function
PDA	Personal digital assistant
PDH	Plesiochronous Digital Hierarchy
PDP	Packet Data Protocol in GPRS context
PDP	Policy Decision Point
PDU	Protocol data unit
PE	Physical entity

PEP	Policy Enforcement Point
PIC	Point in Call
PIM	Platform Independent Model
PIN	Personal identification number
PINT	Phone IP Interworking
PLMN	Public Land Mobile Network
PNNI	Private Network-to-Network Interface Protocol
PNO	Public network operator
PoP	Point of presence
POTS	Plain old telephone service
PPP	Point to Point Protocol
PS	Packet switched
PSM	Platform specific model
PSTN	Public switched telecommunications network.
PSTN/IN	Public switched telecommunications network with Intelligent Network Overlay
Q.931	Synonymous with DSS1
QoS	Quality of service
Radius	Remote Authentication Dial In User Service
RAN	Radio Access Node
RARP	Reverse Address Resolution Protocol
RAS	Registration Admission and Status in H.323 context
RAS	Remote access server
RCF	Resource control function
RCMF	Resource Control and Management Functionality
RFC	Request for Comments
RIP	Routing Information Protocol
RMI	Remote method invocation
RM-ODP	Reference Model of Open Distributed Processing
RNC	Radio Network Controller
ROSE	Remote Operations Service Element
RPC	Remote procedure call
RSVP	Resource Reservation Protocol
RTCP	Real-time Control Protocol
RTP	Real-time Transport Protocol
SA	Service Architecture in TINA
SAAL	Signalling ATM Adaptation Layer
SAP	Service access point
SBB	Service Building Block in JAIN
SC	Service Control in TIPHON
SCCP	Signalling Connection and Control Part
SCF	Service Capability Function in OAS/Parlay Context
SCF	Service Control Function in IN context
SCF	Service Control Functionality in NGN Framework context
SCMF	Switching Control and Management Functionality

SCN	Switched circuit network
SCP	Service Control Point
SCS	Service Capability Server
S-CSCF	Serving-Call Session Control Function
SCTP	Stream Control Transmission Protocol
SDF	Service Data Function
SDH	Synchronous Digital Hierarchy
SDL	Specification and Description Language
SDP	Service delivery platform
SDP	Session Description Protocol in SIP context
SF	Service factory
SFEP	Stream Flow End Point in TINA
SG	Signalling Gateway
SGSN	Serving GPRS Support Node
SIB	Service Independent Building Block
SID	Shared Information and Data
SIP	Session Initiation Protocol
SIPS	Secure SIP
SIP-T	SIP for Telephony
SLC	Subscriber line concentrator
SLCM	Service Lifecycle Manager
SLEE	Service Logic Execution Environment
SLI	Subscriber line interface
SM	Session Manager
SMI	Structure of Management Information
SMS	Short Message Service
SMTP	Simple Mail Transfer Protocol
SNMP	Simple Network Management Protocol
SOA	Service Oriented Architecture
SOAP	Simple Object Access Protocol
SOHO	Small office, home office
SONET	Synchronous Optical Network
SP	Signalling Point
SP	Service Plane in IN context
SPA	Service Provider Access
SPIRITS	Services in PSTN Requesting Internet Services Protocol
SRF	Specialised Resource Function
SRP	Specialised Resource Peripheral
SS	Supplementary services
SS	Session state in IMS context
SS7	Signalling System No. 7
SSF	Service Switching Function
SSM	Service Session Manager in TINA
SSP	Service Switching Point
SSRC	Synchronisation Source
STM-n	Synchronous Transport Module level n
STP	Signal Transfer Point

SUA	SCCP User Adapter for SCTP
Sub	Subscriber
TAPI	Telephony Applications Programming Interface
T-BSCM	Terminating Basic Call State Model
TCAP	Transaction Capabilities Application Part
TCMF	Transmission Control and Management Functionality
TCP	Transmission Control Protocol
TCSM	Terminal Communication Session Manager
TDM	Time Division Multiplexing
TDP	Trigger Detection Point
TDP-N	Trigger Detection Point, Notification Type
TDP-R	Trigger Detection Point, Response Type
TEL URL	Telephony URL
Telco	Telecommunications company
TFD	Technical Functional Domain
TG	Trunking Gateway
THIG	Topology-hiding Interrogating Gateway
TINA	Telecommunications Information Networking Architecture
TIPHON	Telephony and Internet Protocol Harmonisation over Networks
TISPAN	Telecommunications and Internet converged Services and Protocols for Advanced Networking
TMN	Telecommunications Management Network
TOM	Telecommunications Operations Map
TSAP	Transport service access point
UA	User Adapter in SCTP context
UA	User Agent in SIP context
UA	User Agent in TINA context
UAC	User-agent Client
UAP	User Application Part in TINA context
UAS	User-agent Server
UDDI	Universal Description, Discovery and Integration
UDP	User Datagram Protocol
UE	User Equipment
UFM	Unified Functional Methodology
UI	User Interaction
UML	Unified Modelling Language
UMTS	Universal Mobile Telecommunications System
UNI	User-to-Network Interface
URI	Uniform Resource Indentifier
URL	Uniform Resource Locator
USIM	UMTS Subscriber Identification Module
USM	User Session Manager in TINA
UTRAN	Universal Terrestrial Radio Access Network
VAN	Value-added network
VC	Virtual circuit

VHE	Virtual Home Environment
VLR	Visitor Location Register
VLSI	Very large scale integration (integrated circuit)
VoD	Video on demand
VoIP	Voice over the Internet Protocol
VoN	Voice on the [Inter]Net
VPN	Virtual private network

WAN	Wide area network
WAP	Wireless Application Protocol
WiFi	Wireless Fidelity (any short range wireless Ethernet)
WiMax	Worldwide Interoperability for Microwave Access (long range wireless Ethernet)
WIN	Wireless Intelligent Network
WLAN	Wireless local area network
WLL	Wireless local loop
WSDL	Web Services Definition Language
WS-I	Web Services Interoperability
WSP	Web service provider
WWW	World Wide Web

X.25	An ITU-T standard for packet switching
XC	Crossconnect
xDSL	A family of digital subscriber loop standards
XML	Extensible Markup Language
XSD	XML Schema Definition

Principal Graphical Symbols

Circuit Switch or Switching Fabric		SS7 Signal Transfer Point	
Subscriber Line Concentrator (Analogue or ISDN)		Element with SS7 Signalling Point	
IN Service Control Point		IN Service Data Point	
Internet Protocol Router		Home Location Register	
Ethernet Switch		Visitor Location Register	
Ethernet Hub		Mobile Network Base Station System	
ATM Switch			
MPLS Switch		Radio Mast with Air Link	
Call Manager		Element with Physical and Functional Entities with SS7 Signalling Point	
Media Server		Voice Access Gateway	
Conference Bridge		Digital Subscriber Line Access Module	
Media Gateway		Signalling Gateway	
Mobile Telephone		SDH Add-drop Multiplexer	
Personal Digital Assistant		SDH Crossconnect	
ISDN and Plain Phone		Optical Add-drop Multiplexer	
Computers: Desktop and Notebook		Optical Crossconnect	
Server		Wireless Access Point	
Voiceband Modem		Buffer or Queue	

Chapter 1

Setting the Context for Evolution and Convergence of Networks

What do *network*, *evolution* and *convergence* signify in the title of this chapter? *Network* has one of its common meanings, namely, the total set of facilities that an operator needs to provide a telecommunications or information service. *Facilities* are the hardware and software elements and systems that are necessary to provide a service. A *service* at its most general denotes an offering by a provider to customers or end-users. A service may also be offered by one set of facilities to another. *Evolution* captures the continual development of enabling technologies and service offerings that compete for acceptance in the market. As in biological evolution, some technologies are successful because they are adapted to the prevailing conditions while other are not and die out. Similarly, evolution indicates incremental development building on existing successes. *Convergence* identifies a general pattern in the evolutionary process, namely the tendency to bring entities together, for example the coming together of classical telecommunications, the Internet, information technology and broadcasting, the ability to offer multiple services on a single network or the ability to offer the same service via more than one medium.

The starting point of our analysis of evolution and convergence of networks and services is an appreciation of the status of the networks of today, together with the services they offer. Section 1.1 sketches the development of telecommunications – both circuit and packet switched – and the Internet since the 1970s, leading to current networks and services.

Section 1.2 identifies and describes six principal present-day networks with distinctive service offerings: switched circuit networks, both fixed and mobile; the Internet; enterprise networks; packet-switched interconnection networks and leased line services. The capabilities and evolutionary limitations of present networks are analysed.

Section 1.3 then introduces convergence as a theme in seeking to overcome the limitations of present networks and to facilitate new service offerings. Specific instances of convergence are identified from various fields: fixed and mobile networks; telecommunications and the Internet; telecom, information, entertainment and broadcasting services, applications and business models. Convergence is not however a simple phenomenon, but spans technologies, networks, services and business models. A number of general characteristics of convergence are identified.

Convergence is a process rather than an event and we need a concept to describe its desired end point. To this end we introduce the *next generation network* (NGN) in

Network Convergence: Services, Applications, Transport, and Operations Support
Hu Hanrahan © 2007 John Wiley & Sons, Ltd

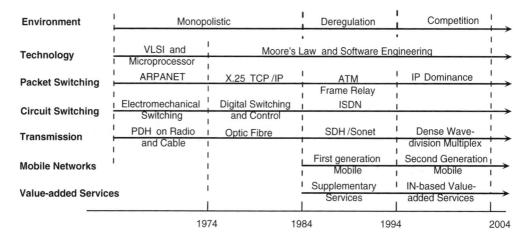

Figure 1.1. Historical developments leading to present state of ICT.

Section 1.4 to represent possible integrated networks that support multiple service offerings. There are many possible NGNs and most must interwork with legacy networks. We therefore use the ITU-T concept of the NGN as embracing the collective improvements to the service provision infrastructures from the base of traditional networks.

1.1   HISTORICAL BACKGROUND TO PRESENT NETWORKS

Telecommunications has a long history. The computer age is also half a century old and has now become the Internet age. While their development was largely independent at first, computing and communications have become inextricably bound and mutually dependent. We seek a unified understanding of convergent information and telecommunications services and the underlying network and software technologies. To visualize the future, we need to understand the historical development leading to present day and emerging technologies. Figure 1.1 identifies several historical threads leading to the present networks. A useful starting point is the mid-1970s. By that stage, large-scale integrated circuits and the microprocessor allowed a change from analogue to digital transmission and switching of information.

The decade 1974–1983 saw several important developments. The already established principles of packet switching led to the development of the X.25 standards. By the end of this period, TCP/IP had been adopted as the basis of the ARPANET, the predecessor of the Internet. In parallel, digital telephone switching developed in two ways: first the incorporation of processor-based control and, second, digital encoding of speech as the basis of switching. The concept of a single network providing both voice and data services, the first multiservice network, was developed into the narrowband Integrated Services Digital Network (N-ISDN) standards. In the transmission area, the first optic fibre cable was deployed. First generation (analogue) mobile networks started operation.

The next decade, 1984–1993, saw increasingly important developments, both technological and regulatory. The break-up of the Bell System was the first step in a worldwide trend toward deregulation and competition in telecommunications. The decade started with one thousand hosts on the Internet. Digital telephone switching penetrated into public and private networks. Packet switching standards expanded to include Asynchronous Transfer Mode (ATM) and Frame Relay. The World Wide Web was launched and the number of hosts on the Internet grew to one million by the end of the period. The GSM second generation mobile networks were standardised and successfully launched. In the PSTN world, the concept of the Intelligent Network was formulated as a means of implementing value-added services in the PSTN and the first standards were developed. In the transmission arena, the Synchronous Digital Hierarchy gave network operators the opportunity to provide readily configured and managed transmission services, both for their own needs and to customers requiring point to point connections.

The recent decade, 1994–2003, saw the launch of commercial Internet service providers in 1995, taking over from government agencies. Web usage overtook other types of Internet services in volume of data transferred. The IN standards developed through two capability sets and became the basis for value-added services in the PSTN. Standards for telephony using Internet Protocol (IP) networks were developed and the concept of a new multiservice network was formulated as the next generation network. The first third generation (3G) mobile network licences were issued but deployment was limited by excessive licence fees and economic downturn. This decade also saw the dot.com boom with unlimited optimism about new Internet-based services; this optimism was soon followed by the March 2000 crash. Optic fibre transmission capacity increased due to both higher speeds of transmission and the use of multiple wavelengths on a single fibre. The growth of the Internet to over 100 million hosts called for rapid increases in core transmission capacity, yet many optic fibres operated at a fraction of their capacity. Internet technologies, for example the use of IP networks and browser-based applications, became the way of delivering IT applications and corporate communications. Interworking between circuit-switched and packet networks was enabled by the development of media and signalling gateways.

This book attempts to create an understanding of the present and emerging network technology as well as its future trajectory. What does the decade 2004–2013 hold? Several trends are already evident. The switched circuit network is likely to start its decline in the volume of traffic carried during the period. Packet switching for various classes of traffic in a single network will take over. The distinction between telecommunications and the Internet will be increasingly blurred. Telecommunication networks are likely to become open to control of services by applications in other service provider domains. New business models will emerge. The variety of terminals and access methods will increase and the distinction between fixed and mobile networks will become less meaningful. As these next generation networks and services develop, the need to interwork with legacy networks remains important. The state of telephone and data networks and the Internet at the end of the decade 1994–2003, described in Section 1.2, is the point of departure of this book.

1.2   DEFINING PRESENT STATE USING REFERENCE MODELS

This book is about a range of future network architectures, collectively called next generation networks and shaped by the process of convergence. We identify a number of present day networks as well as their characteristics and distinctive services that define the point of departure for considering convergence.

1. *Switched circuit networks* (SCN) encompass fixed-line and mobile telephone networks. The *fixed line public switched telecommunications network* (PSTN) and the Integrated Services Digital Network (ISDN) provide basic voice bearer services and value-added services implemented using the Intelligent Network (IN) overlay. With suitable terminals, ISDN also supports video conferencing. Despite providing digital connectivity to the customer premises, the ISDN has made little impact on data services apart from dial-up Internet access. The provider is a telecommunications company (telco).

2. Closely related to fixed networks in (1) are second generation (2G) *mobile telephone networks*. Such networks provide voice and data services to mobile users, together with messaging (SMS and MMS). Mobile networks differ from fixed networks in two main aspects: first, the access method is radio, and second, mobility management is required to keep track of mobile phones. Value-added services are also implemented using the IN. Mobile networks are multiservice networks, supporting voice and various data services through a single air interface.

3. The *Internet*, a worldwide arbitrary interconnection of autonomous networks unified by the IP, supports Internet services: World Wide Web, e-mail, file transfer, transactional services and, increasingly, peer-to-peer services such as file sharing. A user of Internet services is reliant on both an Internet access provider (IAP) for the physical means of connecting to the Internet and an Internet service provider (ISP) for logical access to the Internet, that is the ability to address other parties and be addressed. The ISP and IAP roles may be common or separate. Internet content providers are generally independent of ISPs.

4. *Enterprise networking*, using both telecommunications and data networking technology to create private networks, is devoted to supporting the information and communication requirements of corporations and institutions. Telcos or ISPs provide interconnection between sites and they and other providers may provide completely managed enterprise networks.

5. Telecommunications companies own data networking facilities to provide *switched interconnection services* to support activities such as private and virtual private networking and Internet service provision. Both layer 2 and 3 connectivity is provided, for example Frame Relay, X.25 and, in some cases ATM. Increasingly, IP switched interconnections are offered, usually with virtual private network (VPN) support.

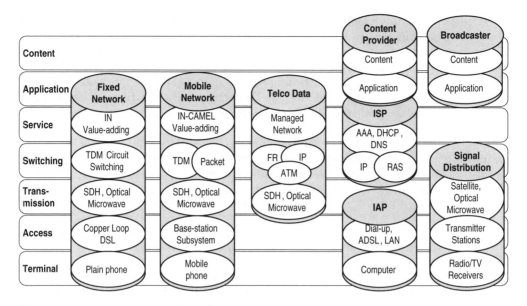

Figure 1.2. Vertically integrated traditional telecommunications and broadcasting businesses.

6. *Leased-line services* provide semi-permanent, nonswitched connections between client sites at specified bit rates in standard multiplexing hierarchies. Telcos are the usual leased line providers.

A seventh category, broadcasting, will become increasingly important as convergence proceeds. Section 1.2.1 examines some current relationships between broadcasting and telecommunications.

1.2.1 SILO MODEL FOR VERTICALLY INTEGRATED NETWORKS

Historically, different telecommunications services, the Internet and broadcasting have been vertically integrated both as businesses and sets of facilities.[1] The vertically integrated nature of these businesses is often depicted by a silo or stovepipe metaphor, as shown in Figure 1.2. Three present telecommunications businesses, fixed telephony, mobile telephony and switched interconnection services, the Internet and broadcasting, are shown. Within each silo, the required facilities are layered to identify commonalities. A set of layers that provide a backdrop for various networks is shown in Figure 1.2.

Section 1.2.2 reviews fixed and mobile switched-circuit networks. Commonalities between fixed and mobile telephone networks occur in the transmission, switching and service layers in Figure 1.2 and are examined in Section 1.2.2. Different technologies are used in the terminal and access layers.

---

[1]We use *facilities* to denote hardware and software required to provide services. *Infrastructure* is a special case of facilities that exist in areas other than the customer's or provider's premises.

Telco-switched interconnection services form a separate silo, marked Telco Data, also relying on common transmission facilities.

Internet service provision falls into three domains. Physical access to the Internet is usually provided by a telco that acts as the Internet access provider. Internet service provision involves service and switching layer concerns. The ISP provides admission control, allocation of network addresses and a connection to the IP network. Content provision on the Internet, often linked to an application, is not necessarily linked to the ISP.

Figure 1.2 shows that the transmission layer exhibits the greatest degree of commonality as all networks require high-volume short and long distance transmission.

Historically, *broadcasting* was vertically integrated. Traditional broadcasting divides naturally into two areas, now operating as different businesses in many cases, as shown in Figure 1.2:

1. Programme and *content assembly and playout* of the broadcast signal into the signal distribution system. This is the *broadcasting* function.

2. *Signal distribution*, comprising the distribution of signals to transmitter stations and the transmitters, provides radio coverage of the reception area(s), analogous to the core and access regions of a telecommunications network. Contact between telecommunications and broadcasting was limited to leasing transmission links from telcos to support signal distribution.

Broadcasting, for most of its history, has been a one-to-many, one-way service but is progressively becoming two-way. Initially, broadcasters used phones and the Short Message Service to allow the audience to respond and react to programme material. Interactive services became possible over broadcast networks when the downlink acquired a data transmission capability that could be complemented by an uplink, for example, by telephone. With digital broadcasting, the availability of a return channel enables interactive services. Early incarnations of such services are essentially Internet-type services, often related to the broadcaster's business.

Broadcasting is also practised over a wired network, for example cable television networks and by audio streaming on the Internet. Cable TV provides a physical access network that is readily adapted to support two-way telephony [155].

1.2.2 PRESENT STATE: FIXED AND MOBILE NETWORKS WITH IN OVERLAY

Switched circuit networks are vertically integrated with fixed and mobile networks constrained to their respective silos by historic regulatory practices, despite having a degree of commonality of technology. Figure 1.3 shows the main architectural features of two circuit-switched networks. We use this reference model for current fixed and mobile networks to distinguish the two networks with respect to terminal capability, access network and mobility management. The model also shows commonality in voiceband switching, call control, signalling network and Intelligent Network overlay.

PSTN users are shown connected via copper pairs with two types of interfaces in the subscriber line concentrator. Subscriber line concentrators may be remote from the exchange or co-located. Analogue subscriber loops terminate on an interface

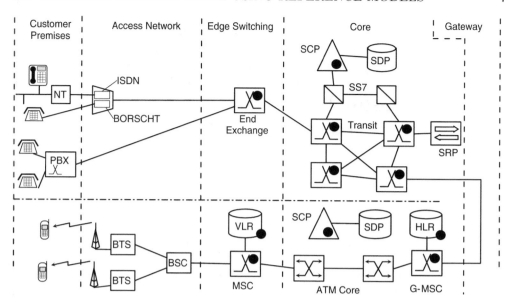

Figure 1.3. Reference network architecture for switched circuit networks.

that performs the BORSCHT functions. ISDN subscribers are connected via digital subscriber loops [100]. Both interfaces deliver time-division multiplexed signals to the end exchange. The end exchange is the network edge element providing the network service access point to the user. Legacy PSTNs have core switching also based on exchanges switching time-division multiplex (TDM)signals. All exchanges use the Signalling System No. 7 network for inter-exchange signalling. In the core, PSTN and ISDN services are supported by the same switches and transmission systems.

The mobile network differs from the fixed line network principally in the cellular wireless access network but also in its ability to track mobile users. *Mobility management* is achieved by three principal means. First, the mobile station assists the network by determining the strongest signals that it receives from time to time and reporting these to the network. Second, using signal strength information, the mobile can be handed over between frequencies on a base station, between base stations or between Mobile-system Switching Centres (MSC). Third, the mobile network has two databases that support tracking the location of the mobile station. The Home Location Register (HLR) is normally associated with the Gateway MSC, that is the MSC that provides interconnection with other networks. A Visitor Location Register (VLR) is associated with every MSC. The HLR contains the permanent subscriber data (subscriber profile) as well as the identity of the VLR to which the mobile station is currently logged. The VLR contains data that supports the management of the subscribers currently in the serving area of the visited MSC.

A common practice in mobile networks shown in Figure 1.3 is to use packet-mode core networks rather than TDM transit switches as in the legacy PSTN. The TDM trunks are adapted to the ATM network using either ATM Adaptation Layer 1 (AAL1)

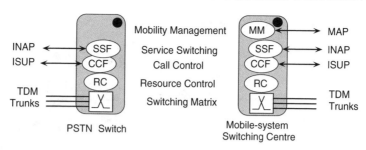

Figure 1.4. Comparison of the functionality of fixed line and mobile telephone network switches.

performing circuit emulation or AAL2 in which 64 kbit/s speech is compressed and packetised before being adapted for transport in ATM cells.

Interconnection of fixed and mobile telephony networks occurs between a transit exchange in the fixed network and a designated MSC, the Gateway MSC (G-MSC) in the mobile network. The Gateway MSC is associated with the Home Location Register and therefore has access to the subscriber profile data and temporary location information. Speech signals are transferred in 64 kbit/s time slots on a TDM link. Signalling interchange between the G-MSC and the transit exchange uses the ISDN User Part ISUP call control messages.

Each network in Figure 1.3 is shown with an Intelligent Network overlay, consisting of a Service Control Point (SCP), a supporting database, the Service Data Point (SDP) and a Specialised Resource Peripheral (SRP) that allows announcements to be played to users and dialled digits to be collected. The principles underlying the Intelligent Network are reviewed in Section 1.2.2. Most principles are common for fixed and mobile networks, except for the latter requiring location and mobile system user information to be available.

Switch Functionality in Fixed and Mobile Networks

Figure 1.4 identifies the essential functions in exchanges for fixed and mobile networks based on circuit switching, for example those shown in Figure 1.3 [91].

The heart of each switch is the actual switching matrix that operates on the time-division multiplex signals. Speech signals are encoded as 64 kbit/s streams and each channel is allocated a time slot on an incoming line to the switch. The switching operation involves reallocating the data bytes representing this speech signal to an allocated (and possibly different) time slot on the desired output line from the switch.

The Call Control Function (CCF) is concerned with making the end-to-end connection. Connections are set up one hop at a time. Each exchange has a first choice and alternate route toward every destination end exchange. An exchange on the path between source and destination signals to its adjacent exchange to agree on a time slot to be allocated. The first choice route is tried first and the alternate route is used if no time slot is available on the first choice. A CCF signals to the CCF of the neighbouring exchange using the ISUP protocol. The Call Control Function must exhibit the external behaviour specified in the ISUP standard.

Control of the switching operation in a particular exchange is essentially the allocation of a time slot on an outgoing trunk. An essential function in the exchange is therefore resource control (RC), the resource being the set of available time slots on each link or trunk group to adjacent exchanges. A standard Circuit Identification Code (CIC) is used in ISUP messages to identify the resource used but, in general, implementations of resource control are proprietary.

Closely linked to the CCF is the Service Switching Function (SSF). The SSF was introduced in the development of the Intelligent Network architecture as a standard way of allowing the CCF process to invoke the assistance of service logic on an external computing platform called the Service Control Point. The SSF is based essentially on the definition of a standardised set of points, called *Detection Points*, in the execution of the CCF logic. Examples of such points are AddressCollected, that is the user has dialled a number that is judged to be complete, and AddressAnalysed, when the CCF logic has completed its analysis of the dialled number to determine whether it represents a routable destination or requires special treatment. At each Detection Point, the CCF reports to the SSF. The SSF tests whether the detection point is enabled to invoke external logic (armed) and whether call-specific conditions are fulfilled that external logic should be invoked. If such conditions are fulfilled, the CCF process waits while the SSF engages in a transaction with the SCF using the Intelligent Network Application Protocol (INAP). A request, usually an INAP InitialDp operation, is sent to the SCF to invoke external logic. The external logic may invoke queries on a database called the Service Data Point or instruct the Specialised Resource Peripheral to prompt the user and collect information. To complete execution of the external logic, the SCP logic sends one or more operations to the SSF to instruct the CCF to resume execution and how to proceed with call processing.

The Mobile-system Switching Centre has an additional function, namely *mobility management*. This distributed function interacts with the HLR and VLR and receives information from the Base Station Subsystem. Mobility-related operations are invoked using the Mobile Applications Part (MAP) protocol in the core network.

Signalling System No. 7

Switched circuit networks are robust, high-availability networks. Robustness and availability are due largely to the use of the Signalling System No. 7 (SS7) as a common channel signalling system that enables switches and nonswitching nodes such as Service Control Points to exchange service control information. Signalling System No. 7 is defined as an architecture and a number of protocols [158].

The SS7 architecture has two types of node. The *Signalling Point* (SP) is the point of access to the SS7 network for the user, for example a switch, SCP, HLR or VLR. The *Signal Transfer Point* (STP) is a high-performance packet switching node. Nodes are connected by bundles of links, allowing load sharing and a means of dealing with link failure.

The principles underlying the architecture of a SS7 system are usually depicted by the diagram in Figure 1.5(a). Two interconnection patterns give the network the ability to deal with link or node failure without degrading its performance. The *quad*

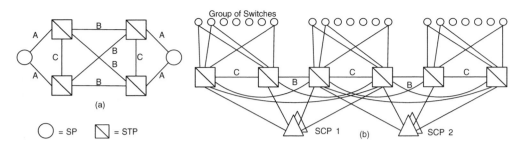

Figure 1.5. (a) Quad structure in Signalling System No. 7 architecture. (b) Typical architecture of an SS7 network.

structure consists of two pairs of STPs connected by four bridge or B-links. Each Signalling Point is connected to a pair of STPs in a quad via two access or A-links. In addition, a cross or C-link joins the two STPs forming a *mated pair*.

An actual SS7 network generally has more than four STPs and supports hundreds or thousands of SPs. Multiple quads must therefore exist. The number of switches, and hence SPs, is often at least an order of magnitude greater than the number of STPs. The number of SCPs in a network is generally small but each SCP requires significant protection from access link or STP node failure as well as failure of its own interfaces. Figure 1.5(b) shows a configuration that addresses these requirements. Groups of switches in a geographical area are connected by A-links to the STPs of a mated pair serving that area. Service Control Points are usually constructed as high-availability computers with duplication of functions. Duplication may include the use of access links to different mated pairs of STPs.

Signalling System No. 7 supports two classes of protocol [189]. The first is concerned with setting up connections between exchanges. In most cases today, the *connection-oriented* application layer protocol supporting inter-CCF signalling is ISUP. The second class of protocol is *transaction-oriented* and is geared to supporting large volumes of database queries or remote operation invocations. Examples of transaction oriented application protocols are INAP and MAP used in mobile networks to query the HLR and VLR. A special application sub-layer, the Transactions Capability Application Part (TCAP), supports transaction-oriented application protocols. TCAP allows multiple applications to have multiple concurrent transactions in progress at any time. TCAP also allows related transactions to be linked.

Both classes of protocol are supported by a robust, high performance protocol stack at OSI-RM layers 1, 2 and 3 called the Message Transfer Part (MTP). The data link layer, MTP Layer 2 (MTP-2), contains protection against frame loss and mis-sequencing, freeing other layers from the need to perform error recovery procedures. The network layer, MTP Layer 3 (MTP-3), is connectionless and uses absolute addresses for Signalling Points called *Point Codes* that are unique to the network.

MTP Layer 3 network functions are supplemented by the Signal Connection and Control Part (SCCP). While SCCP provides connection-oriented services, these are

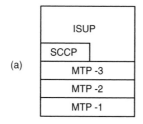

Figure 1.6. (a) Circuit-oriented protocol stack in Signalling System No. 7. (b) Transaction-oriented protocol stack in Signalling System No. 7.

seldom used. Two valuable enhanced addressing modes are provided in SCCP. The first addresses the MTP's limited capacity for identifying upper layer users at each SP. The SCCP sublayer therefore provides a one-byte *Subsystem Number* allowing a number of upper layer users to be connected to a Signalling Point with a single Point Code. The second allows destination Signalling Points to be addressed by means of an E.164 number, called the *Global Title*. One or more of the STPs in the SS7 network must be able to translate the Global Title to a Point Code (and Subsystem Number if required). The Global Title is useful for addressing elements such as SCPs, HLRs and VLRs and in routing ISUP messages to control international calls.

The protocol stacks for the connection-oriented and transaction-oriented classes of protocols are shown in Figure 1.6. The ISUP protocol is shown making use of MTP-3 directly or SCCP with Global Title addressing. Two examples of application layer protocols, namely the CAMEL Application Part (CAP) (the mobile network version of INAP) and MAP, are shown.

The Classical Intelligent Network

The term Intelligent Network (IN) refers specifically to a method of providing value-added services in telephone networks according to the ITU-T's Q.1200–1290 series of Recommendations [93, 151]. In view of the emergence of other types of intelligence in networks, the term *classical Intelligent Network* is used for this architecture. We review the classical Intelligent Network to elucidate its role in fixed and mobile networks and as a baseline for value-added services in next generation networks in later chapters.

We have introduced the physical elements that comprise the classical IN, the SCP, SDP and the SRP, into Figure 1.3. These elements are overlaid on the PSTN or 2G mobile network, using Signalling System No. 7 for message transport. The IN standards are complex and use a framework called the Intelligent Network Conceptual Model (INCM) to provide levels of abstraction.

The IN standards were originally conceived to support an evolutionary progression of services and underlying network capabilities. The IN standards are therefore based on the concept of *capability sets*. A capability set represents a level of functionality available in the IN overlay elements as well as in the switching network. The particular level of capability allows services in a stated range of complexity to be implemented.

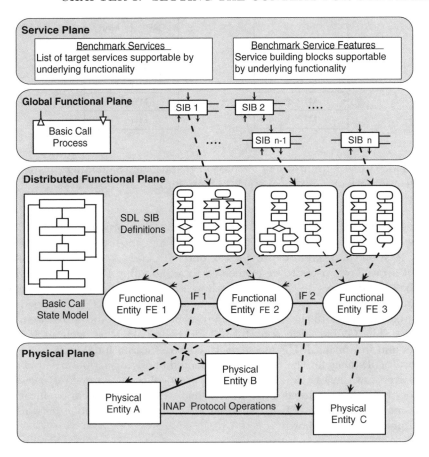

Figure 1.7. The four planes of the Intelligent Network Conceptual Model.

While the number of capability sets was open-ended, only two have been implemented significantly.

The level of capability is indicated by a set of benchmark services, that is target services in a range of complexity that telcos may wish to offer. Listing benchmark services is not standardising services. Since the objective of classical IN is to allow telcos to compete by differentiating their services, services are not standardised.

The first level of abstraction in the INCM is the *Service Plane* shown in Figure 1.7. The Service Plane provides exemplars or benchmarks of services that should be capable of implementation at the stated level of capability of the IN infrastructure. The benchmark services for Capability Set 1 (CS-1) [115] are characterised by only one call process (originating or terminating) interacting with the logic on the SCP and the SCP being invoked only during call setup or clear down, not in the active phase of the call. Examples of such services involve number translation and alternative billing: freephone, split charging, premium rate, abbreviated dialling, and credit card calling.

Voice virtual private networks based on PSTN infrastructure are implemented using the IN architecture.

Capability Set 2 (CS-2) [124] benchmark services include all CS-1 services with the added ability to perform *call party handling* (CPH), that is to manipulate bearer connections in mid-call under user control. Such benchmark services include call waiting and conference calls. Capability Set 2 supports services that require co-operating Service Control Points, for example, in different networks in an international freephone service.

The *Global Functional Plane* (GFP) of the INCM represents the software creation methodology based on defined reusable elements. The method used is based on Service Independent Building Blocks (SIB). The SIB is a scripting type of software methodology. Typical SIBs are User Interaction and Translate Data. The former initiates playing an announcement and optionally collecting digit(s) from the user. The latter initiates a database lookup. Service logic is created by means of a script that chains SIBs and has branching paths depending on conditions encountered during execution of SIBs. The Global Functional plane also contains an abstract description of the call process, the Basic Call Process. Service logic launches from the Basic Call Process and returns to it in standardised ways. While standards bodies defined sets of SIBs for Capability Sets 1 and 2, little benefit occurred from standardisation since each IN equipment vendor developed a proprietary set of SIBs.

The *Distributed Functional Plane* (DFP) of the INCM contains an abstract definition of the functionality that supports the execution of the SIBs, that in turn allows services typified by the benchmark services to be implemented. The logic of each SIB is distributed across different nodes in the IN. For example logic embodied in a SIB called Translate Data starts and ends executing in the SCP but performs the database lookup in the SDP. Abstract data definitions are given and the internal and external behaviour of the SIBS are specified using the Specification and Description Language (SDL) and message sequence charts (MSC). Each description shows the distribution of the SIB logic and the communication between parts. Similar sets of functionality from different SIBs are allocated to abstract elements called *functional entities*. Three principal functional entities are defined. The *Service Control Function* (SCF) provides an implementation-independent definition of the functions that may be performed on the external service platform. The *Specialised Resource Function* (SRF) is an abstraction of functions for interacting with the a caller. The *Service Switching Function* (SSF) contains the functions needed to set and test trigger conditions and to interact with the SSF. Capabilities are therefore defined in terms of functional entities and the information flows between functional entities.

The Distributed Functional Plane specification also defines the Basic Call State Model (BCSM). The BCSM defines standard interface mechanisms for invoking services hosted on a service control point. The next section examines the standardisation approach and how external logic is invoked from the call process.

The *Physical Plane* represents possible physical realisations of the IN using a finite repertoire of physical elements, such as the SCP, SDP and SRP. The Physical Plane defines rules for allowed mappings of the abstract functional entities onto predefined types of physical nodes, for example SCF to SCP. The Physical Plane specification defines the application-layer protocol, INAP, at the various capability sets. Variants on INAP exist. ETSI has defined a reduced form of the ITU-T INAP, called the

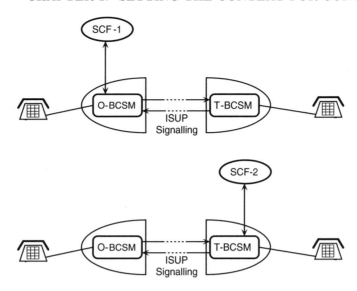

Figure 1.8. Representation of a PSTN call by two half-call models with possible relationships to Service Control Functions.

ETSI Core INAP [56], to ensure interoperability between different implementations. Network-specific billing formats are incorporated into implementations of charging-related operations in INAP.

The GSM mobile network standards define a rich set of switch-based supplementary services. New supplementary services cannot be added speedily or readily differentiated to meet the needs of different operators. GSM networks therefore rely on IN-based services for new features [147]. A leading example of an IN-based service in mobile networks is prepaid calling. The ETSI standards for IN in GSM networks are known as Customised Applications for Mobile network Enhanced Logic (CAMEL) [59]. A version of INAP at CS-1 level, called the CAMEL Application Part, has been defined for supporting services in mobile networks under the CAMEL standard. CAMEL standards allow the invocation by a roaming user of SCP-based logic hosted in the home network.

Call Models and Invocation of IN Services

The Intelligent Network was overlayed on pre-existing digital PSTN switches. These switches conform to external signalling specifications, namely ISUP and Q.931, but implementation detail differs among vendors. The authors of the IN standards were faced with the problem of enabling the different switch vendors to expose consistent interfaces between their uniquely implemented call processes and the Service Switching Function. The call process was abstracted using two *Basic Call State Models* shown in Figure 1.8. The *Originating Basic Call State Model* (O-BCSM) encapsulates the processes associated with the originating side of the call, for example authorising

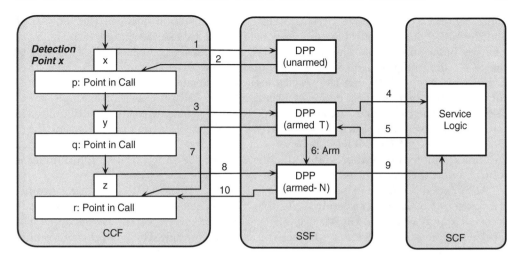

Figure 1.9. Role of detection points, detection point processing and service logic.

the user, collecting digits and routing. The *Terminating Basic Call State Model* (T-BCSM) abstracts the functions at the terminating side of the call including authorising and alerting the called party.

The calling party signals to the O-BSCM. In general, the O-BCSM and the T-BCSM are in different switches and signal to each other through zero or more exchanges using the ISUP protocol. Interaction with a SCP containing service logic is possible from either BCSM, but normally only one at a time.

Figure 1.9 illustrates several concepts in the IN standards. The BCSMs model call control functionality and are located in the CCF. A Basic Call State Model has two building blocks: detection points and points in call. A *Detection Point* (DP) is a stage in the call control process where external logic hosted by the SCF can be invoked if predetermined criteria, the *trigger criteria*, are met. For example, the number translation logic for a freephone service must be invoked if the dialled number has a specified prefix, say 080. At each detection point, the call process halts and sends a notification carrying call parameters to the Service Switching Function as shown in Figure 1.9. In the SSF, *detection point processing* (DPP) determines whether call parameters satisfy the trigger criteria.

The first notification (1 in Figure 1.9) shows the case of a detection point with no criteria set or the trigger parameters not meeting the criteria. Execution of the call process resumes execution where processing was interrupted (2). The call process between this point and the next detection point is encapsulated in a *Point in Call* (PIC). The PIC abstracts this part of the vendor's implementation of the call process. A PIC can receive and emit signalling such as ISUP and Q.931 messages.

The second notification (3 in Figure 1.9) encounters a case where call parameters meet criteria that have been preset by a management function for that detection point. A detection point with a permanent set of criteria is said to be of *Trigger Detection Point* (TDP) type. For example, in an abbreviated dialling service, three dialled digits contain a hash follows by two digits. The translation between the short code and the called party's routable number is held in a database associated with the SCF. The combination of digits #nn meets the trigger conditions. A message (4) is sent to the SCF containing a *Service Key* generated in the DPP that identifies the service logic to be executed. When the service logic has executed, it returns a message (5) containing one or more instructions to the DPP and call process. For example, a DP elsewhere in the call process could be armed to detect some event or condition later in the call, as shown at 6. Such a temporarily armed detection point is called an *Event Detection Point* (EDP). The call process is instructed (7) to resume at a specified DP. For example, in an abbreviated dialling service, the called party's full number is returned and the call process must re-analyse that number.

At 8, we show the call parameters being passed to the DPP for testing against criteria set at 6. If the test result is positive, two actions are possible. As shown at 9, an event report is sent to the service logic and no reply is expected. For example, the service logic may need to know whether the call is answered. The call process resumes execution at the start of the next PIC, as shown at 10. Alternatively, a request may be sent to the service logic that calls for a response. In this case, the call process is not instructed to continue until the response is received.

The O-BSCM and T-BCSM for IN Capability Set 1 are shown in Figure 1.10. The BCSM for second generation GSM networks is a simplified version formed by merging points in call as shown in Figure 1.10. Detection points are eliminated when PICs are merged. For example, *Analysed_Information* is not available in 2G mobile networks.

Example: Call Connection and Value-added Services in the PSTN/IN

This example illustrates two aspects of practice in the PSTN/IN: first, invoking external logic to enhance services (messages 4–12), and second routing connections via a first choice route (messages 13, 14) or an alternate route if no circuit is available on the first choice (messages 15–19).

Three forms of signalling in use in the public switched telecommunications network are illustrated in Figure 1.11:

- *Loop Signalling*: the user and network use loop signalling, represented by pseudo-messages 1–3, 17, 20 and 21.

- *ISUP*: the ISDN User Part Signalling Protocol is used for setting up and clearing down connections between switches. This message sequence chart shows some ISUP messages (13, 14–16, 18, 19, 22, 23).

- *INAP*: control logic on the external Service Control Point (SCP) provides value added services and interacts with one of the switches and the Specialised Resource using the Intelligent Network Application Protocol (4–8, 11, 12)

The numbered messages perform the functions described below.

1–3. Using loop signalling, the caller, A, goes offhook, receives a dialtone and dials a freephone number.

  4. The originating exchange (E1), on analysing the dialled digits, discovers that it cannot route the call to this number. Exchange E1 therefore sends the INAP

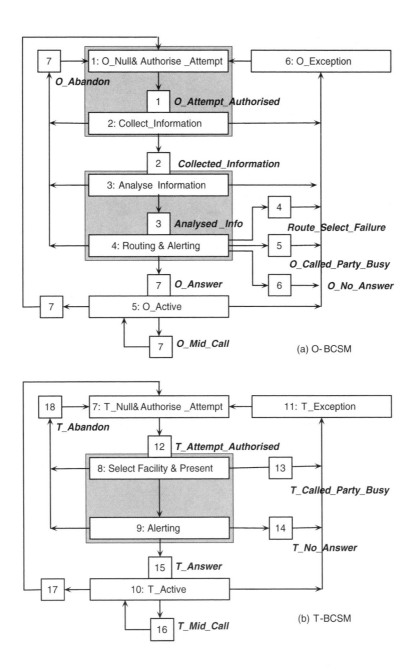

Figure 1.10. Basic Call State Models for Intelligent Network Capability Set 1. The shaded areas show merged points in call for GSM.

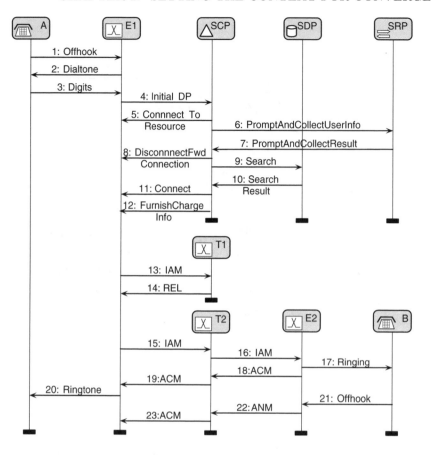

Figure 1.11. Example of a message sequence for PSTN call setup with Intelligent Network service support.

InitialDp message to the SCP requesting it to translate the dialled number to a routable number and to determine the charging details.

5. The SCP needs further information from the user, for example a service option, and requests the switch to connect the caller to an SRP.

6, 7. The SCP instructs the SRP to prompt the caller with a set of options and asks for a digit to be entered. The entered digit is returned to the SCP.

8. The SCP instructs the switch to clear the temporary connection to the SRP.

9, 10. The SCP queries the database (SDP) for the called party routable number corresponding to the freephone number dialled and the choice digit entered by the caller.

11. The SCP instructs the switch to complete connecting the call using the INAP Connect operation, using the routable number as the B-party number.

12. The SCP instructs the switch to mark the billing ticket for reverse charges.

13. Exchange E1 now has a routable number and continues processing the call with the new destination number. E1 sends an ISUP Initial Address Message (IAM) to T1, the transit exchange on the first choice route to the destination, requesting it to reserve a circuit and to take over routing.

14. Here we assume that T1 finds that it cannot route the call onward due to lack of circuits. T1 refuses the request by returning a Release Message (REL).

15. Exchange E1 must use the alternate route to the destination via transit exchange T2. E1 sends the IAM message to T2.

16. T2 has free circuits to the terminating exchange E2 and forwards the IAM to E2.

17. The called party is free and ringing current is applied to the phone.

18, 19. An indication of the ringing condition is returned using the Address Complete Message (ACM).

20. Exchange E1 plays ringing tone to the caller.

21. The called party answers; ringing current is disconnected by E2.

22, 23. The Answer Message (ANM) passes back to the originating exchange; ringing tone is removed by E1; the connection is made.

Limitations of Switched Circuit Networks

Public circuit-switched networks set the standard for network availability, voice quality and grade of service. Switched circuit voice networks have a significant range of value-added services implemented by means of the Intelligent Network. Fixed line switched circuit networks are limited as multiservice networks because their analogue subscriber loops require voiceband data modems to adapt the digital user data to an analogue signal suitable for the voiceband channel. Voiceband modems are limited to speeds of 56 kbit/s. Voiceband access is also inefficient. The voiceband channel occupies a circuit through at least the end exchange before encountering the access element to a packet network.

The N-ISDN, while it sets out to be a multiservice network, offers only a few more real-time services than the PSTN: videoconferencing and high-fidelity audio. The N-ISDN is inherently limited as a data network because of the use of circuit-oriented B-channels that limit dial-up user speeds to 64 or 128 kbit/s and need to be switched through the end exchange in circuit mode.

The Intelligent Network approach to implementing value-added services, while enjoying widespread deployment, has several limitations. First, classical IN is strongly linked to the underlying circuit-switched voice network. IN is thus a vertically

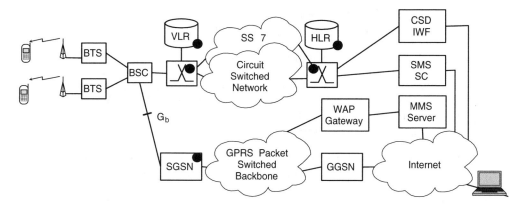

Figure 1.12. Reference network architecture for data services in 2/2.5G mobile networks: GPRS, SMS and MMS.

integrated solution. A single provider offers access, connectivity and the value added service. Second, there has been limited success in opening the Service Control Points in a secure manner to allow requests generated on the Internet to initiate or enhance telecommunication services. Third, the SIB approach to reuse of software suffers from both the dominance of proprietary products and not being an object-oriented paradigm.

1.2.3   PRESENT STATE: DATA SERVICE IN MOBILE NETWORKS

The GSM network is an integrated services digital network since it can support multiple services (voice and various data services) over a single interface. Initially, GSM phones and the network supported only circuit-switched data and the Short Message Service (SMS). Recently, packet mode data services have been introduced in the form of the General Packet Radio Service (GPRS). The GPRS allows the enhancement of messaging in the form of the Multimedia Messaging System (MMS).

Figure 1.12 provides a framework for discussing four data services available on GSM mobile networks. The principal architectural elements of the GSM network supporting circuit-switched services as well as added elements to provide the General Packet Radio Service are shown.

Circuit-switched Data

The mobile station and air interface support a range of synchronous and asynchronous data transfer capabilities using time slots on the air interface. Such data channels are switched through the MSC. GSM standards specify means of interworking with other ISDNs and ITU-T packet standard networks such as X.25. Interworking with the Internet requires an *interworking function* (IWF) for circuit-switched data (CSD). The IWF acts as the Internet service provider's point of presence (PoP). Circuit-switched data rates are generally limited to 9.6 kbit/s. Higher rates are obtained in High Speed Circuit Switched Data services by allocating multiple time slots to the data connection. Circuit-mode data bearers are generally charged on a time basis and such data services are consequently unattractive.

Short Message Service

The GSM Short Message Service provides a means of sending limited length text messages to and from mobile stations. SMS is a point-to-point service that is based on the store and forward principle. An outgoing message from a mobile station is directed to an SMS Service Centre (SMS SC shown in Figure 1.12). The message is stored in the SMS Service Centre and then forwarded to the destination. Service Centre implementations provide an interface to the Internet. Short messages may therefore be sourced and received by Internet hosts.

The Service Centre is associated with an MSC denoted the Interworking MSC. The short message is transferred from the MSC of the serving area to the Interworking MSC over the SS7 network using operations provided in the MAP protocol. In the case of a short message originating in another network, the Gateway MSC for the SMS service must determine routing information from the HLR. The gateway then sends the short message to the visited MSC via the SS7 network.

The interface between the Interworking MSC and the SC is not defined in the GSM standard. Broadly, the chosen protocol stack for the SMS SC must provide services compatible with those defined for the Short Message Relay Layer defined for the air interface. The Short Message Transfer Layer (SM-TL) provides a peer-to-peer service between the Mobile Station (MS) and the Service Centre. The SM-TL supports the delivery of messages from the SC to the MS and submitting short messages from the MS to the SC. The SM-TL tracks and reports on the status or failure of message transfer. The Short Message Relay Layer performs functions including transferring data from and to the MS, acknowledgement of transfer, sending error messages and notifying memory availability in the MS.

General Packet Radio Service

The General Packet Radio Service enhances the GSM network by the addition of a packet mode data interface at the mobile station as well as a packet core network shown in Figure 1.12. The Base Station Controller (BSC) has an added interface (Gb) at which packet traffic from the mobile station is extracted. GPRS data is not switched through the MSC.

The packet mode core network has a number of standardised nodes. The Serving GPRS Support Node (SGSN) provides access for packets from the mobile station to the core network. The SGSN is responsible for the delivery of packets to and from the MSs within its serving area. The SGSN also keeps track of the mobiles within its service area using HLR and local location data. For mobile to mobile data services, packets are routed to the SGSN that serves the base station system to which the mobile station is logged.

The Gateway GPRS Support Node (GGSN) provides an interface between the GPRS packet network and other packet networks, for example the Internet, X.25 networks or private networks. The GGSN acts as a logical interface to external packet data networks and maintains routing information used to tunnel protocol data units (PDU) to and from the SGSN that is currently serving the MS.

The GPRS provides a packet mode access method from a mobile handset to the Internet. The connection can operate on an always-on basis.

While GPRS provides packet mode transport from the MS to the network, the physical channel occupies one or more of the time slots on the air interface as allocated by the network operator. Data rates available on the air interface therefore depend on individual network operator's configuration.

Multimedia Messaging Services

The availability of the GPRS packet mode access and core network transport allows services involving the transfer of longer messages than are permitted in the SMS. The Multimedia Messaging Service (MMS) is built on the GPRS capability. The MMS is essentially a file transfer service, allowing text, image or audio to be transferred. Current incarnations of MMS do not support instant messaging. Like SMS, MMS operates on the store and forward principle.

A simplified form of the MMS system is shown in Figure 1.12. The MMS server, provides storage of messages and general support for the MMS service. The MMS server is located behind an node called the MMS Proxy-relay, not shown in the figure. The Proxy-relay allows the transfer of messages between the mobile client and several types of other nodes in addition to the MMS server, namely, another MMS system, a legacy messaging system such as SMS or a server on the Internet, for example an e-mail server.

The MMS system uses a Wireless Application Protocol (WAP) gateway. The Wireless Session Protocol used between the client and the WAP gateway is capable of starting a session, ending a session, suspending and resuming a session. Within a session, methods, namely Hypertext Transfer Protocol (HTTP) invocations such as GET, may be invoked. HTTP is used between the WAP gateway and the MMS Proxy-Relay, not shown in Figure 1.12.

Limitations of and Opportunities for Mobile Data Services

Second generation mobile networks have successfully provided circuit-mode data services. Principal limitations for the end user are low bit rates and time-based charging.

The introduction of the GPRS packet mode interface in the so-called 2.5G mobile networks provides the capability for end user data services that benefit from always-on connectivity. The data rate limitation remains the allocated time slots in the GSM-air interface. GPRS data services in 2.5G networks compete for bandwidth with voice services. The GPRS standards are however the foundation for the 3G mobile network standards described in Chapter 7, where new air interfaces provide faster data and do not occupy capacity in circuit mode. Subject to low data rates and the limitations of the screen and keypad, Mobile Stations increasingly function as Internet access devices. These limitations are eased in 3G systems.

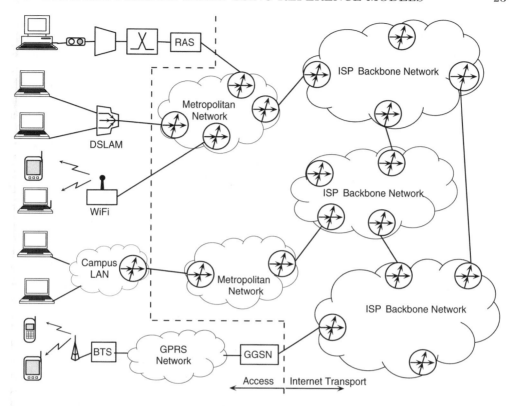

Figure 1.13. Reference network architecture for the Internet.

1.2.4 PRESENT STATE: THE INTERNET

The Internet takes on different meanings when viewed from different perspectives. To end users, it is an arcane means of providing a set of characteristic services including e-mail, Web browsing and transactional services. To network experts, the Internet is a complex set of protocols for providing connectivity and services. The strict meaning of Internet is a set of interconnected autonomous networks. To summarise the present state of the Internet as a basis for discussion of converged networks, we consider the architecture, building blocks, principal protocols and services.

Architecture of the Internet

Perhaps the most striking difference between Internet and telecommunications standards is the paucity of high-level architectural definitions for the Internet. Rather, Internet standardisation is based on the development of specific, focussed protocols as and when needed. Figure 1.13 provides a reference network for discussion of the Internet. The diagram is divided into two main areas. First, a number of methods for accessing the Internet are shown. Second, the concept of the Internet transport network as a set of interconnected networks is shown.

The Internet can be described as an arbitrary set of interconnected networks that provides worldwide packet transport capability. The physical building blocks of the Internet are hosts and routers. *Hosts* are computers that are the senders and recipients of information across the Internet. Hosts may be end-user computers or be involved in the provision of services or the management of the network itself. *Routers* are network elements that steer packets on a hop-by-hop basis from source to destination using a routing protocol called the Internet Protocol. The functions performed by a router are defined in Figure 1.15.

The Internet is made up of networks, formed by interconnecting routers by links or by subnetworks that create virtual links. The networks so formed are interconnected by links between routers at network edges. The individual networks are autonomous and are capable of determining how to route packets The concept of an autonomous system is key to the Internet: an *autonomous system* (AS) is a packet network that has its own routing technology, usually a single routing protocol, is administered by a single authority and interfaces with other autonomous systems using a separate exterior gateway protocol.

The interconnection of autonomous networks is arbitrary in the sense that any network that is connected with another that is part of the Internet is on the Internet. In this sense, connected has two meanings. First, there is a physical connection between two routers, one in each network. Second, routing information is exchanged between the two networks through an exterior routing protocol.

As the Internet has developed, the manner in which networks are structured and interconnected has become less arbitrary. The concept of an ISP network has developed. Large ISPs with national coverage require a network architecture with access, metropolitan and backbone segments as shown in Figure 1.13. ISP networks interconnect at peering points, that is via a link between agreed routers. Interior detail of autonomous systems is not shown in Figure 1.13. In the backbone, the ISP network may have routers only at the edges with full mesh virtual links implemented using an ATM network.

The IP transport network stretches from the first to last router encountered by packets traversing the network. Connection of users to the Internet requires an access network connection that takes on one of several forms. In the access segment, other protocols take over the function of ensuring that datagrams reach the destination hosts.

Access to the Internet

Access to the Internet involves two considerations. First, the end user must have a physical access mechanism and, second, the end user must be identified to the network and other users and must have authority to communicate.

Several forms of *physical access* to the Internet are available currently. Figure 1.13 shows a number of access mechanisms including voiceband dial-up, digital subscriber loop, wireless local area networks, wired local area networks as well as mobile networks via the GPRS. Other access mechanisms not shown include cable modems.

Logical access to the Internet relies on four elements. First, the host requires identification an Internet Protocol Address, a 4-byte number in IP version 4. Second, the user or resource to be used must be identified by a Uniform Resource

Locator (URL), for example for a Web page http://www.ee.wits.ac.za/ and for an e-mail user joe@somewhere.net. The URL must be mapped to the IP address before packets can be routed across the network. Third, the configuration of the host's TCP and IP protocol layers must be matched to the access method. Fourth, the user must, in most cases be authenticated.

Issues of logical and physical access to the Internet are tied up with the question: 'What is an Internet service provider?' The essential elements required by an ISP to provide physical access are:

- Facilities to terminate access networks, for example a remote access server (RAS) for dial-up PSTN voiceband modems or ISDN B-channel connections; digital subscriber line access module or wireless local loop gateway router.

- The ISP must own or have access to a router, called the upstream router, that is connected to one of the networks that forms the Internet.

An ISP provides logical access to users by:

- Getting a user name and password from the user, and querying an authentication, authorization and accounting (AAA) server, for example using the RADIUS protocol, to authenticate the user.

- Using the Dynamic Host Configuration Protocol (DHCP) server to issue configuration data to the host, including an IP address and subnet mask. Except for servers, IP addresses are allocated dynamically. The period of time during which an address is issued is called a lease. A lease may be renewed or given up by the client.

- Permitting the user's packets to be admitted to the Internet.

The host is then able to send and receive packets to and from other hosts on the Internet.

Internet Protocols

Figure 1.14 identifies the principal protocols required in the Internet classified according to OSI-RM layer. A number of *application layer* protocols support characteristic Internet services, for example the Simple Mail Transfer Protocol (SMTP) is the basis for constructing mailer applications.

Several *transport layer* protocols are designed to meet the needs of particular classes of applications in the light of the unreliable nature of the underlying IP network. For example the User Datagram Protocol (UDP) suffices when reliable transport is not essential while the Transmission Control Protocol (TCP) attempts to provide reliable transport for services not having real-time requirements. The Real-time Transport Protocol (RTP) supports the transport of real-time media streams while SCTP is designed to support the transfer of critical messages across IP networks. These protocols are described in Chapter 4.

The *network layer* centres on the packet routing protocols, IPv4 at present with IPv6 for the future. Routing requires the support of protocols for constructing the routing tables, for example OSPF and BGP-4. Translation of addresses used in private

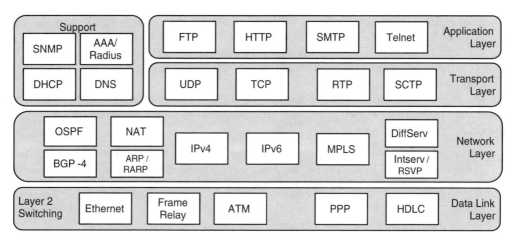

Figure 1.14. Summary of the main Internet protocols arranged by the OSI-RM layers.

domains to addresses that allow end-to-end routing is performed using the Network Address Translation protocol (NAT).

Internet access networks, such as Ethernet LANs, normally use layer 2 (L2) switching and addressing. Translation between IP addresses and L2 addresses is the function of the Address Resolution Protocol (ARP).

Path-based switching is important in the core of IP networks and is supported by the MPLS protocol. As IP networks carry real-time streams, quality of service mechanisms such as DiffServ and IntServ become important.

Packet Switches and Routers

Many packet-switching protocols and specific network architectures exist. Some are based on switching nodes called *routers* while others use *switches*. We explore the commonalities and differences between switches and routers. While implementations vary, all switches and routers conform to the general architecture shown in Figure 1.15.

The *forwarding engine* receives frames (packets) and directs frames to the correct output link, L_{OUT}, toward its final destination, according to the incoming address field, A_{IN}.

The *forwarding table* contains information on the output link to be used for a given incoming address.

The *control protocol* exchanges information and executes algorithms to construct the forwarding table.

Three examples show how the forwarding table is used.

- For the *Internet Protocol*, the table is indexed using the destination address contained in the layer 3 datagram header. The longest match with a table entry is found. The identifier for the output link, L_{OUT}, to be used is retrieved.

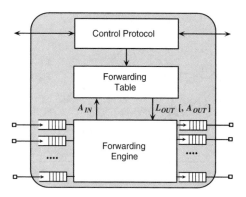

Figure 1.15. Generic architecture for a switch or router.

- In *Frame Relay*, the layer 2 frame contains a field called the DLCI unique to each destination. The DLCI is used as the index into the forwarding table. The table contains the predetermined link toward the destination (L_{OUT}).

- Some protocols, for example *Asynchronous Transfer Mode*, translate the address field. In virtual path (VP) switching, the Virtual Path Indicator (VPI) in the layer 2 cell header is used as A_{IN}. The table lookup gives L_{OUT} and a new value of the VPI, A_{OUT}, to be inserted into the cell header.

We therefore define *router* and *switch* as follows. A *router* is a node for directing packets to the next node toward their destinations using network-wide addresses normally contained in the layer 3 packet header, using information on available routes, their states and costs built up by router-to-router information exchange.

A *switch* is a node for directing packets along a pre-determined path toward their destinations using addresses peculiar to the source-destination pair normally contained in the layer 2 frame header and routing information related to the end-to-end route.

Routing Protocols

The Internet is an interconnection of autonomous networks with no administrative authority above the autonomous systems. How then does the Internet succeed in routing packets from one host to another? The Internet uses the principle of *route discovery* to ensure proper routing in an arbitrary collection of autonomous systems and in the face of possible router or link failure. The functions performed by a router, explained in Figure 1.15, underpin a discussion of route discovery protocols. The essential problem is to construct the forwarding table in each router. In an IP router, a table entry (row) has a number of fields [153]. The Destination Address and Route Mask are used to index into the table. The Route Mask indicates the significant part of the Destination Address to be matched in the search for table rows. The Interface Identifier specifies the link to be used to the next router. The Next Hop Address is the IP address of that router. Depending on the routing protocol, one or more metrics are used to rate this choice of route over other possible routes.

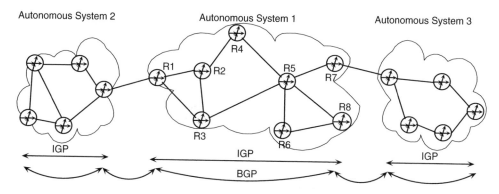

Figure 1.16. Arbitrary autonomous systems in the Internet.

The forwarding table is built up by routers exchanging reachability information by a process termed a routing protocol. Two broad classes of routing protocol exist. An *interior gateway protocol* (IGP) is a mechanism for exchanging routing information among gateways and hosts within an autonomous system. A *border gateway protocol* (BGP) is an inter-domain routing protocol used in IP networks allowing the exchange of network reachability information, including the list of autonomous systems that reachability information traverses, with other BGP-speaking systems.

Two current interior gateway protocols that are Internet standards are Routing Information Protocol version 2 (RIP) and Open Shortest Path First Protocol version 2 (OSPF). These protocols differ in the way they exchange routing information and the metrics used. In RIP, routers within an autonomous system request and supply partial or full routing tables to neighbouring routers. Each router updates its routing table and, for each destination, calculates a single cost metric, the number of hops to the destination. RIP is described as a *distance vector* protocol.

The more recent OSPF is designed for use in large networks. An autonomous network can be split into areas, each of which runs the routing protocol and exchanges information with other areas. Each router broadcasts information about its connectivity with neighbours to all routers in the area in the form of Link State Advertisements (LSA). Link state information includes the interfaces that are attached to the links and metrics. The LSAs are used to build a topological database for the area. Routers within an area generate identical databases. Each router generates a routing table by building a tree of shortest paths from itself as root toward destinations using information in the topological database. Information distributed by OSPF routers includes the state of attached links. Routes that become unavailable are notified rapidly throughout the network and routing tables are updated. Within each area, a router is selected as the Designated Router. The Designated Router generates LSAs for the whole network, thereby reducing the total traffic due to routing protocol messages.

The routers having connections to routers in other autonomous systems are described as border routers; otherwise routers are interior routers. Routers that interchange information via a BGP are called *BGP speakers*. These routers are called

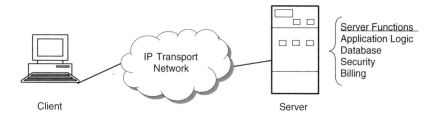

Figure 1.17. Service model for the Internet.

external peers if in different AS, otherwise they are internal peers. BGP routers in the same AS must all present to neighbouring ASs consistent routing information, that is the same reachability information. The border routers in an AS must therefore exchange information. A BGP or an IGP may be used for this purpose. A connection between two AS, that is between two external peers, has both physical and logical aspects. The physical connection between border routers in each AS is a data link layer connection usually referred to as a subnet. The DLL connection may be point-to-point or be supported by a protocol such as Frame Relay. No IP routing is needed between the border routers of two adjacent networks. The logical connection is referred to as a BGP connection.

Current BGPs are the IETF Border Gateway Protocol Version 4 (BGP-4) and the ISO Inter-Domain Routing Protocol (IDRP), a scalable inter-autonomous system routing protocols capable of supporting policy-based routing for TCP/IP internets. The IETF's objective is to promote the use of BGP-4 to support IP version 4 (IPv4). IDRP is seen as a protocol that will support IPv4 as well as the next generation of IP, IPv6.

Internet Services and Service Models

In the Internet, the network provides the basic packet transport service to support packet transfer between any source–destination pair. Network transport is transparent to the end user. The Internet operates on the *end-to-end model*: all intelligence is in the terminal domain and is, in general, under the control of parties other the bearer network provider.

Within the end-to-end model two configurations are common. The *client–server* model is shown in Figure 1.17. Value is created by intelligence in the client interacting with content and service logic in the server as indicated. The server must perform all support roles such as ensuring security and billing. The *peer-to-peer* configuration involves two or more hosts participating in a service where there is no hierarchy, as in the client–server case. All participants can receive or supply comparable services.

The end-to-end model is adequate for simple client–server applications. However, the overhead of providing common support such as security and billing integrated on every server adds cost. Also, complex computer-to-computer interactions in the Internet model are seen to require some common intelligence. The Web services model described in Chapter 3, for example, requires a registry of business services that are available that allows potential users of these services to locate them.

Internet Potential and Limitations

The Internet is characterised by two features. First, the transport network is provided by many providers, each acting locally to their networks. Second, the Internet provides seemingly unlimited opportunity for the creation of new services by a large number of service providers. New services are based on existing protocols and capability added by defining new or alternative protocols as the need arises.

The flexibility of the Internet in allowing new services and protocols to be developed and deployed is also its potential weakness as two examples demonstrate. First, protocols for supporting real-time services such as voice and videoconferencing on the Internet are now well established. The Internet is a best-effort network and provides no inherent guarantees of transport delay and loss. Further, new protocols to compensate for the shortcomings of the best-effort transport network have also been developed but are also best-effort in operation: they allow the specification and monitoring of quality of service (QoS) but cannot offer guarantees. Second, new protocols such as peer-to-peer file sharing achieve high levels of usage and become the dominant form of packet traffic on the Internet, impacting on the performance of other services.

Internet protocols such as voice and videoconferencing are being applied in public and private managed IP network contexts, where carrier-grade performance of the transport network can be ensured. Application of Internet protocols in a managed network environment, that can also be used for real-time and information services, represents an important instance of convergence. However, the public Internet, composed of autonomous networks that in general do not guarantee performance, is not a carrier-grade network.

1.2.5 *PRESENT STATE: ENTERPRISE NETWORKING*

Enterprise networks historically and to a great extent still fall into two separate types: voice and data. Interconnecting private branch exchanges situated at different company sites to form a private telephone network is a long-standing practice. Interconnection between PBXs is typically achieved using time-division multiplexed leased lines. The different sites frequently have a single numbering plan, making the voice network look like a single network. At one or more points, the private network interconnects with the public switched telephone network.

Private data networks, more recently called intranets, consist of local area networks (LAN) at the various corporate sites. Sites are interconnected by physical or virtual links. Frame Relay gives an efficient way in connecting sites in an apparent full mesh. Routing information is configured to give a wide area network (WAN) spanning all the enterprises sites. Off-site connections, for example to electronic data interchange (EDI) networks or the Internet are provided by network operators. The private networks may be physical, that is provided by means of the enterprise's own on-site facilities with interconnection leased from a public network operator. The private networks may be virtual, that is provided on the facilities of one or more service provider but configured by a provider to function as a private network.

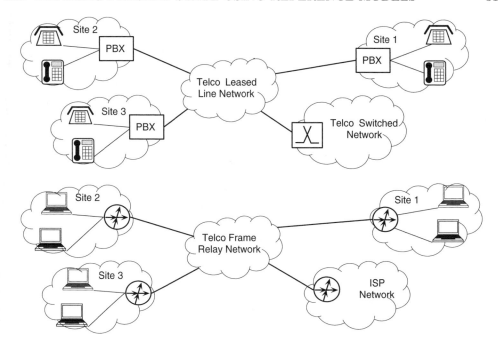

Figure 1.18. Historical types of enterprise networks.

Integrated enterprise networks have developed beyond the separate voice and data networks described above. Enterprise networks have been a development arena for multiservice packet networks based on Internet protocols. For example, the H.323 multimedia conferencing protocols developed originally in the private packet network context. Many equipment vendors offer suites of products that allow integrated enterprise voice and data networks to be constructed. While such networks abound, we treat the integrated enterprise network as a specific instance of a next generation network in Chapter 5.

1.2.6 PRESENT STATE: SWITCHED AND LEASED-LINE SERVICES

Figure 1.2 shows the facilities required by existing vertically integrated service providers. A significant degree of commonality exists at the transmission layer where most networks are dependent on interconnection of switching elements via Synchronous Digital Hierarchy (SDH) and optical equipment. Switching elements support other businesses, for example, virtual private network providers. Telcos provide interconnection services at two general levels. First, the provision of fixed capacity links, and second, switched interconnection using Frame Relay, ATM and more recently the Internet Protocol. While these facilities are used to support the telco's own services, they are made available to other network operators and to specialised end users.

Telcos seek configurability and service assurance in the infrastructure for leased line and switched interconnection services. Historically, SDH has provided a high degree of manageability, supporting configuration of connections at a wide range of speeds, as well as protection switching giving rapid response to fibre and terminal faults [192].

Users of leased line and switched interconnection services demand a wide range of link speeds. At one extreme, a small business may be served by 64 or 128 kbit/s access links. At another extreme an ISP operating Gigabyte backbone routers may require 2.4 Gbit/s links.

Transmission networks based on SDH have great versatility. Managed connections can be provided at rates from 1.544 and 2.048 Mbit/s to 139 MBit/s within a basic STM-1 structure and to higher bit rates by concatenation of the payloads in an STM-N. The practical limit of concatenated STMs is a single stream below 2.4 Gbit/s in a STM-16. At this speed and above, rates match that of a single fibre or wavelength within a fibre.

Historically, transmission systems were the servants of the switching layer and were configured for long periods. Increasingly, the need arises for more rapid configuration of switching and transmission systems and for joint management of the two layers.

1.3   EVOLUTION AND CONVERGENCE

Present vertically integrated networks are successful in their own spheres, for example telephony networks and the Internet. Opportunities for evolution to deliver new or enhanced services are limited within each sphere. Service providers seek new business opportunities, requiring historical boundaries to be breached. In response, technology providers strive to provide appropriate facilities to support new services. This process involves breaking down barriers and developing commonality between the silos of Figure 1.2, as well as bringing previously separate domains together to create new business. Initially, barriers are being broken between pairs of domains: telecommunications, the Internet, information technology (IT), broadcasting and media.

Convergence describes the tendency of such domains to come together in various ways, for example by finding increasing synergy, new applications, using common technology, developing new business models, or through services being offered as packages. Convergence is enabled by technological development and is driven by markets. The most fundamental enabler of convergence is the ability to encode in digital form information embodied in different media: text, data, sound and fixed and moving images. Common approaches to storage, manipulation, co-ordination and transmission of all forms of information are possible. Convergence therefore involves not only broadcasting and telecommunications but also traditionally stand-alone media such as print, film, photography and games. Convergence impacts on all communications, knowledge and transactions-based activities [99].

The process of convergence is viewed from different perspectives. On one hand, advances toward new services have been made pragmatically by creating linkages and synergy between disparate technologies. Alternatively, concepts of infrastructure supporting multiple services have long been advanced, for example the Integrated Services Digital Network, first in its narrowband form and then in the never-deployed broadband form. Convergence proceeds from limited developments to multiservice scenarios. We therefore introduce convergence by presenting a number of instances of limited convergence in Section 1.3.1. From these examples, we list a number of features that characterise in convergent situations. We then introduce the concept of the *next generation network* as a unifying concept for future network facilities in Section 1.4. A number of developments within the next generation network paradigm are identified.

1.3.1 INITIAL CONVERGENCE EXAMPLES

Eight examples typify the pragmatic initial instances of convergence. Most are characterised by the convergence of two existing technologies or the interworking of two existing networks.

1. *Computer–telephony integration (CTI):* Many business applications, for example call centres, rely on the ability to link user information in an IT application to the identity of a telephone caller. At its simplest, CTI may involve passing the PSTN or mobile network Caller Line Identity to the IT application. More complex applications may have other elements such a location-based services as described in example 6. Creation of CTI services is enabled by application programming interfaces (API) that allow applications to invoke telephony-related functions.

2. *Call completion service:* Revenue generated by voice services is lost if a call attempt is not completed, for example due to nonavailability of the called party. Several services provide notification of the uncompleted call to the called part via SMS or e-mail. This service relies on a standard or proprietary interface that allows the telco's IN SCP to initiate the SMS or e-mail to the called party. Data on the mapping of the E.164 number to an e-mail address must be available.

3. *Telco call or message initiation from the Internet:* With more telco subscribers having Internet access as well, a Web-based telephone book application may have the added ability to initiate a voice call using the conventional telephone network by clicking on a link in a browser-based application. This application relies on the ability of the Web application to send a request to the IN SCP of the telco network via a standard or proprietary interface.

4. *Mobile terminal integration:* The mobile telephony industry relies on increasingly powerful handsets for market development. Not only must the handsets support different services, voice, SMS, MMS and general data services, but a number of air interfaces are required, for example all the GSM bands, WiFi and Bluetooth.

5. *Voice network interworking:* A number of available protocols enable the transmission of voice on networks using the Internet Protocol. While their use in the Internet is an example of convergence, a more significant application is the deployment of managed IP networks by telcos to support voice services. These packet voice networks interwork with legacy circuit mode voice networks through gateways. Media gateways allow conversion of voice signals from their circuit mode encoding to a packetised format. Signalling gateways allow SS7 messages to be carried to and from call control servers in the IP network.

6. *Location-based services (LBS):* Second generation mobile networks are capable of generating location information of different precision for a mobile station logged onto the network. Mobile network operators make this information available to Web or IT applications through an API. The application may be offered either by the mobile operator or by a third party operator.

7. *News delivery via multiple media:* Consider a business news and information media company. Information and Communication Technologies (ICT) are used extensively in the gathering, editing and formatting of content. Convergence allows multiple services to be delivered on a common platform as well as particular services to be provided on different platforms. For example, print media retain their traditional form only in the final physical form and diversion to another output medium, for example Web pages on the Internet, is readily achieved.

8. *Interactive broadcasting:* Digital broadcast standards provide capacity on the downlink for both programme material and general information, together with a return channel.

This set of examples is not exhaustive but rather illustrates basic instances of convergence.

1.3.2 FEATURES OF CONVERGENCE

Convergence is a process that manifests its results in different ways in various contexts. Convergence is evidenced by one or more of the following features, identified by the italicized descriptor:

1. Multiple services, new or formerly supported on different networks, are supported by a single set of facilities: the *multiservice* feature.

2. Multiple services are supported on a single terminal: the *multifunction* feature.

3. *Points of integration* occur: diverse equipment and software can work into a single, common standard interface to access supporting facilities.

4. Different infrastructures *interwork* to perform a function or extend functionality.

5. Content of different types can be encoded digitally and transmitted, processed and stored: the *media convergence* or *multimedia* property.

6. The same service or content can be delivered by different types of infrastructures or media: the *versatility* property.

7. Developers are able to exploit multiple services (or types of content) to provide more powerful services (or more complex content) by combining simpler services and content: *composition* of services or content.

8. Abstractions become more than mental tools but are applied to create architectural *layering* to give simple views of hidden detail.

9. Related to (8) is *technology neutrality*: the increasing practice of designing applications to be independent of the implementation technology, allowing multiple incarnations of the application on different platforms.

10. A technological development may be classified as *disruptive* if it is not evolutionary and potentially disrupts markets, businesses or the regulatory regime. By contrast, a technological development that is evolutionary and provides orderly development of markets is termed *sustaining*.

Convergent situations are characterised by one or more of the attributes listed above. Some attributes may be mutually exclusive. For example, *multiservice* networks allow several services on a single network while *versatility* allows a particular service to be delivered over multiple infrastructure types or media. We illustrate and contrast the multiservice property on one hand and the media convergence and versatility properties on another by two examples.

Example 1: The Multiservice Transport Network.

The historical switched circuit network is based on time-division multiplexing 64 kbit/s channels for transmission and switching. The SCN is well suited to telephone-quality speech but is limited as regards data transmission, especially if high rates are required. Packet networks were initially developed to transmit data reliably but without real-time delivery delay constraints. Delivering both data and real-time services on either type of network requires special adaptation, for example data modems in SCN and adding quality of service protocols to packet networks. As data services proliferated, the goal became a network that would carry several classes of user data with performance appropriate to the type of data being transmitted. Such a network is called the *multiservice transport network*.[2] Legacy TDM circuits proved unequal to this task since all channels would be cumbersome and wasteful assemblies of 64 kbit/s channels. Packet networks are suited to carrying data from diverse types of sources provided that mechanisms should be in place to ensure adequate transport quality for each class of user traffic.

Numerous types of access networks serving different types of terminals must be accommodated. A selection of access networks is indicated in Figure 1.19. Access networks can use a variety of layer 1 and 2 protocols. In the multiservice transport network, packets should be transported between endpoints without excessive protocol conversion and adaption. While this common protocol is not used for switching at layer 2 in the access networks, these networks should transport packets transparently.

Realisation of the goal of a multiservice network involves both core network, and access network considerations. Also, all terminals should use the same network layer protocol to give a uniform end-to-end routing method. The Asynchronous Transfer Mode transport network developed in the late 1980s and described in Section 6.2.2 meets all these requirements. Currently, the ubiquitous Internet Protocol is seen as the unifying factor leading to a multiservice transport network. This approach has been adopted in 3G mobile standardisation with IP to the handset. Terminals carry all services using IP as the network layer protocol.

---

[2]The term *multiservice network* may embrace other aspects of service provision such as call and messaging control and end user services not considered here.

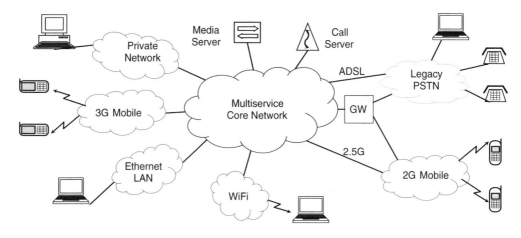

Figure 1.19. Convergence example: the multiservice transport network.

A multiservice network must meet the following requirements. The network:

- uses a single protocol for end-to-end routing (with layer 2 segments in the access or core carrying packets transparently);

- accommodates different traffic classes and ensures quality of service set by policy or agreement across different administrative domains;

- provides carrier-grade availability since some services such as telephony must give such guarantees;

- allows different access networks to be used;

- does not constrain the types of terminals that can be used, preferably using the same layer 3 protocols;

The multiservice core network must interwork with legacy networks through gateways (GW) as shown in Figure 1.19. Legacy networks may also support packet access to the multiservice network, for example ADSL in fixed line networks and GPRS in 2.5G mobile networks.

The physical point of integration is the use of a common network transport for all classes of traffic arising from different end user services. Logically, integration occurs due to the use of a single network layer protocol, the Internet Protocol.

Example 2: Content Convergence.

Figure 1.20 shows some of the essential processes in a modern news firm. Material is gathered by reporters, interviewers, cameramen, researchers and others. It is gathered in different forms: speech, still images, moving images, text and as data. Information in all these media is stored for use in an editorial process that produces various forms of

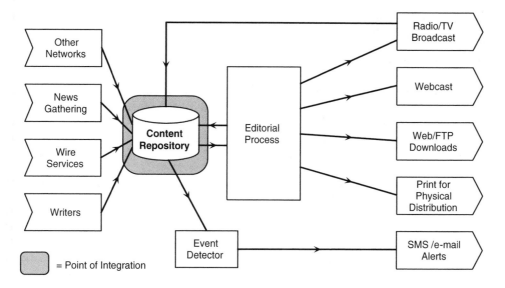

Figure 1.20. A news service built on content convergence and versatility of delivery.

news outputs. Voice and video are recorded digitally and stored in a content repository along with encoded images, text and data. The repository also contains previously disseminated radio, TV and print material. All material is indexed for ease of retrieval.

The common content repository supports an editorial process in which material for dissemination is prepared. Multiple channels are used to distribute content to end users ranging from real-time sound and TV broadcasting, webcasting, Web pages and printed newspapers and magazines. Also, subscribers may receive alerts of specified events as information is placed in the content repository, for example 'Market falls 10%!' or 'Bulls win!'

This example demonstrates three indicators of convergence. The *point of integration* in this scenario is the common storage of all types of media, that is the *convergence of content* into a single logical repository. The ability to provide information in suitably processed formats to different groups of news consumers is enabled by the common information resource but demonstrates *versatility* in distribution.

1.4 THE NEXT GENERATION NETWORK CONCEPT

Convergence is a concept that attempts to capture current and future developments in which benefits are obtained by using common facilities for new activities or being able to deliver services over different media. Convergence is a *process* with no predefined long-term endpoint.

Convergence is one of several patterns observed in the evolution of ICT and its applications. A second pattern of technology development was identified by Ferguson in the context of the development of transmission systems [94]. Three phases are observed, which we illustrate by means of a topical example.

1. Initially, an opportunity for a new service is recognised but the existing facilities do not support the new requirement. For example, the Internet was well established as a best-effort, non real-time network supporting messaging and file transfer-type services. Addition of the ability to transfer voice signals, *voice on the Net* (VoN), was seen as a potential Internet service as an alternative to telco voice services but could not be implemented with existing file-oriented application and transport protocols.

2. The next phase in this pattern of development is to adapt the existing facilities to accommodate the demands of the new service. For VoN, this involved the development of low bit rate packet speech encoders, the Real-time Transport Protocol (RTP) and measures to improve the quality of packet transport including the Resource Reservation Protocol (RSVP). While these developments enabled the transmission of voice on the Internet, the quality, availability and dependability of speech services could not be assured. As long as the Internet remains a best-effort network with end-to-end connectivity provided by autonomous systems, voice on the Internet has no service guarantees.

3. The third stage in the pattern of development is to specify, design and implement infrastructure that provides proper support for the service. Continuing the VoN example, the standards and protocols developed for the Internet have been supplemented by network protocols that attempt to assure quality of service. These protocols are applied in managed networks, giving voice services that are indistinguishable from traditional telco services in quality and availability. The managed network potentially also provides Internet services but protects the real-time services from excessive Internet traffic by segregating packets belonging to different services into classes.

Many developments follow this pattern. Significant instances of convergence occur in the second phase but may not be sustainable in the long term. In this book, we are concerned principally with the third stage, namely the creation of proper support for convergent services. Two standards bodies, the ITU-T and ETSI have both encapsulated this requirement in the concept of the *next generation network*. The ITU-T defines the NGN as a concept embracing the collective improvements being made to the service provision infrastructures building on the base of the three traditional networks: the fixed PSTN/ISDN, second generation mobile networks, and the Internet [37]. These enhancements are bringing about the convergence, firstly, of these networks through flexible interworking arrangements and, secondly, of services. ETSI's concept of the NGN focusses on the means managing the complexity of new, converged networks through the use of a number of separations into layers, planes and open interfaces to support evolution.

The *next generation network* concept, concentrating on the technology and services aspect of convergence, encapsulates the *product* of the convergence process as well as ways of abstracting and modelling the complex networks and software systems. The long-term objectives of the NGN are:

- convergence of infrastructure for telecommunications and information services;

- convergence of applications composed of IT and telecommunications aspects;

- support for new and evolving business models, including those that open the market to new service providers; and

- support for multiple services include both real-time and information services, either singly or in combination, to form multimedia services.

Like convergence, the NGN is an evolving concept: there is no single NGN. Rather, the term captures the movement or development process to one or more future networks that exhibit convergence. Several developments have been or are candidates for consideration as next generation networks:

- IP telephony, including H.323 and SIP;

- packet mode networks based largely on ITU-T protocols: ATM, BICC;

- the ETSI Telecommunications and Internet Protocol Harmonisation over Networks (TIPHON) initiative;

- integrated enterprise networks;

- the 3GPP third generation mobile network;

- The 3GPP2 all-IP third generation mobile network; and

- Fourth generation networks (4G).

Subsequent chapters examine specific aspects of the NGN and these candidates in particular, not as isolated cases but with emphasis on unification of the description methodology. These NGN solutions must be analysed within a common framework and tested against a number of general requirements: flexibility, availability, dependability, scalability, accountability, manageability, interoperability, reusability, portability and the ability to federate across different administrative domains.[3]

1.5 CONCLUSION

In this chapter, we establish the current state of a number of networks and the services that they provide, principally fixed and mobile switched circuit networks, mobile data services and the Internet. These networks are for the most part vertically integrated but have mutual dependencies in the access and core transmission areas. We have not considered the present state of operations support systems (OSS), including management signalling, service support data, accounting and billing. These issues are taken up in Chapter 9.

The concept of convergence is established as a general pattern in the evolution of networks and services. Examples of convergence cited are generally incremental, with a limited number of networks, services or technologies coming together. The result of large-scale and long-term convergence is captured in the concept of the

---

[3]Key terms are defined in the Glossary.

next generation network. Subsequent chapters focus on individual candidate next generation networks as well as unifying principles. The number of candidate networks is large, all are complex and all require significant common support, for example in the form of operations support systems and common underlying transport infrastructure. We therefore need to establish a framework for analysing and describing convergent situations and next generation networks that effectively controls the complexity of the study. This framework is developed in Chapter 2.

Chapter 2

A Framework for Examining Next Generation Networks

Chapter 1 introduced convergence as a two-sided coin. One view shows multiple services, including services currently delivered on a variety of networks as well as new services, moving to a common delivery network. This view of convergence centres on *multiservice networks*. The next generation network concept introduced in Chapter 1 describes such networks. The opposite view of the coin reflects the delivery of particular content and applications via multiple delivery methods, for example networks, broadcast and print. This *content delivery versatility* arises from the ability to encode all types of information in digital form. Actual instances of convergence range between the two views.

While convergence is intended to bring greater utility to the end user, the service provider is often faced with greater complexity, particularly in the migration process from current to new technologies. Chapter 1 therefore recognises the need for a framework to manage the complexity and diversity of NGNs. In this chapter, we seek an understanding of the characteristics of communication systems that render them complex. Building on accepted methods for dealing with complexity, we establish a framework that supports the understanding of diverse emerging and legacy communication systems.

Section 2.1 examines the characteristics of evolving networks that increase complexity. Section 2.2 reviews the characteristics of complex systems and methods of managing complexity specific to telecommunications. The NGN Framework is presented in Section 2.3. This framework covers the physical and functional architecture of NGNs and their building blocks. Section 2.4 tests and illustrates the NGN Framework by using a number of applications.

2.1 CHARACTERISTICS OF EVOLVING NETWORKS

Network evolution follows several paths, some of which converge, while others become dead-ends. Over time, several NGNs are possible. The evolution process and paths followed are influenced by several factors, described below.

Despite convergence, underlying facilities do not necessarily become simpler. Each offered service requires underlying facilities. Each NGN attempts to be an effective combination of facilities to deliver new services. Convergence allows an increased number of services to be supported by common facilities. However, the total set of facilities is not common and, paradoxically, we are faced with a *proliferation of facilities* illustrated in Figure 2.1. Core networks show greatest convergence while

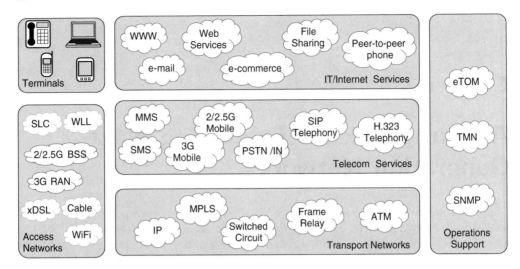

Figure 2.1. Proliferation of networks, access methods, telecommunications services, terminals and IT/Internet services.

access networks and terminal types currently increase in number. In the case of service platforms, spanning the telecommunications and Internet domains, some 30 architectural milestones are identified in Figure 2.2 between the start of digital telephone switching and near-term NGN service architectures. Several approaches to operations support systems exist. With an increased supply of building blocks, protocols and service architectures, the number of potential NGN architectures proliferate.

Each proposed NGN architecture or contributing standard competes with other contenders for acceptance. Each proposal faces *uncertainty of adoption*. The long-term, fully converged NGN cannot therefore be defined; rather a set of imperfect NGNs is identified. The problem is similar to that encountered in the development of the classical Intelligent Network. The IN standards embody the principle of evolution through a number of capability sets, each allowing more complex value-added services and possibly moving to a Broadband-ISDN environment [113]. The idea of a *long-term target IN* was frequently used but, being far ahead of present capability sets, always remained hypothetical.

The telecommunications, Internet and information technology communities are prolific producers of standards for networks and methods of delivering services and applications. *Multiple standards paradigms* exist. Telecommunications standards have traditionally been systems-oriented: families of standards start with the broad architectural principles then work toward increasing levels of detail. Internet standards, by contrast, having established general principles, consist of a large number of stand-alone or loosely-coupled specifications, generated incrementally as required. Information technology standards are concerned with diverse topics, including software engineering, programming languages and databases. With convergence, it becomes necessary to analyse, design and integrate systems based on standards from the three paradigms with their different emphases on architectures, processes, protocols and design and management tools.

Several stakeholders have *varied perspectives* on NGN architectures and the resulting services. Researchers generate and prove new ideas. Standards bodies attempt to capture promising ideas in standards. Equipment and software vendors seek to develop new marketable products for service providers and end users. Service providers such as telcos constantly evaluate technology developments to formulate a technology adoption strategy. Corporate end users seek effective choice, deployment and utilisation of ICT facilities. Regulators must create an environment that fosters competition and end user protection in a rapidly changing technological environment. All parties must be conscious of legacy networks and services. Each of these interests leads to diverse perspectives of each technology; complexity is often dealt with by ignoring other relevant aspects. Not only do individuals and organisations vary in their propensity to adopt new technologies early or late, but their perspective on an emerging technology changes over time in the often protracted period between conception and adoption. The Gartner Hype Cycle [98] models this behaviour using the concept of visibility to describe the initial enthusiasm before the technology is proven, followed inevitably by disillusionment before the technology matures and is adopted. We revisit the issue of hype in Chapter 10. Practitioners therefore need support in dealing with complexity. Students of emerging ICT systems must build a fundamentals-based understanding of emerging networks as a basis for becoming a researcher, facilities designer, service provider or a regulator.

We therefore seek to assist several audiences to deal with the proliferation of complex networks, elements and functions. We have in mind ICT system architects, standards writers and users, analysts, designers, operators and students. We unify viewpoints and paradigms to support the understanding of varied NGNs. The approach has two thrusts: systems and software. The latter occupies Chapter 3, where we review the methods of analysis, design and description of software processes and protocols. In this chapter, we develop a framework for the analysis and design of next generation networks that can be applied across a number of types of legacy and emerging networks. We first examine the proliferation of facilities illustrated in Figure 2.1 and service architectures.

2.1.1 *PROLIFERATION OF TERMINALS AND ACCESS NETWORKS*

The number of types of terminals in use has expanded beyond the traditional fixed line phone and fax as well as the ISDN phone and videoconference terminal. Second-generation mobile phones, originally supporting circuit-switched voice and data, have expanded in capability to 2.5G packet-mode GPRS data. Third generation mobile phones offer enhanced data services such as videophone and high-speed Internet access. Personal digital assistants with 2.5G or wireless LAN connectivity primarily offer data services and, in some cases, voice as well. Personal computers are capable of having terminal functionality added for many services, for example to support a soft phone or videoconferencing. Devices connected to a cable modem or the set-top unit in Digital Video Broadcast have both broadcast reception and interactive capability.

Similarly, the number of access network technologies has increased. The PSTN's original twisted pair copper loop supports voiceband signals, narrowband ISDN and various higher speed digital subscriber loops such as ADSL, the last providing access to a packet-mode core. Numerous wireless access mechanisms are available. For example, [176] describes 19 wireless access technologies at various stages of maturity from conception to deployment. Mobile networks support both 2/2.5G and 3G wireless access with full terminal mobility management. Wired Ethernet local area networks have increased in speed and are capable of being extended to high-speed wide area networks. Wireless Ethernet LANs provide both flexibility in fixed installations as well as support for mobile data users. Cable TV and Digital Video Broadcast [180] are doubling as broadcast and interactive service access networks.

The increase in terminal options and access network types results in some instances of convergence identified below. These instances are limited by associations between terminals and access network, terminal and service (including support functions such as mobility management), and access network and core network. An example illustrates the effect of these associations. A third generation mobile phone has a standard 3G air interface and is also fitted with WiFi (Ethernet wireless LAN) and Bluetooth. The standard 3G interface, the access network and the core network together support realtime services (voice and video), messaging (SMS and MMS) and Internet services (e-mail, Web browsing, transactional services and file transfer) limited only by the screen and keypad of the phone. This *multifunction terminal* represents the convergence of several services to the user and a unified, standards-based approach to facilities provision to the telco, that is technology convergence. The WiFi air interface is generally best effort and therefore suited to data rather than real-time services. The mobility management approach is different to that of the 3G network and handover from one type of air interface to another is problematic. Also, the probability of there being WLAN coverage in a place where there is no 3G coverage is small. Bluetooth is a short-range cable replacement technology and does not enhance services or the basic 3G wireless access mechanism. Thus, the basic 3G standard provides clear benefits of convergence while the addition of Bluetooth and WiFi interfaces may not.

The fourth generation (4G) network concept recognises that the range of access networks is likely to increase rather than diminish. The 4G network therefore envisages multiple, heterogeneous access networks but with mechanisms for seamless access to an IP core network. In the access area, convergence is likely to take the form of interworking mechanisms to give seamless roaming rather than merging of actual access networks. Terminals may in fact become more complex, for example with multiple radio frequency front ends: mobile network, wireless LAN and Bluetooth. Seamlessness from the user point of view is embodied in session control and mobility management as well as compatible quality of service for different access mechanisms.

2.1.2 CORE NETWORK TYPES

While core networks show least proliferation, several types of switching network are available: circuit-switched, best effort IP, both versions 4 and 6, MPLS, Frame Relay and ATM. The circuit-switched network, with its limited bearer channel bit rate and capacity to support voiceband services mainly requires consideration in the area of interworking with packet networks. The circuit-switched network is vertically integrated with its access mechanisms, call control and IN service control. Packet *core networks* by contrast are inherently multiservice networks and several packet-mode access networks can be accommodated, but each type is best suited to particular suites of services.

Managed multiservice networks that are capable of supporting both real-time and information services with quality of service appropriate to each service are required for next generation networks. Such networks must support end-to-end quality of service requested for the specific application. The core and access networks must therefore work together to fulfill this goal. The multiservice core network must interwork with legacy networks. Managed multiservice networks must also support a best effort class of service but must protect QoS-assured classes from excessive best effort traffic.

Internet Protocol networks divide into unmanaged and managed categories. The underlying transport network of the Internet is unmanaged and offers no performance guarantees. Managed IP networks generally use both a quality of service mechanism such as DiffServ and MPLS virtual circuit-based switching.

Asynchronous Transfer Mode networks are capable of filling the managed multiservice core network role, but are being supplanted by MPLS networks. Frame Relay, while managed, is intended to provide the core connections between local area networks to create wide area networks.

Migration toward a managed multiservice core network with compatible access networks is an important focus of convergence that is outside the scope of this book.

2.1.3 EVOLUTION OF SERVICE ARCHITECTURES

A *service architecture* is an arrangement of computing elements, control logic, data and communications protocols intended to support a service or category of services. In the service and application areas, we are concerned with the generic control of connections and the logic to created value-added services. The telecom service and IT applications areas shown in Figure 2.1 have received significant attention and standardisation effort in the quest for new services, multiservice architectures and the convergence of telecommunications and IT services.

Figure 2.2 shows the proliferation of approaches to service architectures and traces the evolution toward the goal of a converged service delivery platform. Two starting points are identified: the processor-controlled PSTN switch-based services and basic Internet services. Figure 2.2 shows initial silo-like development of the two streams with increasing crossover of development paths.

PSTN Era Service Evolution

In the PSTN area, the concept of the narrowband Integrated Service Digital Network was built on the capability to switch digitally encoded signals. The narrowband ISDN supported switch-based supplementary services (SS). The N-ISDN in turn gave rise to second generation mobile networks (GSM and IS-54). The classical Intelligent Network evolved to overcome the inflexibility of supplementary services in the ISDN by locating service logic on a Service Control Point. The Intelligent Network standards developed through two capability sets (CS-1 and CS-2) in the switched circuit network environment. Intelligent network standards were specialised for mobile networks in the

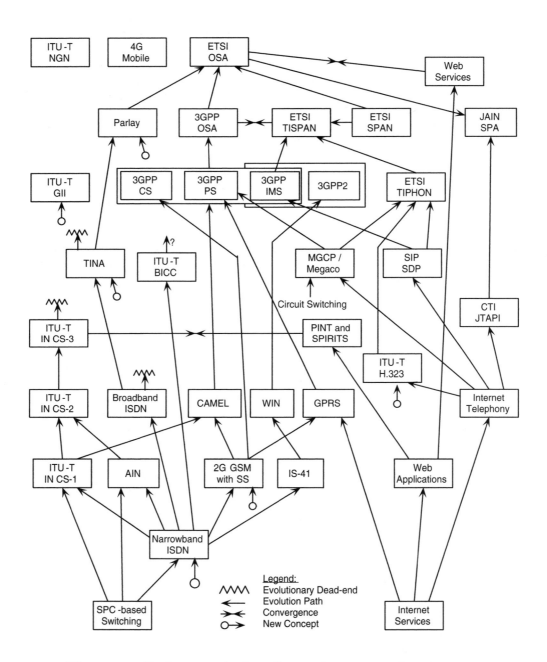

Figure 2.2. Evolution paths for call control, services and applications.

CAMEL and WIN standards. The N-ISDN concept was extended to the Broadband ISDN based on a QoS-enabled multiservice network using Asynchronous Transfer Mode. Further development of the IN capability sets did not progress beyond CS-3. Both B-ISDN and IN reached evolutionary dead ends.

The availability of the GPRS in mobile networks provides mobile phones with always-on packet data communications. While mobile phones previously had circuit-mode data connections, GPRS transformed the mobile phone into an Internet terminal. Thus, an important instance of convergence occurred: the mobile phone became a seamless access method to the Internet.

Building Services on Internet Standards

Building on basic Internet services, browser-based applications became widespread. Two initiatives sought to supplement phone services using Internet capabilities and Internet applications with phone services. The PINT standard [170] allows an Internet client to make a request to a PSTN Service Control Point, for example to initiate a voice call. PINT was seen as part of the suite of protocols for IN CS-3, opening up an Internet–telco convergence possibility. A complementary protocol called SPIRITS [103] aims to allow notifications to be sent from the telco Service Control Point to an Internet client, for example to provide missed call notification.

Internet telephony resulted from three developments: the ability to encode speech in packet form at low bit rates, call session control protocols and a special transport protocol, and the Real-time Transport Protocol (RTP). The last monitors and provides information to control real-time streams in packet networks. The H.323 multimedia communications suite started as a LAN-based real-time communication system but developed for use as a general packet telephony standard. The Session Initiation Protocol (SIP), with its companion Session Description Protocol (SDP), is a competing multimedia session control protocol. Several standards emerged for integration of computers and telephony, including JTAPI, which provide applications developers with an API that allows control of telephone calls from an application.

With the emergence of packet telephony standards, the need to interconnect with circuit-switched networks was satisfied by media gateways. Protocols developed for controlling media gateways are MGCP and Megaco, discussed in Chapter 4.

The ETSI TIPHON standards define a generic approach to packet telephony, allowing individual protocols such as members of the H.323 suite, SIP, RTP and Megaco to play defined roles in a unified architecture.

With the emergence of packet networks as potential bearer networks for telephone services, the ITU-T extended the widely used ISUP call signalling protocol to BICC, the Bearer-Independent Call Control standard.

Convergent Service Architectures

During the 1990s, the TINA initiative defined a technology-neutral architecture for the control and management of services and network connections, using the then new concepts of distributed object computing. TINA failed to gain acceptance for reasons given in Chapter 6.

The second generation mobile network concept, with GPRS as the packet-mode connection, was developed into the 3G mobile standards. A circuit-switched (CS) mode was standardised to allow access via 2G Base Station Subsystems but the envisaged transport network is packet switched (PS). Valued-added services based on CAMEL are included in the 3G standards for compatibility. The IP Multimedia System (IMS), that allows users to access applications using SIP signalling, is part of the 3G standards. While the 3GPP standards evolved out of the GSM 2G standards, the 3GPP2 initiative aimed directly at an *all-IP* set of standards, where all IP means using standards of the IETF.

TINA introduced the concept of a *3rd party service provider* that could, by using standard interfaces, offer services to supplement to those of the *retailer*, that is the party to whom the end user subscribes. With TINA's failure to gain acceptance, the Parlay initiative adopted an approach embodying simplified concepts from TINA to define an open API that could be used to give an application provider the ability to make use of network connection, messaging and data, such as mobile network databases. The 3G standards developed a similar concept called the Open Service Access (OSA). Parlay, OSA and the ETSI SPAN are harmonised and are now published as ETSI standards. In parallel, the JAIN initiative sought to define an open network interface for applications on the Java platform. JAIN adopted the Parlay definition for the open interface. A single interface is therefore standardised to allow applications, that may be developed and hosted in an IT environment, to invoke connections, messaging and other services of the telco network while protecting the network security and integrity. The OSA/Parlay concept and standards are described in Chapter 8.

In the Internet area, the concept of a Web service had become established. A Web Service is a software programme with a well-described interface that is published, and can be discovered and utilised by another software application. OSA/Parlay provides information technology with an important convergence opportunity with telecommunications by making services available as Web Services.

Other work in progress includes ETSI's TISPAN initiative, attempting to harmonise TIPHON and OSA/Parlay, the ITU-T NGN initiative and the 4G mobile initiative reviewed in Chapter 10.

2.2   DEALING WITH COMPLEXITY

Present and emerging networks are complex systems and form part of a complex evolutionary process. *Complex system* means one that is typically built from a large number of entities that are highly interactive [179].

The complexity of ICT systems lies partly in the physical architecture but more so in the control and management processes and protocols. As many of the entities are software processes, their individual behaviour may exhibit many responses to different combinations and sequences of inputs. The evolutionary process also exhibits characteristics of a complex system, particularly as there are many human and organisational roleplayers: decisions are not made on technological grounds alone.

It is argued that, for proper understanding, analysis and prediction, complex systems require a description that is at least as complex as the system [179]. Simplified descriptions do not allow the system to be understood or to predict its behaviour. This observation is true in systems with complex technology, dynamic change and

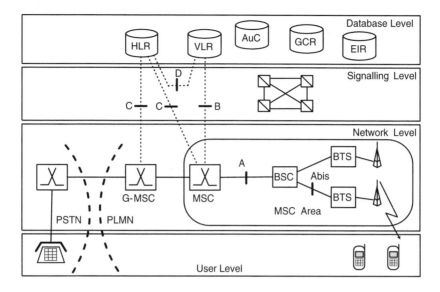

Figure 2.3. Use of levels for describing a GSM system. Adapted from [57].

human and organisational factors. The development of ICT systems has long adopted the principle of separation of a complex system into subsystems with defined interfaces. While the separations are not always complete, understanding, analysis and design of complex ICT systems is successfully pursued through a series of subsystems and reference points that can be adequately described. The framework presented in Section 2.3 provides high-level organising principles for the division of ICT systems into sub-systems and locates the reference points in a consistent way.

After dividing the system, two common techniques are used for managing complexity: modelling and abstraction. *Modelling* is a formal way of describing the system or object of interest that ensures a consistent description for a specific purpose. *Abstraction* hides unnecessary detail of the system or object of interest, exposing only the detail required for the purpose at hand.

Abstraction is used in the well-known OSI-reference model for data communications systems. The entire set of functions required to support two applications to communicate over a particular type of network is divided into seven groups. These groups are arranged in a sequence where the functions of one group serve only its neighbour. The groups are depicted as layers with each layer serving that immediately above it. An interface, called a *service access point*, is defined at layer n and its operations may be invoked by layer $n+1$. Communications protocol definitions provide a detailed description of the behaviour of a particular layer, usually in the form of message sequence charts but also supplemented by state transition diagrams or SDL diagrams of the internal logic. The implementation of the layer is thus *abstracted* and is described by means of a *model*.

A second example of separation of functions shown in Figure 2.3 is drawn from the GSM standards [57]. The levels shown demarcate concerns of databases, signalling,

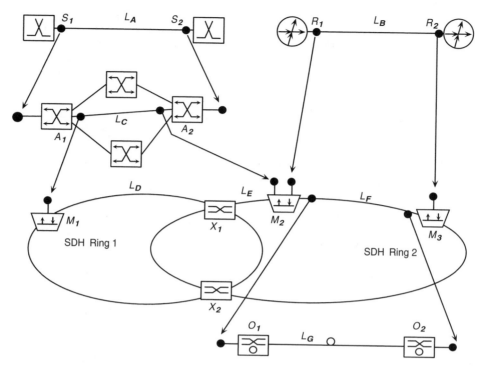

Figure 2.4. Illustrating the process of abstraction applied to physical components, namely links, in circuit- and packet-switched networks.

the network and user terminals. A number of reference points marked B, C and D are defined between the network and database levels. The reference points provide a means of interacting with the databases without knowledge of their detailed implementation.

A third example of abstraction is shown in Figure 2.4. Two constructs often found in network diagrams are shown. Two telephone switches S_1 and S_2 are connected by link L_A. In another network, routers R_1 and R_2 are connected by link L_B. The link construct (a line) is an abstraction for the detail of the underlying network.

Link L_A is implemented by a combination of ATM switches, SDH nodes, optical multiplexers and optic fibres. Link L_A is a virtual link created by means of an ATM switching network. The time-division multiplex frames are switched via link L_C. This link in turn is configured in two SDH rings. The TDM frame is multiplexed into the STM-n at add-drop multiplexer M_1, switched to the second SDH ring by crossconnect X_1, and extracted at add-drop multiplexer M_2. All this detail is hidden in link L_A.

The link L_B between routers R_1 and R_2 is implemented in SDH ring 2, using add-drop multiplexers M_2 and M_3 and link L_F. Link L_F is implemented as a wavelength on physical fibre link L_G, terminated in optical crossconnects O_1 and O_2.

The abstractions in Figure 2.4 hide detail of the physical elements that make up the links L_A and L_B. Where elements have software-based processes, functionality may also be abstracted. The switches S_1 and S_2 connected by link L_A are abstractions

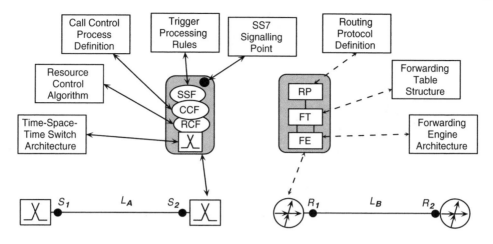

Figure 2.5. Illustrating the process of functional abstraction applied to physical components, namely switches and routers, in circuit and packet switched networks.

for the functions shown in Figure 2.5. The switch is expanded to show the switching matrix, the resource control function (RCF), the Call Control Function (CCF) and the Service Switching Function (SSF). The ability of the call process in the switch to communicate over Signalling System No. 7 is indicated by the presence of a Signalling Point.

Figure 2.5 also shows router R_1 expanded using the router model presented in Chapter 1.

2.3 A FRAMEWORK FOR MODELLING AND ABSTRACTING EVOLVING NETWORKS

Numerous modelling and abstraction methods are in use across the broad field of ICT. The range of systems to be described extends from largely content- and application-based to largely network-oriented. Systems span terminals, access and core networks, and operate under a number of business models. To promote a consistent approach across diverse systems in a range of NGN contexts, we seek a neutral framework onto which different models and abstractions can be inserted. The framework must accommodate emerging as well as legacy networks. The framework must satisfy the needs of a range of stakeholders and it should be possible to use an appropriate part of the framework within a given problem. We refer to resulting the framework as the *NGN Framework*.

The logic underlying the NGN Framework is shown in Figure 2.6. An ICT system is viewed as made up from building blocks that are structured to create the system The building blocks that constitute a system are classified as *physical entities* (PE) and *functional entities* (FE). These concepts are adapted from the corresponding elements in the IN Conceptual Model, described in Section 1.7. Physical and functional entities are different types of views on system constituents.

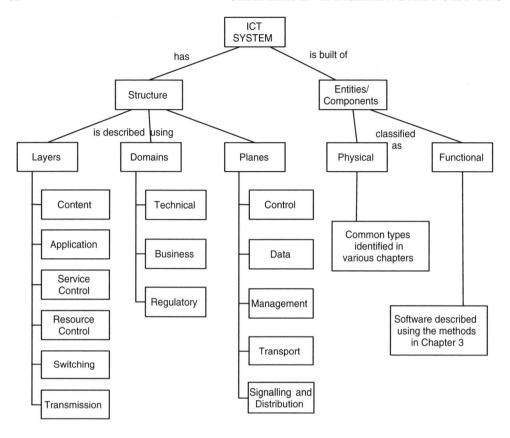

Figure 2.6. Classification of methods used for describing an ICT system.

Physical entities are generally stereotypes of physical building blocks, for example telephone switches, routers, links, multiplexers and crossconnects encountered in the example in Figure 2.4. Other physical entities are defined in later chapters. Information flows between physical entities as streams of user data and control and management messages. Interfaces between physical entities may be defined using concrete protocols, metaprotocols, implementation-independent formats, reference points or application programming interfaces (API).

Functional entities by contrast are abstract descriptions of functions performed, for example call control in a PSTN switch and constructing a forwarding table in an IP router. Functional entities encapsulate one or more related operations. In the IN Conceptual Model, FEs interact to control services. Communication is defined by *information flows*, that is, abstract descriptions of protocol requirements. In the NGN framework, FEs may be involved in control and management and their interactions may, as in the case of the PEs, be defined in abstract or concrete forms. Functional entities are abstractions for software processes, exposing only an external view of functionality. Only the identity, broad functions and interfaces or reference points are normally shown. The detailed internal processes are described by means of the methods presented in Chapter 3.

Mapping of functional entities onto physical entities is in general less restrictive than dictated in the INCM rules. For example in Figure 2.5, a telephone switch is recognised as a physical entity. The logic to set up and clear down calls is grouped in the Call Control Function entity. Information flows between Call Control FEs in different switches may be defined in abstract terms. Actual information flows between such physical entities use the ISUP signalling protocol. The unstandardised resource control functional entity is recognised but no information flows have been defined: these are internal to the switch. The switching matrix, controlled by the RCF, is concerned with user stream data and is not therefore represented as a functional entity.

The framework is concerned with the *structure* of the ICT system, that is the way the building blocks are arranged and grouped to reduce the complexity of the description. Figure 2.6 classifies the separation methods used to describe ICT systems: layers, domains and planes. These methods are not new but are adapted to the NGN context. For example, use of *layers* goes beyond that of the OSI-RM, to describe any hardware and software functional groups with client–server type relationships. Across each layer, a number of *domains* may be identified, corresponding to related functions or administrative responsibilities. Generally, but not always, peer-to-peer or user-to-provider relationships exist between domains within a layer. Concerns cut across layers and domains and are described using *planes*. In addition, business models may be mapped by assigning functions and activities of business entities to layers and domains in the overall model. Detailed specifications of layers, domains and planes follow.

2.3.1 LAYERING

Layering is a method of partitioning an ICT system that identifies separations of functionality into subsystem where any two adjacent sections have a client–server relationship. One subsystem, usually depicted as the lower layer, provides services that can be requested by the other, depicted as the upper layer. Each layer has characteristic functionality that is not generally found in another layer. Functional entities do not straddle layers while physical entities may do so. Inter-layer relationships are represented by reference points, applications programming interfaces or protocol service access points.

Layering Systems in Use

A number of layering schemes are in use, some specific, others general. Four schemes are shown in Figure 2.7. The OSI seven-layer model shown in (a) is a familiar means of abstracting the complexity of data communications protocols. While successful in that role, it does not address the complexities of service and management architectures and is not used in the NGN Framework layering. Layering in the TIPHON IP telephony architecture is shown in Figure 2.7(b) [60]. Control functions for setting up the call association (CC), specifying the end-to-end bearer connection (BC) requirements, controlling the media flows (MC) and the actual network connection are separated

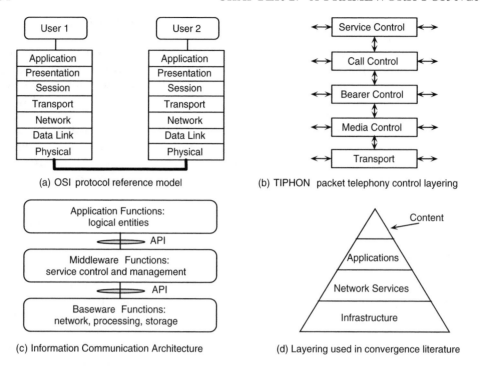

Figure 2.7. Four established methods of using layering to manage complexity.

into client–server relationships. The Service Control (SC) function serves the Call
Control function, departing from the graphical convention of showing the client layer
uppermost. Peer-to-peer interactions take place within each layer.

Figure 2.7(c) shows the layering in the ITU-T Global Information Infrastructure
(GII) Recommendation, intended to guide the development of information communication networks and services [129]. The GII attempts to provide a generalised framework
by identifying three principal layers. The uppermost layer is concerned with the
logical functions that create applications. These applications may call on network
connections and services such as messaging. The middle layer provides the method of
access for applications to invoke network functionality. The lowest layer represents the
underlying functionality: network connectivity, information processing and storage of
information. This form of layering conforms to the client–server definition. Using only
three layers provides too coarse a level of abstraction and too few reference points to
successfully divide most systems.

The four-layer model shown in Figure 2.7(d) arises in the literature concerned
with the content-oriented services and regulatory processes [25]. The *infrastructure
layer* represents the core switching, transmission and access required to transport
information between end-user and service provider premises. The *network services
layer* provides generic functions required to support communications. These functions
include setting up, modifying and clearing connections, mobility management, as well

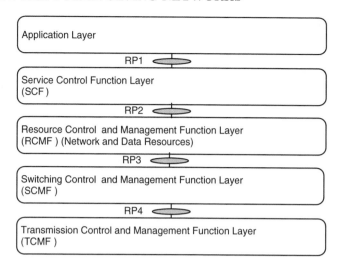

Figure 2.8. *Layers* used for partitioning ICT systems with reference points.

as circuit (TDM) and packet switches. The infrastructure layer provides facilities to support bearer connections as requested by the network services layer. The *application layer* is the locus of the value-adding logic. This logic may enhance a communications service or may support access to or enhance the value of content. The *content layer* is the locus of information owned by various parties and accessed using communications. While this layered model has the reach to describe a variety of ICT systems in telecommunications and broadcasting, the network services and infrastructure layers are too coarse for abstraction and modelling of NGNs.

Other layering systems are possible, for example that used in Figure 1.2. The access and terminal layers do not follow the client–server model. For example the access network may contain transmission and switching elements while the terminal may contain application and service logic.

Defining a Set of Layers

We require a layering system that is generic and has a sufficient number of abstraction layers to model a range of systems, including telecommunications, broadcasting and related information technology applications. The NGN Framework layers are developed from those in Figure 2.7(d) by splitting the network services and infrastructure layers and treating content as a resource. The layers must be chosen to provide the opportunity to define meaningful reference points. Network services in Figure 2.7(d) breaks down into a user-focussed service control layer and a network- and content-focussed resource control layer. The infrastructure layer breaks into two layers using the well-established client–server relationship between switching and transmission that exist in the infrastructure layer [192]. The layering system must accommodate important inter-layer interfaces already standardised, including the

OSA/Parlay API described in Chapter 8, as well as allowing further interfaces to be defined in future.

We present a set of five layers in Figure 2.8 giving a division of functionality that satisfies the requirements on the framework. Pairs of layers have client–server relationships with the lower layer serving the upper layer with one minor anomaly. Reference points such as RP1–RP4 shown in Figure 2.8 may be defined. The relationship between content and application layers may have different models: content serving application or application serving content. We view content as a serving role. The layers of the NGN Framework are defined as follows.

1. *Application Layer*: the locus of ICT application logic that may be in the telco or in an application service provider (ASP) domain.

2. *Service Control Functionality (SCF) Layer*: the locus of generic, stable and robust functionality to support realtime and information services using network connectivity. This layer is accessible to the Application layer through an open, secure API.

3. *Resource Control and Management Functionality (RCMF) Layer*: is the locus of functions that allow fulfillment of requests from the SCF layer for stream flows, messaging and access to network data in the transport layer. Functionality in this layer may relate to the *control* of resources for individual service instances or the broader *management* of resources. Resource control functionality (RCF) is a subset of the RCMF layer functions. Chapter 9 examines management functions and their relationship to control functions.

4. *Switching Control and Management Functionality (SCMF) Layer*: the part of the transport layer concerned with routing of flows at the packet level and making physical connection to resources and gateways. Aggregate flow, per flow and best effort mechanisms are in general supported. This layer may similarly have service instance related control functions or broader management functions.

5. *Transmission Control and Management Functionality (TCMF) Layer*: provides the means of carrying high volumes of packets as well as TDM streams between elements such as switches. Control and management of this layer does not take into account individual flows. Functions in this layer are predominantly management functions.

Content is not treated as a layer above the Application Layer, as suggested in Figure 2.7(d). The method of incorporating content in the layered model is introduced after detailed descriptions of the five layers.

The emphasis of the Application, Service Control and Resource Control layers falls on their respective functionality. The physical entities, usually servers, on which the functionality is implemented are often of secondary importance. In the Switching and Transmission layers, physical entities are often distinctive while functionality may be largely of a management nature. The following sections enlarge on the purpose and characteristics of individual layers shown in Figure 2.8.

Transmission Layer

The purpose of the *Transmission Control and Management Functionality Layer*, or simply the Transmission Layer, is to provide high-volume, high-speed transport of multiplexed user data, both time-division and packet multiplexed, between interfaces on switching elements such as PSTN switches and concentrators, routers, switches and media gateways. The TCMF layer also supports the provisioning of leased lines and unmanaged services such as dark fibre.

Typical physical entities found in the TCMF Layer include SDH add-drop multiplexers (ADM) and crossconnects (CC), dense wavelength division multiplex (DWDM) ADMs and CCs, and fibre level crossconnects. Physical elements such as crossconnects are described by information models, object-oriented definitions of the data that specifies the configuration and state of the entity.

Functionality in the TCMF Layer is largely management-related; on-demand connections are a rarity in existing networks. Typical functions are fault, configuration, accounting, performance and security (FCAPS) management [111]. Switching control concepts including Generalised MPLS are of increasing interest to the control of TCMF layer elements and networks.

The TCMF is frequently divided into sublayers, for example an electrical multiplexing layer and an optical multiplexing layer.

As in the case of other layers, the elements in the TCMF may present a vertical interface to the switching layer and horizontal interfaces to network management systems and to peer networks.

Switching Layer

The *Switching Control and Management Functionality Layer*, or simply the Switching Layer, is concerned with making connections in a connection-oriented switching network, transfer of packets in a connectionless network, making circuit switch connections, and the specialised control of connections to media processing elements such as media gateways. Networks may be structured as subnetworks and tandem networks. Such packet networks may be controlled and managed using per class aggregate flow or per flow, or may have private virtual circuit connections provisioned.

The PSTN switch and IP router shown in Figure 2.9 are physical entities present in the SCMF layer. Other physical entities include MPLS switch-routers, ATM and Frame Relay switches and connections to the media processing functions of media gateways and interactive voice response units. Physical entities found in the switching layer straddle higher layers as well.

Functional entities include some recognised in standards, such as the PSTN switch CCF, while others such as the resource control function (RCF) reflected in Figure 2.9 are not standardised but rather emphasise the need for this function. The detailed forwarding function in the router may, if required, be captured in a functional entity. Management-oriented functionality in the FCAPS areas may be present.

The SCMF may present an interface to the RCF layer for exchange of control information. Horizontal interfaces to management systems are important. Interfaces may require adaptation to the internal control functions of the switch.

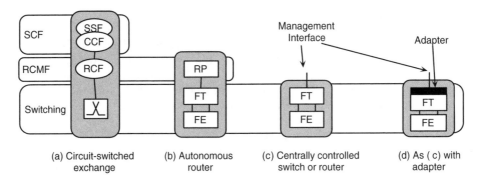

Figure 2.9. Illustrating aspects of the definitions of SCF, RCMF and Switching layers.

The restriction of FEs to a layer while PEs may span layers is illustrated in Figure 2.9. In the case of the circuit-oriented switch shown in Figure 2.9(a) the SSF and CCF are classified into the SCF Layer while the resource control function (RCF), concerned with the allocation of time slots, is not separable from the switching matrix in proprietary exchange implementations. In the case of an IP router shown in Figure 2.9(b), the routing protocol falls into the Resource Control Layer while the forwarding table and the forwarding engine belong in the Switching Layer. It is generally possible to implement centralised forwarding control by writing to the forwarding table of switches and routers using a management interface as shown in Figure 2.9 (c) and (d). In this mode, the management interface is presented to the Resource Control Function Layer with or without an adapter. The SCMF may thus contain OSS-related functions described in Chapter 9. The SCMF, in the ideal case, presents a technology-independent interface to the RCMF layer. Element adapters are in general required to interface the SCMF to actual network elements. The switching layer should have an API for control and management of network resources to meet service needs.

Resource Control and Management Layer

The *Resource Control and Management Functionality Layer* expresses the need for a locus of functionality that deals with the detailed methods of accessing connection, messaging and network data resources. Several types of resources are of significance: time slots in circuit-oriented elements, bandwidth in packet-switched elements, addresses, and network information such as user profiles and locations. The layer also handles inter-domain interactions. Typical functions include connection admission control and gateway-based routing. An important function of the RCMF layer is finding routes in switching networks to meet QoS requirements. Types of connections range from real-time streams to file transfer and messaging.

Physical entities implementing RCMF layer functions are essentially servers or are parts of entities that span this and other layers, for example a media gateway.

The control of the gateway lies in this layer while that actual media processing and connections are in the Switching Layer.

Specific functions that are placed in the RCMF layer are the Media Gateway Control Function and the control of interactive voice response units.

Few standards expose a resource control interface to support the Services layer other than [35, 97]. The network interface is often left as an implementation issue. A long sought after goal is a technology-independent API to provide a standard way for the SCF Layer to access network resources.

Service Control Function Layer

The *Service Control Functionality Layer*, or simply the Services layer, contains functionality that is generic to a number of services and applications including call and session control for end users and applications, messaging support for applications and secure access to network data. These services may be invoked by network access signalling or by an application making a call on the upper interface of the SCF layer. These services are stable in that their logic does not evolve, and they are robust, predictable implementations.

As in the case of the RCMF layer, the SCF functions are essentially distributed computing processes and are implemented on servers. Stereotype PEs are therefore uncommon. An exception is the legacy PSTN switch where the switch spans the SCF, RCMF and SCMF layers, as shown in Figure 2.9(a). The legacy PSTN CCF is an example of an SCF layer functional entity.

The concept of the SCF layer is drawn from the Open Service Access (OSA)/Parlay architecture [66]. A typifying set of SCF functions that suggest functional entities is defined in the OSA/Parlay standards. Functions include multi-party, multimedia call control with conference support, data session control, mobility management, messaging and user interaction control. The generic nature of the SCF supports a range of possible applications. This functionality is, in the ideal case, independent of the transport network. The SCF layer presents an interface to serve the Application layer. In the best scenario, the interface is an open standard application programming interface (API) such as OSA/Parlay. Call control functionality in other architectures, for example the H.323 Gatekeeper, is mapped into this layer.

The upper interface of the service layer in switched circuit networks is the Service Switching Function INAP interface. In managed IP voice networks, the interface is a soft service switching point (SoftSSP) enabling call-related events in IP telephony to trigger application logic. The SCF layer relies on a number of network databases. The model envisages that these databases interfaces are located at the upper edge of the Resource Control and Management layer.

Application Layer

The *Application Layer* is the locus of computing applications that may be enhanced by network services including call control, messaging and access to network data. Applications may support the enhancement and distribution of content and provide access to content for network end users. IN-style applications may enhance basic communications. Applications can be provided by a variety of parties for several purposes.

Table 2.1. NGN Framework layers and sub-layers, with examples

Layer	Sub-layer and Function	Example
Content	Archive	Image/video library
	Interactive	Web
	Streaming	Video-on-demand
Application	User AAA	Radius, Diameter
	Application logic	Enterprise application
	Reusable logic	EJB, CORBA Beans
Service	User AAA	TINA Access Session
Control	Application AAA	Parlay Framework
	Feature server	IN Service Control Point
	Call/session	TIPHON CC
	Data access	Parlay Mobility SCF
Resource	Quality of service	RSVP, DiffServ
Control	Endpoint addressing	DNS, DHCP, ARP
	Bearer control	TIPHON BC
	Media control	TIPHON MC,
	Gateway location	
	Routing: autonomous	IP OSPF, BGP-4
	Routing: central	PNNI
	Network state aggregation	
	Connection admission control	
Switching	Flow-based routing	IP
Control	Aggregate flow routing	MPLS
	Packet filter	Firewall
Transmission	Electrical multiplex	SDH
Control	Wavelength multiplex	DWDM
	Fibre switching	

Such application servers may be in the telco domain or in a trusted external application service provider (ASP) domain. The nature of applications also varies. An application may for example be invoked in the Internet domain and result in a multiparty conference being set up using the public telephone network. An application may be invoked as a result of an event occurring in the network. For example, the call control logic in the SCF layer detects that the number dialled requires special processing and invokes an application inside the secure telco domain that contains the necessary logic. An application may support access to content, implement value-adding processes and deliver the content over a communication network.

The application layer contains computing processes and therefore has servers as the main type of PE. Several types of servers are distinguished. An *application server* (AS) hosts logic implementing a variety of applications that may be invoked directly by a user or as a result of a network event. Applications are software-engineered for high availability and dependability. Applications execute in a container

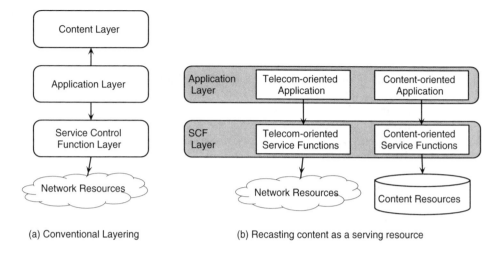

Figure 2.10. Recasting content as a serving resource.

such as the JAIN Service Logic Execution Environment (JSLEE), providing lifecycle support and delivery of notifications [164]. A *feature server* (FS) is a special case of an application server that implements functions similar to switch-based services or classical IN functions [151] invoked in response to events occurring in the network, for example address translation and call screening. A *Web service provider* (WSP) is an application server hosting services that can be discovered and invoked by another computer using the Web Services model [101]. External invocations are simple and the complex logic to implement the application executes in the application server.

Because of the diversity of providers and applications, identification of generally accepted functional entities has not taken place. Rather, several software reuse paradigms have been pursued, ranging from the SIB methodology in classical IN, Enterprise Java Beans, CORBA Beans, and Web Services. Over time, some may gain recognition as functional entities.

Content in the Layered Model

Any *content layer* is potentially diverse due to the wide scope of electronic media, based on the ability to encode, store and process media in digital formats. Content includes information of all types: voice, image, video, text and general data. The information is encoded in digital form, is indexed, may be real-time or stored, and is capable of transformation into other formats and being composed for delivery.

The layers of the NGN Framework shown in Figure 2.8 have client–server relationships: in each adjacent pair of layers, the lower layer serves the upper. The client–server model implied in Figure 2.7(d) between the content and application Layers is not meaningful. Digitally encoded content is passive data that requires an external agency to process and provide to an end user. Each type of content needs particular methods to be applied. Prompting a telephone user to make a choice by

pressing a digit requires message selection, playout and capturing the DTMF digit as
well as procedures for misuse, for example using timeouts and repeating the prompt.
Similarly, a video on demand service requires access control and VCR-type control
over streaming the video signal. Content is therefore considered to be a resource
that requires the control of an application, supported if necessary by content-oriented
service control functionality.

Content is regarded as a serving resource, similar to a transport network being
a serving resource. Thus, rather than using the layering suggested in convergence
literature shown in Figure 2.10(a), content resources are aligned with network resources
in Figure 2.10(b). The NGN Framework thus accommodates telecommunications-
oriented applications, for example GUI-based control of a multiparty conference,
similarly to a content-oriented application, for example downloading ringtones to
mobile telephones or an image library.

The broadcasting and news media industries have a strong content focus. Similarly,
the Internet has been the main means of accessing public-domain content. With
convergence, significant opportunities exist for developing content-related applications
that exploit communications and interactivity. The telecommunications-oriented
notions of physical and functional entities are new to the field of content production,
processing, storage and dissemination but are essential to describing converged
facilities.

2.3.2 DOMAINS

A domain is part of an ICT system demarcated to contain similar technical functions,
business interests or regulatory concern. We define a *functional domain* as one in
which a related set of distinctive technical functions is performed in one or more layer.
The functions may be technical, business or regulatory. Business, administrative and
regulatory domains in general span one or more functional domain. A domain may
cut across some or all of the layers. For example, the business domain of a full service
telco could span all layers, while an application service provider's business domain is
restricted to the application layer. Figure 2.11 shows functional domains as vertical
divisions. Note that a particular domain may not span all layers. For example, the
access network domain may straddle only the RCF, SCMF and TCMF layers.

Technical Functional Domains

Five types of *technical functional domains* (TFD) are identified below; most of the
domains are commonly used in describing telecommunications systems. The *Customer
Premises* TFD accommodates terminals and, if present, customer premises networks
which must interwork with the telco network infrastructure via an access network.

The *Access Network* TFD represents the circuit- or packet-mode transport from
the CPE to the edge of the network, supported by transmission and resource control.
Routing in the access network normally is based on layer 2 addresses and therefore has
local significance. For example, in an access network based on Ethernet, network-wide
addresses are resolved into Ethernet MAC addresses.

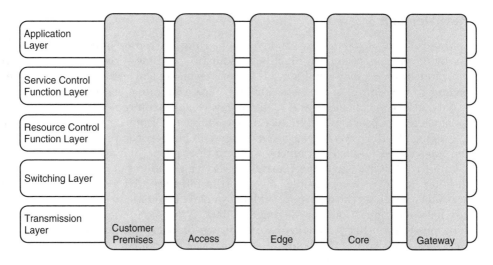

Figure 2.11. *Layers* and *functional domains* used for partitioning resources and functional areas of facilities.

The *Edge* TFD is the point at which network-wide switching or routing begins and the user accesses network services. In the PSTN case, we regard the end exchange as an edge element. In a packet network the edge elements could be paired IP routers.

The *Core Network* TFD provides edge-to-edge transport, using one of a number of switching/routing paradigms. The core functional domain is not restricted to the Transmission and Switching layers. For example, service control functionality within a telco domain is a core domain function. In fixed networks, core network functions have originating, transit and terminating roles. Mobile networks distinguish additional network roles including home network, visited network and serving network.

The *Gateway* TFD accommodates elements between two networks performing a transport, signalling or media adaptation function. A gateway may have admission control and may generate accounting information. Inter-network gateways appear between core networks, for example the Gateway MSC in a GSM network. A media gateway (MG) adapts the bearer traffic between switched circuit and packet networks and maintains connections between packet- and circuit-mode bearers. A signalling gateway adapts control and management messages from one protocol stack to another. Service Control Points are joined by an interworking function. A residential gateway (RG) may occur between the CPE and the access network. In general, an access gateway occurs where an access network meets the Edge.

Technical functional domains are often grouped into *administrative domains*. For example a telephone connection may pass through an originating network, one or more transit networks and a terminating network. Each network is the responsibility of a particular telco and is referred to as an administrative domain.

Business Domains

A *business functional domain* is usually mapped onto a representation of the ICT system by layer and technical functional domain. Business domains have been vertically integrated as shown in Figure 1.2. Both technological and regulatory changes are leading to new forms of business domains. For example, a service provider may invest only in the Application and Content layers and acquire connectivity from a traditional telco to provide IT-type services that benefit from network connectivity and messaging. With convergence, the traditional divide between telecommunications, information services and entertainment starts to disappear. Similarly, the end-to-end paradigm of the Internet and the centralised service paradigm of telco networks are not the only models. For example, opening the interface of the SCF layer to the Application layer allows third party providers from an IT domain to create applications that use network connections, messaging and data.

Many possible business domains can be drawn on a grid of layers and technical domains. In most cases, the boundaries of a business domain will correspond to inter-layer and inter-domain boundaries. The Edge and Gateway domains ensure that physical entities do not straddle business domain boundaries.

Regulatory Domains

We define a *regulatory domain* as the area in which a regulator applies a particular set of policies and regulations. Previously, these domains were readily demarcated by the silo model shown in Figure 1.2. A broadcast regulator treated the broadcaster and the signal distributor separately. A telecommunications regulator may have treated fixed and mobile networks separately. In general, new types of regulatory domains can be drawn on the backdrop of the layer-technical domain grid using the interlayer and inter-technical domain boundaries.

2.3.3 PLANES

Layers and domains give an orthogonal mapping of functional and physical entities in the NGN Framework. A physical entity may straddle layers and is generally confined to a technical domain. Functional entities are restricted to one layer and one technical domain. Several concerns exist that involve functional entities acting in concert across layers and technical domain boundaries. Operations support and business support, loosely termed management, falls into this category. Similarly, an application may request a connection with a required quality of service. Fulfilling this request requires entities from the Application to Switching layers to work in concert. Thus, QoS considerations cut across layers and domains.

The concept of a *plane* is used to capture a crosscutting concern. A plane is defined by selecting entities that relate to the particular concern from the two dimensional field of layers and technical domains. The term plane is used in the ISDN standards in the form shown in Figure 2.12 to group concerns of a particular kind. In ISDN, connection control and the transfer of data user-to-user are the task of separate processes each communicating over its own protocol stack. The control application and its supporting protocol stack are termed the *Control Plane*. Similarly, the user application and its information transfer protocol stack form the *User Plane*.

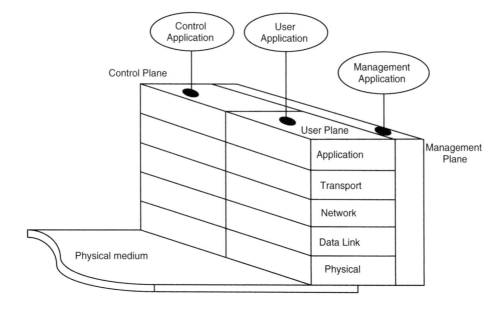

Figure 2.12. Protocol stacks showing ISDN-style Control, User and Management Planes.

The Operations, Administration and Maintenance System (OAM) in ISDN must interact with all protocol layers to ensure proper operation. The *Management Plane* is depicted as cutting across all layers of both protocol stacks.

The IN Conceptual Model shown in Figure 1.7 uses four planes to separate concerns. For example, the Global Functional Plane is concerned with reusable software elements that can be chained together to create the service logic and how that logic interacts with the Basic Call Process. The Distributed Functional Plane is concerned with functional entities while the Physical Plane defines the physical entities and the application layer protocol.

The NGN is in general complex and the ISDN and IN notions of planes are limiting. We therefore adapt their definitions of planes by first redefining planes such as control, user and management and then regarding the list of planes as open ended. For example, a QoS plane could be proposed.

The following definitions of principal planes separates the invocations between control and management processes, and the underlying mechanism for transporting the invocations. For example OSI application layer invocations or method calls on an interface are distinguished from the underlying mechanism, for example a TCP/IP stack or a CORBA distributed processing environment.

Planes likely to be of use in the NGN are: control plane, data plane, management plane, and the signalling and distribution plane. Other planes may be used when needed. For example, a packet communications systems is depicted in [108] using transport, access and applications planes.

Control Plane

The *Control Plane* reflects all actions required to initiate, control and terminate calls and services. These actions are viewed in the NGN Framework at the application level. For example, the exchange of ISUP and INAP messages shown in Figure 1.11 required to set up a freephone call is a control plane process. Reference points and interfaces defined in API form belong to the control plane. Unlike ISDN, the control plane does not concern itself with the underlying message transfer. Information on how the ISUP and INAP messages are transported or how API method calls and responses are passed becomes a signalling and distribution plane concern.

Data Plane

The *Data Plane* encompasses all data resources needed to deliver a service. The data level of the GSM model shown in Figure 2.3 is a data plane. Data supports both control of services and management of resources. In the NGN, each layer and domain requires characteristic data. For example, at the switching layer, forwarding tables in switches and routers form an important data resource. Subscriber profiles are important data for the SCF layer.

Management Plane

Management is regarded as a local operation in the ISDN model but is a global concern in general. The *Management Plane* encompasses all concerns of business and operations support. For example, in Figure 2.1, management concerns range from the configuration and monitoring of core network facilities to care of customers. Management concerns cut across layers. The management plane and its relationship to other planes are explored in Chapter 9.

Signalling and Distribution Plane

The *Signalling and Distribution Plane* (S&DP) reflects the fact that any ICT system is essentially distributed across space and is controlled and managed by processes hosted on different computing nodes communicating over a network. Signalling is not allocated to an explicit layer: signalling cuts across layers.

The signalling and distribution plane reflects the mechanisms used to transport service- and management-related operations, responses and notifications. The plane accommodates communications based on protocol stacks as well as distributed computing using mechanisms such as CORBA. With this definition, the Signalling and Distribution Plane in a PSTN is the SS7 network. To answer the question of how ISUP and INAP messages shown in Figure 1.11 are transported, the S&DP contains the detailed protocol stacks shown in Figure 1.6. In a CORBA-based system the S&DP is concerned with the ORB, its services and the underlying transport protocols.

The definition of APIs at various inter-layer interfaces defines *application layer signalling*. Support for application layer signalling takes on many forms: OSI protocols, SS7 and distributed object technology.

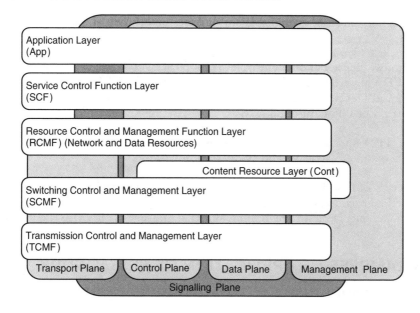

Figure 2.13. *Layers* and *planes* used for partitioning resources and functions.

Other Planes

The planes defined above are those found in common use in telecommunications standards. Other planes may prove useful. For example, all QoS issues may be represented on a plane spanning from the Application to the Transport layers. Similarly, usage accounting may gather data in the transport network, add fields to the data record in the service layer before finally storing the record for off-line processing. Thus, billing and accounting may be a useful plane. Similarly, the concerns of the ISDN User Plane shown in Figure 2.12 may be better expressed by a plane concerned with transporting streamed data.

2.3.4 SUMMARY OF THE NGN FRAMEWORK

Figures 2.8, 2.11 and 2.13 show the various organising elements of the NGN Framework: layers, domains and planes used in combination. We summarise the NGN Framework as follows:

1. The five layers provide the basic horizontal organisation of the framework. Layers have a client (upper)–server (lower) relationship. It is recommended that the layering scheme be followed with possible variations. Sub-layers may be introduced provided that they also have a client–server relationship. Only those layers relevant to a particular ICT system need be used. Layers may be merged if there is no benefit in exposing the interlayer interface.

2. Physical entities represent the typical physical building blocks of an ICT system. A PE may straddle layers.

3. Functional entities represent grouped functionality that resides within a single layer.

4. Technical functional domains reflect groups of related functionality across layers. A TFD may cross one or more layer. The choice of TFDs, CPE, Access, Edge, Core and Gateway, avoids physical and functional entities falling on a business or regulatory domain boundary.

5. Inter-TFD and inter-layer boundaries are useful in drawing the boundaries of business, administrative and regulatory domains.

6. Selection of functional entities across layers and technical domains that relate to a particular concern is a plane. The following planes are recommended: Control, Management, Data and Signalling and Distribution. Otherwise the list of planes is open-ended.

2.4 EXAMPLES OF APPLICATION OF FRAMEWORK

The following examples serve to illustrate the use of the NGN Framework and to demonstrate that it covers many existing and emerging architectures and is relevant to problems of convergence.

2.4.1 LEGACY NETWORKS ELEMENTS IN THE FRAMEWORK

While the NGN Framework is intended to support next generation networks, the ability to represent legacy networks supports studies of the interworking of new and legacy networks. In this example, we illustrate how the NGN Framework layers accommodate key elements from the public switched telecommunications network with an Intelligent Network overlay.

The mapping of legacy network PEs, namely the PSTN switch and the Intelligent Network Service Control Point and Specialised Resource Peripheral, onto the NGN Framework is shown in Figure 2.14. The exchange straddles three layers. The TDM switch falls in the Switching layer.

Functional entities occupy one layer only. The Resource Control Function Layer is not explicit in the switch construction. The implementation-dependent resource control is shown as a (non-standard) functional entity marked RCF. The RCF allocates timeslots from a pool of free timeslots on the circuit group leading to the adjacent exchange. The standards-based Call Control Function (CCF) and Service Switching Function (SSF) fall into the SCF layer. The IN Service Control Function (SCF), as the host for programmable logic, and its associated Service Data Function (SDF) are placed in the Application Layer. The exchange periphery is shown straddling the switching layer and the SCF layer. The subscriber line interfaces (SLI) generally terminate on a local switching function that allocates time slots to active interfaces. The Call Control Agent Function (CCAF), recognised in the ISDN standards, contains end-user interaction functions.

The Specialised Resource Peripheral contains physical elements for terminating bearer connections, receiving DTMF digits and playing announcements. These elements are allocated to the switching layer. The logic required to control these

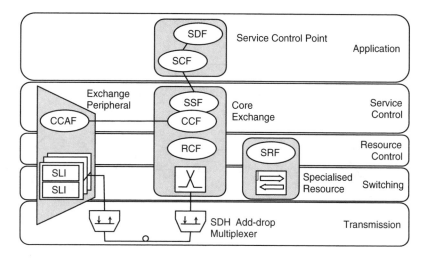

Figure 2.14. Mapping of legacy PSTN/IN physical and functional entities onto NGN Framework.

resources is represented by the Specialised Resource Function (SRF) in the resource control layer.

We also show a possible transmission layer configuration. The exchange periphery is assumed to be located remotely and requires a TDM transmission system link. The link is formed by add–drop multiplexing onto a SDH frame.

This example illustrates that the framework accommodates the principal building blocks of the legacy switched circuit network.

2.4.2 FROM CIRCUIT SWITCH TO SOFTSWITCH

A network becomes intelligent when service control is separated from switching. The pre-IN telephone network has call and service control locked to the switching function in the same physical entity, as shown in Figure 2.14. The addition of the external Service Control Point gives the PSTN programmable intelligence. As networks move toward packet-based transport of real time signals, the meaning of *connection* changes to indicate the logical association between call parties. Connection control is divorced from the switches and is located on a server, known as a *softswitch*. We use the framework to examine the changes that take place in the transition from traditional circuit switching to packet switching with softswitch control.

Figure 2.15 illustrates the effect of the call control process being linked intimately with the switching function in a circuit-switched networks. We consider a call that must be routed across three networks and show part of the signalling sequence. The Call Control Function of every switch on the path is involved in signalling to route the call through predefined switched that have free time-slots. After the initial request from an ISDN phone (Q.931 Setup message), the first CCF issues an Initial Address Message (IAM), corresponding to message 15 in Figure 1.11. Provided that each

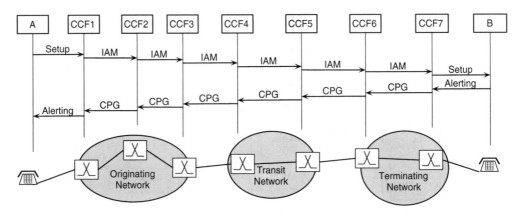

Figure 2.15. Illustrating the effect of locking call control to switching node in a circuit-switched network.

adjacent switch can route the call, the IAM is forwarded to the next exchange. When the IAM arrives at the terminating exchange, the B-party, also assumed to be an ISDN phone, is informed of the incoming call (Setup). On receiving an Alerting message from the terminal, an ISUP Call Progress Message (CPG) is passed switch-by-switch back to the originating exchange. The physical switchpath is connected at each switch.

   This example shows how call control signalling involves every switch along the route. We now contrast the corresponding call setup operation in a packet network, using a hypothetical protocol. We generalise such networks in Figure 2.16 to establish the main features. For convenience, we consider a voice network.

   The principal components of a packet voice network are shown in Figure 2.16 on a background of three NGN Framework layers. The access networks are not shown. The call is routed through three networks in three administrative domains.

   Each core network provides packet transport with QoS for speech traffic. The nature of the packet network is such that a packet containing a segment of encoded speech, once offered to the first router, finds its way through the network according to the routing protocol used. Unlike the switched circuit case, it is not necessary to set up the route on a router-by-router basis for every call. Routes determined by the routing protocol or configured by a management system are used.

   In a switched circuit network, call control involves both maintaining the logical association between the call parties and setting up the dedicated channel between the two users. In a packet telephony network, call control is concerned with the association of the users. Control of the call is vested in an element called a *call manager* (CM), a server hosting the call control programme. Since the call control functions do not include the detailed routing of packets, the call manager functions include admission of the user, setting up and controlling the association, invoking service features and generating billing information. In general, each network has a call manager.

   The call manager may have inbuilt IN-like service features or have access to a *feature server* (FS) as shown.

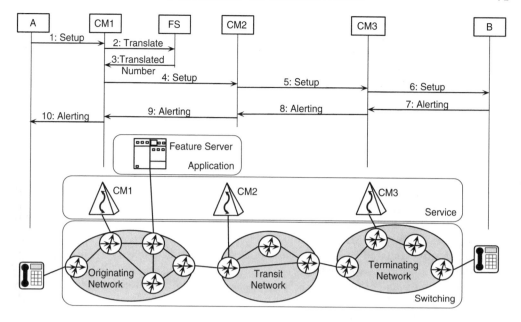

Figure 2.16. Separation of softswitch-based call control signalling and media transfer in packet network.

Other functions not shown in Figure 2.16 but considered in Chapter 4 are packet-level admission control and the negotiation of media between the parties.

Two types of call control signalling are required in the packet telephony environment: access signalling between the terminal and the call manager; and call manager-to-call manager Signalling. In this example we use Q.931 messages from the H.323 standard described in Chapter 4. The initial call setup signalling is as follows:

1: The terminal A, an IP phone, requests the call manager (CM1) of the originating network to set up the call. The called party number is contained in the Setup message.

2, 3: The call manager needs to have the address translated and invokes the feature server (FS).

4: Call manager 1 determines that the call must be routed via a transit network and sends a Setup message to its call manager, CM2.

5: CM2 determines that the call can be admitted to its network and forwards the Setup to the call manager, CM3, of the terminating network.

6, 7: CM3 informs the called party of the incoming call using a Setup message; the called party accepts the invitation by returning an Alerting message.

8–10: Notification that the B-party phone is ringing is returned by the Alerting messages.

Once the call has been set up and answered at the terminating side, each terminal sends real-time streams as required to the other party without the involvement of the call manager. The transport mechanisms in the access and core networks deal with the streaming. The path followed by the speech packets depends on the routing protocols used in the packet networks and does not follow the routing of messages via the call managers.

2.4.3 CONCEPTUALISING CONVERGENCE IN LAYERED MODEL

Figure 2.1 shows a pre-convergence situation. The number of possible terminals has proliferated: plain phones, mobile IP phones, PDAs, computers. Similarly, the range of access network technologies has increased. Some support access to multiple services while others do not. Several types of switching network are available: circuit switched, best effort IP, MPLS, Frame Relay and ATM. Service control mechanisms have proliferated beyond the PSTN call control. Numerous software technologies are available for developing applications.

Services and applications is an area of considerable potential convergence. Legacy service control in switched circuit networks is based in the switches and in external IN platforms. The telephony network is a *closed network*: only the telco can offer services. The end-to-end model of the Internet is open since all intelligence resides in end stations and the network provides only packet transport. Anyone can provide services. Open network or Open Service Access (OSA) models for allowing applications to access network functionality are embodied in standards such as OSA/Parlay [159] and JAIN [30]. Similarly, the Parlay X standard [101] seeks to allow the creation of Web Services that can invoke well-defined communications functions. These developments create a convergence between IT applications, Web services and telecommunications networks.

Figure 2.17 shows the effect of the open service interface in creating a point of integration (1) at which diverse applications, implemented on a number of platforms, access generic services in a standardised way.

The service layer functionality is not bound to particular transport entities or physical location such as an exchange building (central office) and is in general, a distributed computing application. Technological elements of the service layer divide into two sub-layers. First, switching and routing equipment provides per call connections for services such as telephony. Per-packet routing is provided for data services. Second, service control includes control of calls, closely linked to the time-division multiplex switching function in circuit-switched telephony and decoupled in managed packet voice services. Present day voice value added services such as number translation (freephone, call forwarding) and flexible charging are regarded as part of the network service functions.

The service layer is the natural home of ISP and VPN providers. Virtual private networks (VPN) provide managed IP networks for enterprises spanning a number of remote sites. A key building block of both the VPN and the ISP network is the IP router. ISPs are therefore in a position to be VPN service providers.

The second point of integration shown at (2) in Figure 2.17 is intended to allow generic service facilities to control heterogeneous networks – a goal still to be attained.

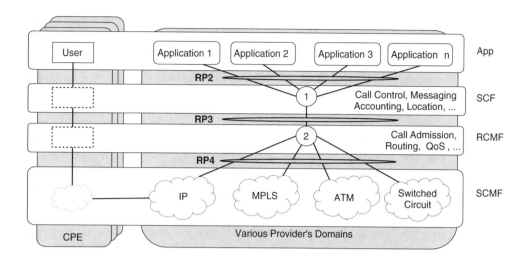

Figure 2.17. Points of integration and diversity of access networks.

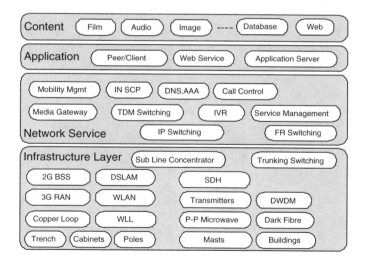

Figure 2.18. Horizontal model for convergent environment with typical elements assigned to layers.

2.4.4   DRAWING THE BOUNDARY OF THE REGULATED DOMAINS

Regulators are faced with the task of drawing the boundaries of regulated domains. It is generally accepted that regulation is necessary for a limited number of purposes. One of these is to regulate the activities of operators who require facilities that cross public or private land, other than the customer premises.

An important distinction in modern telecommunications regulation is between infrastructure and other facilities as shown in Figure 2.7. The upper boundary of the infrastructure layer is chosen to align with a possible boundary between horizontal licensing domains. The infrastructure of a communications provider is any facility that exists on, under or over land, water or buildings that are not the premises of the provider for the purposes of providing communication between the premises of customers or providers. Technological elements includes in infrastructure therefore include works such as trenches, ducts, cabinets, masts, optic fibre cables, copper cables, microwave links, radio base stations, optical multiplexers, SDH multiplexers and switching required for trunking.

This argument seems to draw the boundary above the transmission domain. A number of reasons indicate that the switching layer should also be included. Regulators must enforce interconnection agreements between operators; interconnection involves individual calls or sessions and therefore the switching layer. Constructing the interconnection network may be inefficient if switching is excluded.

2.4.5   DIGITAL VIDEO BROADCASTING IN THE FRAMEWORK

Legacy television and sound broadcasting systems have data capacity in the downlink direction in addition to the normal broadcast channel. Services have been built on this capability: the Radio Data Service in FM radio systems and text-based information systems using spare lines in the television frame. In the latter case, the return channel for interactive services was provided via the public telephone network using a dial-up modem.

The ability to encode television images offers several opportunities in addition to stable, quality reception. At one extreme, the distribution of high-quality television signals over terrestrial broadcasting networks in addition to cable and satellite systems becomes possible. By contrast, digitally encoded standard definition television signals require less channel bandwidth, allowing transmission over telecommunication systems to fixed and mobile terminals. Similarly, effective distribution of CD-quality audio is possible.

Digital Video Broadcasting (DVB) encompasses standards for encoding, multiplexing and distributing digitally encoded signals. Several distribution modes are described in various standards: satellite (with master antenna and cable distribution), terrestrial radio, cable and multichannel multipoint distribution systems (MMDS). Digital video broadcast distribution networks offer significant broadcast downlink capacity in the undirectional *broadcast channels*. The DVB standards specify several methods of providing bidirection interaction channels that can be used with the downlink channel to create services.

The architectural reference model of a DVB system [180] is shown in Figure 2.19 mapped onto the layers of the NGN Framework. The reference model is recast using

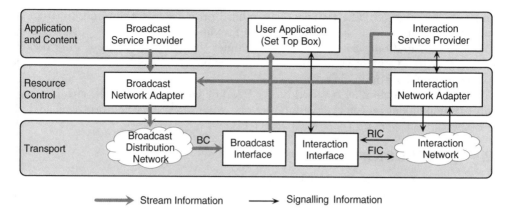

Figure 2.19. Reference model for Digital Video Broadcast systems overlayed on the NGN Framework. The interaction channel is carried a separate network.

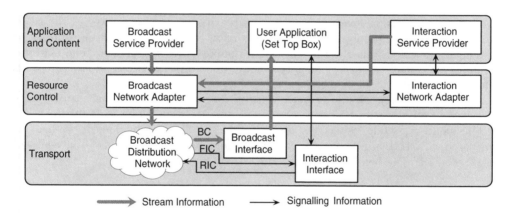

Figure 2.20. Reference model for Digital Video Broadcast systems overlayed on the NGN Framework. The interaction channel is carried in the same network as the broadcast channel.

the layer and domains of the NGN Framework. The broadcast service and interactive service provider roles are shown as distinct applications.

The interaction channel in the reference architecture consists of a forward interaction path and a return interaction path. The interaction channel may be implemented in a standardised way with the aid of an existing telecommunications network: PSTN, ISDN, GSM or DECT. Either the forward or both interaction channels may be implemented via the broadcasting distribution system. The satellite system standards provide forward and return interaction paths by satellite. Similarly, DVB based on Local Microwave Distributions Systems (LMDS) provides a return point-to-point channel. The DVB cable standard provides upstream and downstream interactive channels. The interaction service provider application may insert content or control information into the broadcast channel (by a mechanism not fully defined in Figure 2.20). Control signalling may be exchanged between the interactive and broadcast provider application.

The interaction channel is based on TCP/IP or UDP/IP protocols with low layer protocols dependent on the type of interaction network. Application protocols include DVB-specific and general types such as HTTP.

2.5 CONCLUSION

After arguing that a NGN is a complex system, a framework for mitigating the complexity is presented. This NGN framework uses a *layering* system as it primary organising principle with *layers* and domains providing further demarcations. A number of functional domains may be superimposed on the layers to give a two-dimensional background for describing a particular NGN. The long established notion of a *plane* as a means of concentrating on a particular class of concerns may also be used. Possible planes extend beyond the control, management and user planes of the ISDN.

The notion of using layers to control complexity is not new; however, the set of layers in the NGN Framework is selected to allow various NGNs to be represented in sufficient detail. Similarly, the functional domains are well known. The Gateway domain is specifically identified for interdomain boundaries between administrative or technology domains.

The NGN Framework allows both *physical entities* and *functional entities* to be represented and also supports the mapping of functional onto physical entities.

This chapter, while using the abstract concept of functional entities, has not considered how software processes and communications protocols are analysed and described. That role belongs to Chapter 3.

Chapter 3

Software Methodologies for Converged Networks and Services

Chapter 2 developed the NGN Framework for describing the architectures of ICT systems. Descriptions contain physical entities and functional entities, organised by layers, domains and planes as required. Functional entities are abstractions for software processes. Software for ICT systems is complex, serves diverse needs, and has demanding performance requirements. Such software is inherently distributed, executing on different computing nodes in response to messages flowing between the nodes. Software considerations range from the design of complex programmes for applications, services and operations support to communications protocols.

Since the 1970s, the telecommunications, Internet and information technology communities have been active in developing software-intensive systems and standards. All three communities have faced increasing complexity of facilities and applications. The three communities, with their particular objectives, developed distinctive methodologies for software systems. With convergence of network and applications, the body of knowledge and practice of software design and development in telecommunications, the Internet and information technology also experience convergence.

The student, analyst, designer and operator of next-generation ICT systems require an understanding of the methods of analysis, description, design and, at a broad level, implementation of complex, distributed software systems. This chapter reviews frameworks, methodologies and standards for specifying, designing and standardising software processes and communications protocols. We identify a combination of methodologies drawn from telecommunications, the Internet and information technology for use in the converged network, service and application environment.

Section 3.1 examines the origins and development of various software methodologies in telecommunications, the Internet and information technology. Section 3.2 examines the problem of analysis, modelling and description of software processes and protocols in complex systems. Section 3.3 reviews two frameworks for software system analysis, modelling, design and implementation. The remainder of the chapter examines important notations that support analysis and design of software processes and protocols, organised as follows. Section 3.4 examines methods and notations for determining the requirements for a system. Section 3.5 reviews methods from the various domains for defining static information in the form of objects. Dynamic modelling notations are introduced in Section 3.6. Section 3.7 reviews notations for expressing software components and interfaces. Section 3.8 reviews methods for

Table 3.1. Comparison of software methodologies

Domain	Context	Process	Main Notations
Telecom	Network, service and management architectures	From architecture, to functions and protocols	SDL, MSC, ASN.1, GDMO
Internet	Protocols to support transport and applications	Add new protocols as required	RFC822, text-based protocols, ASN.1
IT	Varied computing and database applications	From use case to executable code	UML, XML

supporting inherently distributed software systems. Section 3.9 closes the chapter by establishing an analysis, description and system design paradigm that draws on existing methods and is geared to complex situations.

3.1   DEVELOPMENT OF SOFTWARE METHODOLOGIES FOR ICT

Table 3.1 summarises the development of methodologies for software systems and protocols in telecommunications, the Internet and in information technology. Methodologies are characterised by the context, the processes used and the principal notations used for expressing the results of analysis and design.

In the *telecommunications* domain, the development of electronic telephone exchanges over three decades ago exposed the for well-engineered software for systems with five-nines availability. Simultaneously, signalling moved from physical signalling to protocol stacks. The need arose to specify software functionality and standardise communications protocols and the behaviour and interaction of software elements in particular. Two important methods were developed, the Specification and Description Language (SDL) for defining processes and interactions with other processes and the message sequence chart (MSC) for defining the flow of messages in the various sequences of successful and unsuccessful operation of distributed processes. Both SDL and MSC developed in power as the systems to be specified and designed became more complex. SDL acquired object-oriented definition of its elements in 1992. MSC has conventions for clearly distinguishing the sequences that ensue from alternative events and decision branches. Many telecommunications signalling standards use MSCs for describing the external interactions between communicating entities and SDL for defining the internal behaviour of those entities.

The *Internet* standards process focusses on the development of protocols to support transport of packets and applications in the Internet. The IP network and TCP transport layer protocols, together with routing and address resolution protocols are well established. Traditional applications rely on protocols that support remote login,

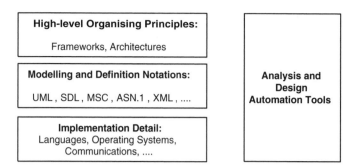

Figure 3.1. An organisation of software concerns in complex ICT systems.

e-mail and Web browsing. Recently, attention has focussed on the transport of real-time signals such as audio and video in the Internet. New network, transport and application layer protocols have been developed. The need to interconnect with the existing telephone network requires new protocols. In summary, the Internet methodology is essentially the definition of protocols, supporting services and management information.

In the *information technology* area, the range of possible types of software application is large. The 1990s saw the development of modelling approaches to the analysis, specification and design of object-oriented software. By 1997, a number of approaches to object-oriented software had been unified to produce the Unified Modelling Language (UML). While standardisation continues to develop, UML has widespread acceptance for the analysis of software system requirements and the design of software systems. Unlike SDL, UML is not focussed on particular application areas such as telecommunications. Rather, UML provides a number of notations based on object-oriented concepts for expressing different views of any software system and gives the designer flexibility to emphasise views for different purposes.

Apart from specific areas of overlap, software methodologies have existed in the telecommunications, Internet and IT silos. With convergence of networks, services and applications, software requirements and methodologies increase in commonality. For example, some IT applications are enhanced by being able to access network connections seamlessly. Similarly, basic communications may be enhanced by IT applications. Internet and telecommunications protocols must co-exist and interwork. The power of object-oriented definition enhances telecommunications standards. Applications become inherently distributed. As systems become more complex, the need increases for frameworks to guide analysis and design and for software architectures to structure systems.

3.2 SOFTWARE PROCESSES IN THE NGN FRAMEWORK

Software concerns in ICT systems range across applications, service control, network management and communications protocols. Figure 3.1 classifies the concerns of software system analysts, designers and implementers. At the highest level, *frameworks and architectures* provide set of concepts, principles and rules that guide analysis and design of complex systems. Frameworks also support learning about and understanding such systems. Next, we require ways of describing the outcome of analyses and designs in suitable notations. The methodology should separate implementation detail from the analysis and design process. Performing analysis and design and presenting the workproduct requires agreed *notations* that are frequently rooted in a modelling approach. Once the design is complete, *implementation concerns* such as programming languages, operating systems and communications protocols arise. Running across these three levels of activity, analysts, designers and implementers require the support of *information representation and automation tools*.

This chapter is concerned mainly with the first two sets of concerns but also demonstrates how high-level organisation methods keep implementation detail out of the analysis and design phases.

3.2.1   *SOFTWARE ANALYSIS AND DESIGN REQUIREMENTS*

The analysis and design of software for control and management in telecommunications systems has been the preserve of a limited number of persons employed mainly by telecommunications equipment vendors. By contrast, a significantly larger number of programmers work in the traditional IT context, producing applications for end-users. Their applications often rely on data communications networks: dedicated networks, intranets and the Internet. Programmes rely on communications interfaces built into the operating systems.

With next generation networks, the possibility arises of applications based on more intelligent interaction between applications, content, realtime connectivity, messaging and network data. Thus, several needs arise in software engineering. First, software systems become more complex and analysis, specification, description and design methods become important. Second, the previously distinct software engineering paradigms of telecommunications and information technology must converge. This chapter addresses these two needs. Third, the complexity of the underlying network facilities must be hidden from the applications developer and this is the subject of Chapter 8.

3.2.2   *FUNCTIONAL ENTITIES AS SOFTWARE ELEMENTS*

The application, service control and management processes in an ICT system are grouped in functional entities with defined relationships. The NGN Framework developed in Chapter 2 locates functional entities by layer and domain or groups FEs for a specific purpose using planes. Figure 3.2 shows the FEs of a hypothetical system mapped onto physical entities and located in layers and domains. Functional entities interact through reference points, interfaces or protocol service access points (SAP). Vertical (interlayer) reference points may be defined in standards or remain undefined if internal to a physical entity. Horizontal reference points are shown and may be standardised. A functional entity has one or more relationship with other functional entities.

Functional entities are, with few exceptions, *communicating finite state machines*. A state is explicitly identified by a name and is characterised by the internal condition of the entity. That condition is reflected in the values of key items of data. The process of moving from one state to another is called a *transition*. A transition is initiated by

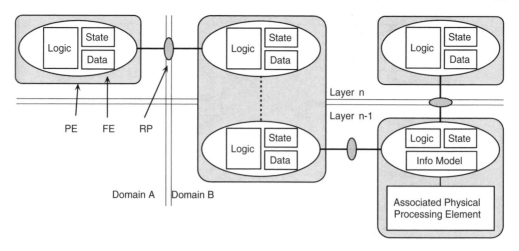

Figure 3.2. Functional entities in NGN Framework are locus of software processes.

an external event such as the arrival of a message or an internal event, for example the expiry of a timer. A functional entity therefore contains logic and data, has state and presents interfaces to other FEs. Interfaces are vertical, that is to an adjacent layer, or interlayer, as shown in Figure 3.2.

The state generally reflects how the history of previous actions affects the response of the entity to a present stimulus. The set of states, together with their linking transitions form the *logic* of the entity. The logic receives input messages from other FEs via defined relationships, executes internal operations and issues messages to other FEs. The response of the FE to a particular message depends on the state of the FE when the message is received.

The finite state machine has the property that internal data may not be changed by an external entity except as a result of a message in a defined relationship. The finite state machine is described as *encapsulated*.

The full definition of a functional entity has both external and internal views. The general method of defining the external view in the form of a protocol or interface specification is summarised in Figure 3.3. Each protocol or interface is intended for use in a specific *environment*. For example, the INAP protocol is intended for application only in the IN overlay to a PSTN or mobile network.

The specification of a protocol has four main parts. First the *message vocabulary* defines the allowed set of messages in a rigorous but human readable format. The *message services* define the actions that a message invokes on the part of the serving entity. In the INAP example, an InitialDP message will invoke the service logic specified by a ServiceKey parameters carried in the message. In the TCP protocol, the receipt of a message with the SYN bit set results in the receiving entity starting a synchronisation process. The *procedures* specify allowed and meaningful sequences of messages both for normal successful execution and for handling incorrect requests (exceptions) and failures and errors occurring during processing. The *encoding rules* specify how the

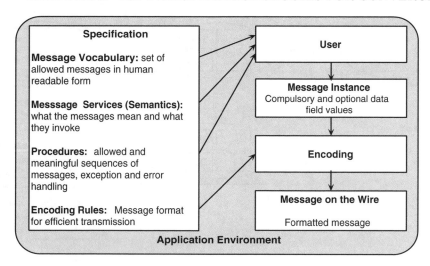

Figure 3.3. Essentials of the specification and use of a communications protocol.

message, originally specified in a human readable form, and hence not compact, is formatted for efficient transmission.

Figure 3.3 shows how the specification influences the implementation of the protocol. The user process generates a message instance with compulsory operation and data fields as well as optional fields. The message is generated in encoded format and passed to the protocol stack for transmission.

Many protocols have only an external definition, leaving the internal detail to the implementer. Others specify some internal detail, for example a state diagram in the TCP specification [173]. In critical cases the internal logic and data supporting the message services and procedures are specified in detail, for example in the Q.931 protocol specification [110].

3.2.3   PHYSICAL ENTITIES IN THE FRAMEWORK

Physical entities are described by the functional entities they host. A physical entity may however contain an element that performs physical operations on user data or streams, for example a switching matrix, a media gateway, a digit receiver, a tone and announcement generator or a forwarding engine in a router. A physical processing entity associated with a functional entity is shown in Figure 3.2. Such physical processing elements are described by an information model, that is structured data that represents the element and its operational state. For example a switching matrix would be described by the current connections between inputs and outputs and the level of traffic being carried. The information model is often implemented as a *management information base* (MIB).

Other functional entities, hosted in the same or a different PE, manipulate the data in the information model and respond to changes at the physical level, for example to notify the onset of an overload condition. The information model is therefore regarded as a functional entity that may have functions for accessing and manipulating the data and issuing notifications of events monitored in the entity.

The RM-ODP introduced in Section 3.3.1, contemplates a different type of interface, the *stream interface* to show that an entity is a source or sink of user data [123]. The stream interface has a number of properties, for example a network address, port number, media carried and the required quality of service. A stream interface therefore has an information model and may have a closely associated functional entity.

3.2.4 SPECIFICATION AND DESCRIPTION REQUIREMENTS

We need to define messages, parameters and supporting data structures as well as the internal processes of functional entities. A special form of data is management information that describes the configuration of physical entities and their operational state. The RM-ODP notion of an *object* as a rigorous, formal representation of a real-world entity is used. The formal definition may be done using an object-oriented method with its powerful concepts of inheritance and containment. Object-oriented representation offers the ability to abstract the internal detail of the object behind an interface and to encapsulate the object's data in such a way as to ensure that the consequences of external interactions are confined to the object.

We recognise, however, that other methods are in widespread use, for example descriptions and interactions using documents. The XML method of definition of objects has an inherent tree structure. Scripting method are used for programming operational procedures and the script may be composed of suitably defined objects.

3.2.5 MODELLING

A *model* is a description of an entity or system that is sufficiently detailed for the purpose at hand. This representation has static and dynamic facets. System dynamics usually follow a state transition type of process. The question arises: what is the correct level of detail? The answer is tied to the management of complexity. A complex system requires a model at least as complex as the system itself for effective analysis, design and operation [178]. We partition ICT systems into layers and domains, capture related functionality in functional entities and use reference points to describe their interaction. A reference point gives a degree of abstraction. Simplification of the abstracted system may be inadequate for two main reasons. First, separation of concerns across a reference point may not be complete: the abstraction is then defined as *leaky*. For example, at a vertical reference point, an application layer protocol may require knowledge of the underlying transport mechanism in order to supply parameters for a message. Second, it seldom occurs that there is a single, isolated functional entity behind the reference point. The FE has relationships to other functional entities that may interact with other FEs in turn and with the external, uncontrolled environment. Put simply, the behavioural complexity of part of a system viewed via a reference point is potentially greater than the complexity of the functional entity in isolation.

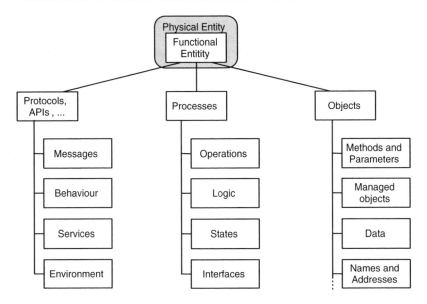

Figure 3.4. Classification of methods used for describing protocols, processes and objects in an ICT system.

3.3   HIGH-LEVEL ANALYSIS AND DESIGN METHODS

Future ICT systems support services and applications that may rely on complex logic, allow access to content and exploit network connectivity and messaging. Analysis, modelling and design methods must therefore support high-level as well as detailed design. Two such methods are the Reference Model of Open Distributed Processing (RM-ODP) [123] and the Model Driven Architecture (MDA) [157].

RM-ODP and MDA have similar objectives in supporting the design of distributed software systems. Their respective emphases differ and are complementary in many respects. RM-ODP provides a formal basis for the fundamental concepts and architecture of distributed systems. MDA seeks to support the portability, interoperability and reusability of software, emphasises modelling and is linked to notations that allow the system designer to apply the MDA methodology.

We review both RM-ODP and MDA and identify their common features. We then unify these frameworks into a common process and identify the role of modelling notations in analysis and design of ICT systems.

3.3.1   THE REFERENCE MODEL OF OPEN DISTRIBUTED PROCESSING

The RM-ODP standard is a response to the growing importance of standards-based support for distributed processing, that is Open Distributed Processing (ODP). RM-ODP addresses the need for a coordinating framework for the standardization of ODP systems. The resulting architecture supports distribution, interworking and

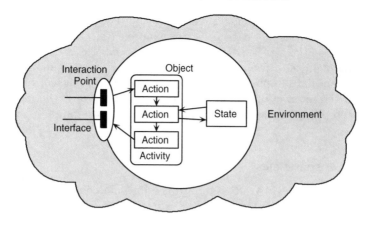

Figure 3.5. Characteristics of an object defined in RM-ODP.

portability. *Distribution* is more than locating the constituent software objects of
a system on separate computing nodes but includes the enabling communication
support mechanisms. *Portability* is a property of an object that, by virtue of the
compliance of its interfaces to an agreed specification, allows it to be applied in different
configurations. *Interoperability* is the ability of objects that are implemented using
different technologies to interact to meet specified objectives

RM-ODP's envisaged users were seen to include "standards writers and architects
of ODP systems" [123]. For example, the TINA Architecture, described in Chapter 6,
follows RM-ODP principles. The set of viewpoints defined in RM-ODP however, serves
as a framework for the analysis and design of applications in distributed systems not
unlike that of the MDA.

While RM-ODP is strongly focussed on standards and architectures, MDA seeks
to support the development of applications by achieving portability, interoperability
and reusability of software components. *Reusability* is the property of an already
implemented software object that allows it to used as part of a new object or system
without redesign.

The RM-ODP standards are based on precise concepts and formal definitions. The
standard first presents a characterisation of concepts used in ODP systems [117].

Modelling concepts are based on the use of an *object* as a representation for an
entity and what it does. The object exists in an environment that is not part of
the model. The object interacts with other objects in the environment via defined
interaction points as shown in Figure 3.5. An *interface* is a defined abstraction for a
subset of an interaction point. The object performs internal *actions*. A sequence of
actions is an *activity*. The object has state, a condition represented by key internal
data, influenced by past interactions and determining future responses to actions and
interaction. The object exhibits *encapsulation*: state changes occur only in response
to interactions or actions. The object, through its interfaces, performs *functions* and
offers *services*.

Specification concepts are defined precisely for both objects and behaviours. *Types* are used to classify entities into categories, for example objects, interfaces and actions. A *class* is a set of entities of a particular type. A *template* is a specification of the common features of any entities belonging to a type. The template may be for an object or an action. A special template is the *interface signature*, a set of action templates associated with the interactions of an interface. *Instantiation* is the production of an entity from a template: an instance is an object in its initial state or the occurrence of an action.

Distribution Transparencies

The second thrust of the RM-ODP standard is a formal definition of the properties of systems or objects. An important property is *distribution transparency*. Transparency differs in emphasis from abstraction in that it hides behaviour rather than system detail. Distribution transparency is the ability to free a user of the system from the need to deal with an aspect of the behaviour of the distributed system. Some support mechanism provides services that mask particular aspects of the behaviour of the distributed system from the specified user. A system may exhibit combinations of different kinds of transparencies.

- *Access transparency* enables an object to invoke the services of another where the objects use different data representation and invocation mechanisms. Interworking between differently implemented objects is supported.

- *Location transparency* frees an invoking object from the need to know the location of an invoked object when identifying and binding to interfaces: naming information is sufficient to invoke the services of an object.

- *Relocation transparency* allows an invoking object to continue to invoke an interface to which it is bound when that interface has been relocated.

- *Migration transparency* enables a system to change the location of an object without compromising the ability of an invoking object to access its services. Load balancing and reduced latency are thus achieved.

- *Replication transparency:* allows an invoking object to invoke an interface that has been replicated without knowledge of the details of the replicated interfaces. The object of replication is to enhance performance and availability.

- *Failure transparency* enables an object to continue operating if other interacting objects and even itself fail and have to be recovered. This property supports fault tolerance.

- *Persistence transparency* frees an object from dealing with the deactivation and reactivation of other objects (or itself) to give the effect of persistence when the system cannot ensure continuity.

- *Transaction transparency* frees an invoking object from the need to coordination of activities among interacting objects to achieve consistency.

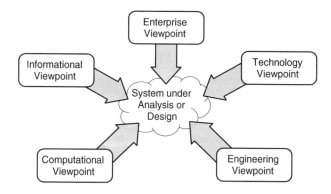

Figure 3.6. RM-ODP Viewpoints giving different but consistent views on the same system.

The mechanisms for achieving several of these transparencies are incorporated in the CORBA environment, described in Section 3.8.4.

RM-ODP Viewpoints

The third thrust of the RM-ODP standard is the definition of a framework for the specification and support of an ODP system [118]. The specification is based on five viewpoints on the system and its environment. Each viewpoint is a way of describing or modelling a system using an abstraction based on selected concepts and principles. Each viewpoint is expressed in a viewpoint language, a formalism suited to the viewpoint. The five viewpoints on an ODP system and its environment are:

- *Enterprise viewpoint* is essentially a requirements analysis, focussing on the purpose, scope and policies for that system without regard to how the requirements are met.

- *Information viewpoint* focuses on the semantics of information and information processing.

- *Computational viewpoint* enables distribution through functional decomposition of the system into objects which interact at interfaces.

- *Engineering viewpoint* focuses on the mechanisms and functions required to support interaction between distributed objects in the system.

- *Technology viewpoint* focuses on the choice of technology in the system.

Support functions for ODP systems are management and co-ordination, repository and security.

In the convergence environment, interest falls more frequently on the development of ICT applications rather than new architectures. The five viewpoints of RM-ODP provide a useful basis for the analysis and design of complex applications.

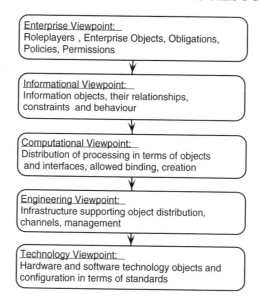

Figure 3.7. Process for distributed application analysis and design based on RM-ODP.

The process of analysis and design suggested by the viewpoints is shown in Figure 3.7. The application requirements are first established using the enterprise viewpoint. The designer answers questions including: 'Who are the roleplayers? What are their obligations? What are they permitted to do and prohibited from doing?' The designer moves to the informational viewpoint where the information and behaviour required to specify the system are detailed using a formal representation. The computational viewpoint then considers the packaging of the information objects into components with interfaces. The design completed thusfar is independent of the implementation detail. The engineering viewpoint then deals with the methods of supporting the working objects in a distributed environment.

Every new application requires the enterprise, informational and computational viewpoint steps. The engineering and technology viewpoint considerations may be reusable from application to application.

3.3.2 MODEL DRIVEN ARCHITECTURE

The point of departure of the Model Driven Architecture is the principle that best practice in software design and development separates the analysis of system requirements and specification of the system from the manner of its implementation in a specific computing and communication environment. The computing environment is characterised by languages and operating systems. The communications environment is characterised by transport protocols and supporting mechanisms. The concept of a *platform* captures considerations of implementation languages, operating systems, communications protocols or distribution support such as CORBA. The platform is a set of defined interfaces to the infrastructure underlying a computing application and their patterns of use [157]. The platform may be generic, specific to technologies or specific to individual vendor's implementations.

The description *model-driven* emphasises the intention that the lifecycle of a software product should be supported by models. Types of models are defined through three MDA viewpoints. The *computationally independent viewpoint* focusses on the requirements on a system and its environment without regard to how the system will be structured and implemented. The *platform independent viewpoint* is a way of describing a system that concentrates on the operation of the system without considering the implementation platform. The system could be implemented on different platforms. The *platform specific viewpoint* is a way of viewing a system that adds platform specific considerations to the platform independent view of the system.

The three viewpoints lead to three types of models for the system: computationally independent model (CIM), platform independent model (PIM), and platform specific model (PSM). Each model gives a view on the system: a representation of a system from the corresponding viewpoint.

- *The computationally independent model* is essentially a requirements analysis considering the system and its environment and has similar objectives to the RM-ODP enterprise viewpoint. The MDA does not specify how the requirements analysis should be done. The RM-ODP enterprise language could be used, wholly or in part.

- *The platform independent model* of the system is expressed in technology neutral or platform independent terms. The objective is to permit implementations on different platforms. Development of the PIM involves identifying the objects that make up the system and the way that they interact to perform processing. These objects must then be packaged into components with interfaces. The PIM therefore has significant commonality with the informational and computational viewpoints of the RM-ODP.

- *The platform specific model* is concerned with implementation issues similar to the concerns of the engineering and technology viewpoints of the RM-ODP. The RM-ODP engineering viewpoint is concerned with support for a distributed processing environment and defines a generic platform in MDA terms. The RM-ODP technology viewpoint is concerned with the choice of technology, its configuration and testing.

The Model Driven Architecture is the highest level organisation in the OMG standards. Other important standards are the Object Management Architecture (OMA) and the Unified Modelling Language (UML). The OMA has two thrusts. The *Object Model* defines concepts and semantics for objects, including requests, types, interfaces, operations, parameters, returned results, exceptions and object lifecycles. The object definition methodology of the OMA is basic to the practical definition of classes using the modelling notation of UML.

The second part of the OMA is its *Reference Model* that defines the components, interfaces and protocols that make up the OMA. Within the reference model, the Object Request Broker (ORB) is defined as to support the various distribution

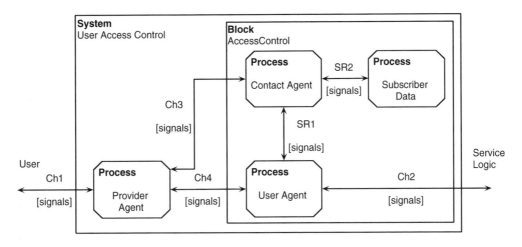

Figure 3.8. System, block and process level organisation in SDL.

transparencies, principally access and location transparency formulated in the RM-ODP. The OMG-standardised ORB, the Common Object Request Broker (CORBA), is reviewed in Section 3.8.4.

3.3.3 SDL AND MSC

The early recognition that well-engineered software was essential to ensure highly available and dependable telecommunications equipment prompted an industry-led development of a specification and description method for state-based communicating processes, the Specification and Description Language (SDL) [53]. Responsibility for SDL was taken over in 1976 by the CCITT, the predecessor of the ITU-T. Initially SDL was a graphical support tool that described internal behaviour of processes. Over time, SDL was enhanced by the addition of a data definition capability and system structuring conventions. In 1988, a formal definition of semantics was added and SDL attained the status of a formal specification and description method. In 1992, object-oriented definitions were introduced for the constructs that make up an SDL description of a system as well as data defined by SDL users. While SDL produces a model for the internal behaviour of the system, it is supported by tools for simulating the system and for producing code in programming languages. SDL is the accepted method in telecommunications standards organisations as well as industry for providing a rigorous definition of the internal behaviour of the processes that make up a software system

The SDL describes the internal behaviour of a *communicating, finite state machine*. Alongside SDL, the ITU-T also maintains a Recommendation for Message Sequence Charts (MSC) [121]. The MSC provides a method of describing the externally observable behaviour of a system via messages passing between entities that make up the system.

The starting point of an SDL description is a definition of the system boundary and the way that the system communicates with its environment, as shown in Figure 3.8.

The system may be subdivided internally at the block and process level. The *process* level contains the detailed description of the functioning of the system. The *block* level is a method of grouping processes but does not add to the functional description. Processes communicate by passing *signals* via identified routes. When a route is internal to a block it is called a *signal route*. If a route crosses a block or system boundaries it is called a *channel.*

SDL initially was concerned only with instances of systems, blocks and processes created to describe particular systems. The graphical notation was, and still may be, used to describe actual systems and processes. The addition of object-orientation to SDL gives a level of abstraction. For example a type is an abstraction for a set of instances with similar properties. Thus, SDL elements such as system, block and process have type definitions.

Data definitions in SDL follow object-oriented principles. A user-defined data type is called a *sort* in SDL. Since SDL is implementation-independent, sorts do not follow computing ideas but mathematical notions. Sorts may be defined in SDL's own notation or in other object-oriented methods such as UML and ASN.1. Conversion rules are defined in standards. SDL process definitions therefore allow the definition of data that is used in the logic depicted by the block.

SDL is technology neutral: the description language does not correspond to a programming language. Practical tools exist for conversion of SDL definitions to programme code.

While SDL has traditionally been regarded as a system design methodology, its starting point is the definition of the system with blocks, processes and signal routes. As the signal routes correspond to the interfaces by which processes communicate, the system diagram has features of the computational viewpoint. The subsequent definition of data and processes has features of the informational viewpoint.

Examples are used to illustrate the modelling and descriptive notations reviewed in this chapter. The first example illustrates the partitioning of a system for description using SDL.

Example 1: User Access Control System.

The first example we use to illustrate the system breakdown in the SDL analysis and design method is a User Access Control System (UACS). This is a simplified form of the TINA Retailer Access Session described in [35]. The UACS is intended to be used to allow a user programme to access service logic. The UACS allows the user to make initial contact, authenticate, and to start and stop a selected service. Figure 3.8 shows the design after initial steps described elsewhere in this chapter at a stage when the constituent processes have been identified.

A process called the Provider Agent supports all interactions with the User process, which is outside the system. In the provider domain, Access Control block provides a Contact Agent process that handles initial contacts and authentication. The User Agent has the user profile data and methods for starting and stopping services. The User Agent communicates with the Service Logic, also outside the system being designed, through defined signals.

The processes communicate via signal routes or channels. The permitted signals are defined for each signal route and channel.

A second example illustrates the use of SDL to define the detailed internal process in a communications protocol. The SCTP protocol is a transport protocol described in Section 4.5.2. We use the procedure for establishing an association between the two transport layer entities as an example to illustrate a number of analysis and design notations.

Example 2: Establishing an Association in the SCTP Protocol.

We consider a SCTP endpoint. A single process is designed that implements both the initiating client and serving roles.

The process performs the following actions. The endpoints are called A and Z. User A initiates the association using an INITIATE primitive. The SCTP-A entity prepares data in a Task Control Block (TCB) and issues an INIT message. SCTP-Z responds with a INIT ACK carrying its TCB. SCTP-A then responds with a COOKIE ECHO message carrying state information in the form of a cookie. STCP-Z then issues a COOKIE ACK. The two ends generate and use a Verification Tag as a security measure. When this four-way handshake is complete the connection is established and the upper layer entities are informed by a COMMUNICATION UP primitive.

The detailed process at a SCTP protocol entity is shown in Figure 3.9. This SDL diagram applies to the processes at both endpoints. The states and transitions traversed depend on whether the end is the initiating end or not. Both ends start in the Closed state. The initiating end A leaves this state on receipt of the upper layer primitive ASSOCIATE. End Z leaves the Closed state on receiving the COOKIE ECHO message from A. The SDL diagram shows the various TCB creation processes, the generation of Verification Tags and checking received tags.

3.3.4 *A LEGACY SYSTEM METHOD: UFM*

The Unified Functional Methodology (UFM) was an initiative arising out of the ISDN standards intended to characterize requirements for services and network capabilities in future networks [126]. Envisaged networks included future mobile networks, the B-ISDN as well as the network management systems based on the Telecommunications Management Network (TMN) [204]. The methodology is specific to that network evolution thread. The UFM defines a four-plane model. This model is the basis of the IN Conceptual Model shown in Figure 1.7, defined to specify services and to define the network capabilities needed to support the services. The software reuse and service creation method in the UFM are based on a scripting paradigm. Functional entities are particular to the class of network considered. Thus, much of the UFM is not beneficial in the NGN context.

The UFM recommendation however defines a conceptual basis for functional entities and relationships that is both useful in the NGN Framework and is adaptable to the object-oriented paradigm. In particular, the UFM Recommendation provides an adaptation of the message sequence chart reviewed in Section 3.6.2 that allows external behavioural descriptions to be related to internal process definitions using SDL. The UFM, while oriented to switched circuit networks, also proposes a decomposition of functionality in terminals and network nodes into a number of layered entities: user application function, service control function, service data function, session

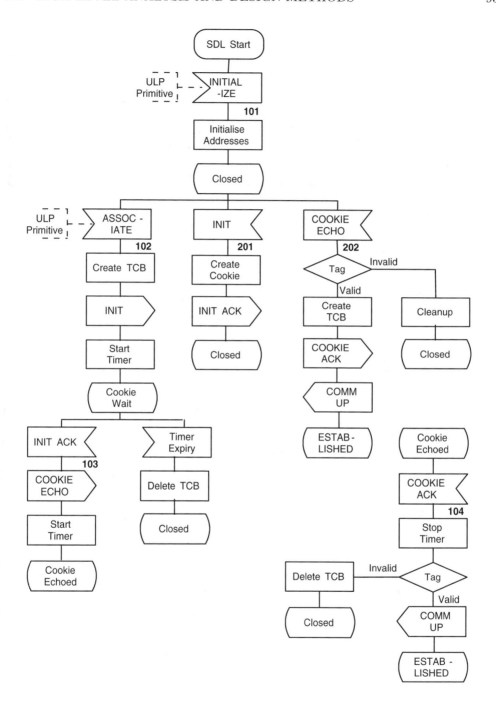

Figure 3.9. Association setup logic for the SCTP protocol expressed in SDL.

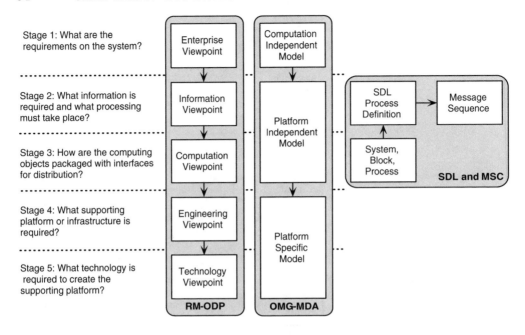

Figure 3.10. Comparison of three system design methods: RM-ODP, MDA and SDL (with MSC for checking the output).

management function, resource management function, call control function, connection management function and fabric management function.

3.3.5   GENERALISING HIGH-LEVEL METHODS

Figure 3.10 compares RM-ODP, the OMG's MDA and traditional software design in telecommunications based on SDL. The comparison is based on the five best-practice stages in system analysis and design answering the following questions:

- *Stage 1:* what are the system requirements, who are the roleplayers and what are their obligations?

- *Stage 2:* what information structures and processes are required to fulfill the system's required functions? A formal object-oriented definition is essential.

- *Stage 3:* how are the objects implementing the processes and data to be packaged into components and what are the interfaces?

- *Stage 4:* what supporting infrastructure such as languages, operating systems and distributed processing support are required?

- *Stage 5:* what actual technologies are to be used?

The five RM-ODP viewpoints address the five questions. The models of the MDA map onto the five questions. The RM-ODP enterprise viewpoint and the CIM both

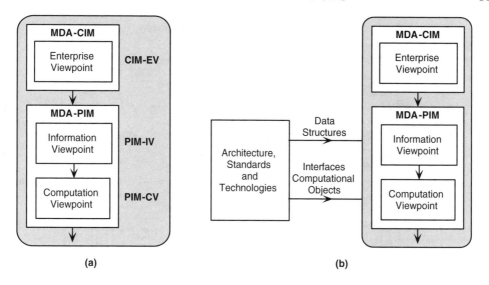

Figure 3.11. (a) Generalisation of RM-ODP and MDA processes for distributed system design. (b) Process with constraints placed by software architecture or standard.

fulfill Stage 1 of the design process, namely requirements specification. The platform independent model corresponds to the informational and computational viewpoints. The platform specific model starts with engineering viewpoint considerations. We therefore generalise the analysis and design framework as shown in Figure 3.11(a).

Figure 3.12 shows how various notations support the use of the framework in analysis and design. The notations are not restricted to a single source, for example UML. Notations are chosen as required.

3.3.6 ROLE OF SOFTWARE ARCHITECTURES

The process specified by the five stages listed above is often not linear, nor does it contain all five stages if there are preconditions on the design. For example, a constraint in the form of technology used or a software architecture may restrict the freedom of the designer. A modification of the process taking the constraint of a software architecture into account is shown in Figure 3.11(b). Data structures and objects already defined in the architecture must be imported into the design. Similarly, part or all of the computational model may be predefined. Constraints may arise in other ways. For example, if the Java programming language is prescribed, the platform-specific model is predefined.

3.3.7 SOFTWARE MODELLING NOTATIONS

We are concerned with analysing, describing and designing the software constituents of ICT systems. To do this, we need to consider both the internal processes of the functional entities and the protocols or interfaces at which external behaviour can be

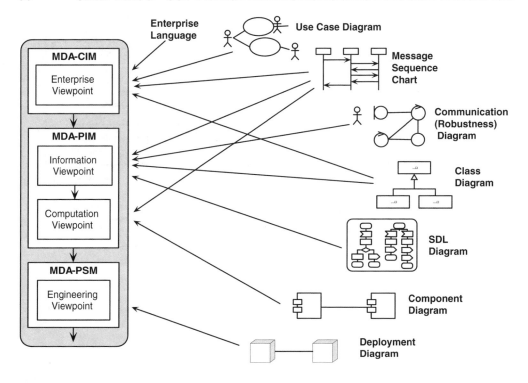

Figure 3.12. Generalised analysis, modelling and design system based on RM-ODP and MDA with tools useful at various stages.

observed. We require the ability to define objects. Figure 3.4 summarises the types of entities that we must describe and the types of objects that are encountered.

Analysis and design of complex software systems has two main forms of support: a framework and modelling notations. The two frameworks reviewed above have different associated notations. The RM-ODP has strong conceptual aspects but is not linked seamlessly to common modelling notations. In contrast, MDA is based on a limited but sufficient set of concepts and is well supported by the UML set of diagrams. The UML notations describe the structure of the software system and the behaviour of the system and assist in the management of the models, for example by defining packages.

Table 3.2 lists modelling notations drawn from a number of sources. At each stage of the analysis and design process, different notations may be required. We therefore regard the set of notations as a resource available to the analyst or designer to be used as required, as shown in Figure 3.12. The following sections enlarge on the listed notations.

A complex system requires several models for effective description. Models may focus on one aspect such as the overall architecture, business roles and requirements,

Table 3.2. Types and example of notations for software analysis and design

View	Type	Body	Modelling Notation
CIM–EV	Static	RM-ODP	Enterprise Language
		OMG-UML	Object diagram (enterprise objects)
	Dynamic	OMG-UML	Use case diagram
		OMG-UML	Message sequence chart
PI–IV	Static	OMG-UML	Class and object diagrams
		ISO/ITU	ASN.1
		IETF	Extensible text-based protocols: RFC822
		ITU-T	ASN.1 definition of managed objects
		OMG-OMA	Interface Definition Language
		W3C	XML Document Type Definition/Schema
	Dynamic	OMG-UML	Sequence diagram
		ITU-T	Message sequence chart
		OMG-UML	Activity diagram
		OMG-UML	Communication/robustness diagram
		OMG-UML	State diagram
		ITU-T	Call state model
		ITU-T	SDL diagram/textual definition
		ITU-T	Connection view (IN-CS-2/Megaco)
		ITU-T	UFM Information flow diagram
PIM–CV	Static	OMG-UML	Component diagram
		OMG-UML	Interface class diagram
		OMG-OMA	Interface Definition Language
PSM–EngV	Static	OMG-UML	Deployment diagram
		ITU-T	RM-ODP engineering viewpoint

information, processing, states, connections or communications protocols. Part of the software framework is a systematic way of choosing and relating the different models.

This review of modelling notations is not intended to be an in-depth tutorial. Rather, we seek to highlight the range of notations that model different aspects of a software system. The concepts underlying the notation and graphical constructions are examined. Example 1, introduced in Figure 3.8, and example 2 are used to illustrate a number of notations.

All modelling notations reviewed in this chapter are described in relation to the generalised process resulting from blending RM-ODP and MDA principles. Figure 3.13 relates the Specification and Description Language to the generalised analysis and design framework. Figure 3.10 reflects the historical mode of use of SDL where the division of the system into processes is the first step. For example, in the definition of the ISUP process, the Call Control Function is defined and the ISUP protocol is the concrete implementation of information flows between adjacent CCF entities.

Figure 3.10 suggests that SDL also has a role as a supporting notation for the generalised analysis and design framework. In this role, the technology neutrality of SDL makes it suitable for describing internal behaviour in the PIM–IV and PIM–CV phases. Translation to the platform specific view is achieved by various mappings.

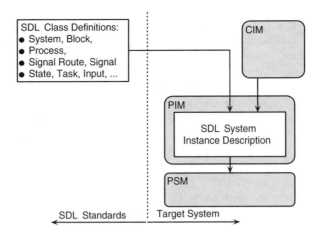

Figure 3.13. Relationship between SDL standard and system instance specification.

Subsequent sections review a number of notations for describing software processes and protocols.

3.4   ENTERPRISE AND BUSINESS MODELLING NOTATION

Enterprise and business modelling notations support Stage 1 of the analysis corresponding to the enterprise viewpoint or the computationally independent model.

3.4.1   ENTERPRISE LANGUAGE

The RM-ODP Enterprise Language is concerned with establishing the requirements on the system in its environment in terms of its purpose, scope and policies [118]. The system is viewed as a *community* of enterprise objects. *Enterprise objects* are things that play roles in the system. Enterprise objects that have a single authority in common may be classified as belonging to a *domain*. Domains may share resources in an agreed way. Such a community of domains constitutes a *federation*.

The enterprise language is applied at the system level and to each enterprise object. Questions asked at the system level are of the form: 'What are the roles of the system? What activities does the system perform? What policies are applied? What contract exists between the system and its environment?' Similar questions are posed about the enterprise objects. Roles of enterprise objects are defined by answering a number of questions:

- *Permissions*: what may the enterprise object do?

- *Obligations*: what must the enterprise object do?

- *Prohibitions*: what must the enterprise object not do?

- *Behaviour*: what observable activities characterise the enterprise object?

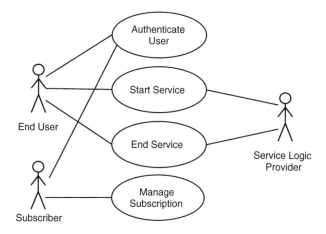

Figure 3.14. A use case diagram showing the external actors and use cases for the User Access Control System.

The Enterprise View is formulated using natural language but benefits from object-oriented descriptions of the enterprise objects.

Example 1 continued: UACS Enterprise Description.

The UACS example has the following enterprise language description.

The entities external to the system are the Subscriber, the User and a Service Logic Provider. The system performs the following roles: to control access by Subscribers and Users; to allow a subscriber to manage users; to initiate and end services provided by another system, the Service Logic Provider.

Two environmental contracts must exist. The end user must have a subscription that is registered in a database accessible by the UACS. The UACS must have an contract with one or more service provider. The service provider could for example be a PSTN that allows access to the SCP for initiating calls.

The enterprise objects making up the community are the End User, Subscriber, the Service Logic Provider and the Access Control System.

3.4.2 UML USE CASE DIAGRAMS

A use case diagram is an UML notation that presents high-level system requirements by indicating related sets of actions and their relationship to external roleplayers. The notion of an *actor*, shown by a stick man in Figure 3.14, is a person or system external to the system being specified. A *use case* is a sequence of actions that performs some function required of the system. Use cases may include or extend other use cases. The *use case diagram* shows relationships between actors and use cases.

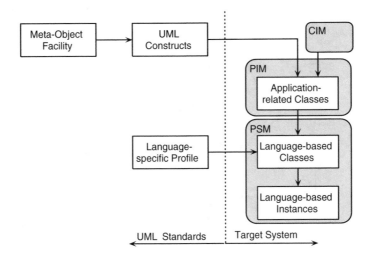

Figure 3.15. Structure of UML standards and resulting development process for classes.

Example 1 continued: Use Cases for the UACS.

The enterprise description of the UACS is now accorded a level of formality through a use case diagram. Figure 3.14 identifies three parties external to the system. The Subscriber has a subscription that permits listed services to be accessed. The subscriber may enrol End Users (including the subscriber) who are permitted to make use of services after authentication. The Service Logic Provider is an external computer that runs the programmes that control the requested services, for example an on-demand conference call.

The activities to be performed by the system are shown as use cases. The actors that may participate in each use case are shown by association lines. Use case Authenticate User allows a subscriber or end user to be authenticated before participating in other user cases. In the Manage Subscription use case, the authenticated Subscriber performs actions such as enrolling or removing end users, checking account balances and changing user profiles. Start Service allows an End User to query available services, select a service and initiate the service. End Service allows the End User to stop a service that is running.

3.5   OBJECT AND DATA DEFINITION LANGUAGES

3.5.1   *UML CLASS AND OBJECT DIAGRAMS*

In the discussion of the RM-ODP we introduced the object as a representation of a real-life entity. Some of the properties of the object were summarised. While the use of the RM-ODP viewpoints and their harmonisation with the Model Drive Architecture are advocated in this Chapter as an analysis and design framework, the Object Model of the Object Management Architecture (OMA) is favoured for defining detail. The Object Management Architecture [167] provides a set of concepts and definitions that support rigorous specification of objects and classes. The Unified Modelling Language (UML) provides the notation for expressing object definitions.

Figure 3.15 illustrates aspects of UML itself and the method of defining classes. UML has constructs for defining its various notations including use cases, classes, interactions and sequence diagrams. The Meta-object Facility supports the definition of standard UML constructs, for example the class diagram, as well as special constructs that may be needed for other purposes. UML also has profiles for mapping classes from its technology-neutral form to language-specific forms. To the right of the dotted line, we show the three steps of the MDA process. Having defined the system requirements in the Computationally Independent View, classes for the application are defined using UML technology-neutral notation in the Platform Independent View. The language-specific profile then allows the classes to be mapped into the target language form, for example Java or C++.

The OMA model considers the interaction of a client with an object. An *object* is an identifiable, encapsulated entity that provides services that can be requested by a client. A *client* is any entity that requests the services provided by an object. The client's understanding of the object and its services is the *object semantics*.

The services of an object are actions that can be performed by the object on receiving a request. A *request* is an event initiated by a client that addresses the target object, specifies an operation to be performed and may have parameters. An *operation* is an identifier for the requested service. Request *parameters* are values associated with a request that are passed to (*in-parameters*) or returned from (*out-parameters*) an object or passed in both direction (*in-out parameters*). A *value* is an actual parameter in a request. The parameters are viewed as having a *signature*: the legitimate values of parameters required in a request and returned, together with the exceptions that can arise. A request can return a *result value*: a single value returned on execution of a service, together with the out-parameters. An *exception* is like a request but consists of an identifier plus parameters returned to the client by an object if an abnormal condition occurs after requesting a service.

Requests are generally identified by the operation part. Related operations are frequently grouped in interfaces. An *interface* is therefore a set of operations that a client may request on an object.

The *implementation* of an object is the executable code and data that perform the services offered by the object, subject to lifecycle constraints. The implementation contains methods: the executable code that performs the service requested by an operation. An operation is therefore often referred to as a method call.

Example 1 continued: Classes and Objects for the User Access Control System.

The enterprise viewpoint and use cases point to a number of consideration for the PIM–IV stage of design. An important consideration is that the subscriber or user should not have to know the location of the provider. The design used in the TINA Architecture is helpful. One or more objects in the end user equipment hide the details of how the provider is accessed. Three objects appear necessary: User Authentication, User Access and User Service Control.

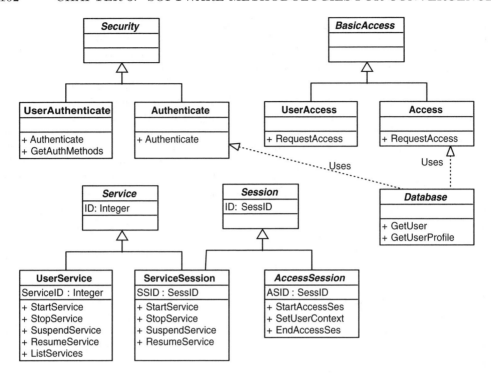

Figure 3.16. Example of class definition in the PIM–EV stage for the User Access Control System.

These objects invoke corresponding objects in the provider domain, **Authenticate**, **Access** and **Service**. While the objects in the user domain relay requests and responses with minimal processing, those in the provider domain perform necessary processing. The definition of classes from which these objects can be derived can therefore exploit inheritance. The **Authenticate** and **Access** objects require database support and a **Database** object is therefore required.

Figure 3.16 defines classes for objects for the User Access Control System. The instances of objects and their relationships are defined in Figure 3.24.

3.5.2 ABSTRACT SYNTAX NOTATION ONE

The Open Systems Interconnection movement gave rise to Abstract Syntax Notation One (ASN.1), a formal method of defining objects, used for defining both communications protocols and managed objects.

Abstract Syntax Notation One is a method of defining *data types* applied mainly in data communications, telecommunications and network management and is a joint ITU-T and ISO standard [116]. ASN.1 defines a set of rules for describing the structure of objects, independent of how they are used by an application or transferred between

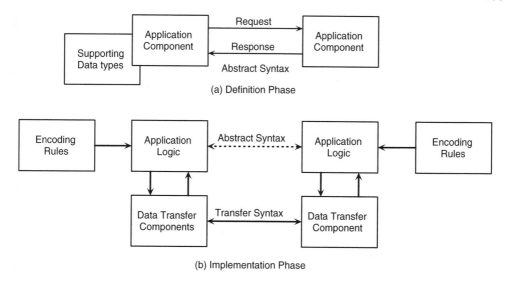

Figure 3.17.   Abstract Syntax Notation One definition approach to (a) object definition and (b) transfer of object instances.

applications. ASN.1 makes use of inheritance, allowing definitions from one standard to be used in another.

ASN.1 has two principal areas of application: first, the formal definition of telecommunications protocols such as INAP, CAP and Megaco, and second, the definition of management information. Both applications are essentially data type definition: the former results in types that describe protocol requests and responses while the latter defines network management information as types.

Using ASN.1 to Define Protocols

Figure 3.17 illustrates the two phases in the life of a protocol or data structure. During the definition phase, data types and *application components* are defined. Application components are the *operations* and *requests* that may be exchanged between two applications, expressed in the abstract syntax of ASN.1. Data and components are expressed in high-level language format for human readability.

For the implementation phase, the actual components transferred between applications are encoded in a compact form using an agreed encoding as shown in Figure 3.17(b). The transfer syntax defines the format for representing object instances for transfer between applications over a network, typically as a stream of octets. Encoding rules define the mapping of objects, defined using an abstract syntax, into a transfer syntax for transmission over a network. The Basic Encoding Rule (BER) is commonly used.

ASN.1 defines a number of basic types and allows the development of application specific types. Data types are based on the ASN.1 primitive types INTEGER, OCTET

STRING, OBJECT IDENTIFIER and NULL or a ASN.1 SEQUENCEs. A SEQUENCE takes the form:

SEQUENCE { <type1>, <type2>, ... <typeN}

where each type is a primitive type. A SEQUENCE thus generates a list. Tables may be created as a SEQUENCE of lists.

Simple types for specific applications are defined in various standards. For example, types specific to TCP/IP networks are defined:

IpAddress:	32-bit IPv4 internet address as a four-octet string
Counter:	Non-negative 32-bit integer which may only increase
Gauge:	Non-negative 32-bit integer which may increase or decrease
TimeTicks:	Non-negative 32-bit integer counting time in 10 ms steps

More complex structures such as protocol messages and managed objects are defined using ASN.1 macros.

A basic type of particular importance in ASN.1 is the *object identifier*. An object identifier is a series of integers that provides a globally unique identification of the object. Object identifiers are arranged in a hierarchical tree structure, starting from a root. The tree grows through the top level standards bodies ccitt, iso and joint-iso-ccitt, each branching out until the actual object is identified. For example, the definition of ETSI Core INAP operations is located by the path:

Core-INAP-CS1-Operations {ccitt(0) identified-organization(4) etsi(0) inDomain(1) in-network(1) modules(0) cs1-operations(0) version1(0)}

The InitiateCallAttempt operation defined below is one of the set of operations so located.

We illustrate the use of ASN.1 by examples drawn from the ETSI Core INAP specification. An example of a protocol operation used by a Service Control Point to initiate a two-party call in a telephone network is:

```
InitiateCallAttempt ::= OPERATION
ARGUMENT
    InitiateCallAttemptArg
ERRORS {
    MissingParameter,
    SystemFailure,
    TaskRefused,
    UnexpectedComponentSequence,
    UnexpectedDataValue,
    UnexpectedParameter
}
```

This operation is issued by a Service Control Point to a PSTN switch to initiate a two-party call. The data type InitiateCallAttemptArg is defined as follows.

```
InitiateCallAttemptArg ::= SEQUENCE {
    destinationRoutingAddress      [0]   DestinationRoutingAddress,
    alertingPattern                [1]   AlertingPattern         OPTIONAL,
    extensions                     [4]   SEQUENCE SIZE(1..numOfExtensions)
                                         OF ExtensionField       OPTIONAL,
    serviceInteractionIndicators [29]  ServiceInteractionIndicators
                                                                 OPTIONAL,
    callingPartyNumber             [30] CallingPartyNumber   OPTIONAL
}
```

The tags, for example [0] and [1], are used in encoding the operation for transmission. The data types such as DestinationRoutingAddress are defined by:

```
DestinationRoutingAddress ::= SEQUENCE SIZE (1) OF CalledPartyNumber
CalledPartyNumber ::= OCTET STRING (SIZE(minCalledPartyNumberLength ..!
                      maxCalledPartyNumberLength))!
```

Structure of Management Information

The IETF defines a network management framework for managed objects as abstract representations for elements that must be managed in the Internet. The concepts, principles and basic definitions are contained in the Structure of Management Information (SMI). Type of objects for specific management purposes belong to a structure called the Management Information Base (MIB). Section 9.2.2 enlarges on the SMI and MIB and the use of the SNMP to perform management operations on elements described by the defined objects.

The Structure of Managed Information describes the common data structures to be held in the MIB and a means of identifying values. Data structures are defined as objects, each object having a unique OBJECT IDENTIFIER. Uniqueness of OBJECT IDENTIFIERs is achieved by an administrative registration process. The process spans a number of standards bodies including ISO, ITU-T, ETSI and the IETF.

Managed objects have a trail to the root of the registration process defined by the number assigned to the organisation and to the grouping of functions. For example an managed object in the IETF MIB grouping would have a registration $1.3.1.6.2.6.x.y$ where x and y indicate groups not shown on the tree in Figure 3.18. Similarly, a managed object in GSM, an ETSI standard, is registered as $1.4.0.x$.

The managed objects are defined in a restricted form of ASN.1. Actual data values are transmitted across the network using the Basic Encoding Rules of ASN.1.

An example of object definition of types peculiar to TCP/IP networks follows. An object called ifAdminStatus holding status information about an interface is defined as:

```
ifAdminStatus  OBJECT-TYPE
     SYNTAX    INTEGER  {
                  up(1),         - - ready to pass packets
                  down(2),
                  testing(3)     - - in some test mode
               }
     ACCESS    read-write
```

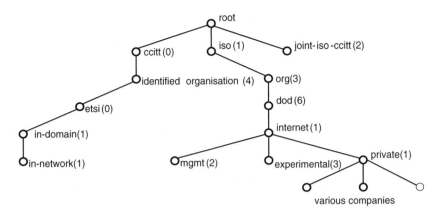

Figure 3.18. Part of the object identifier tree in ASN.1.

```
STATUS        mandatory
DESCRIPTION
              "The desired state of the interface. The testing state (3)
              indicates that no operational packets can be passed."
:: =  { ifEntry 7 }
```

This object is registered as object 7 in the group `ifEntry` defined in the MIB-II specification in RFC1213.

3.5.3 INTERFACE DEFINITION LANGUAGE

Interface Definition Language (IDL) is an object-oriented method of defining interfaces and the operations supported. The IDL was defined to support interoperability of client and server object implementations in different languages running on different platforms. The IDL language is declarative, that is it supports the definition of interfaces and does not itself produce executable code. IDL has a strongly typed syntax that has mappings to most of the common programming languages. For example, the interface definition can be translated into the header for a class in C++.

The use of IDL in a heterogeneous language environment is shown in Figure 3.19. The process for generating the client and server implementations is shown. Assume the client object is developed using language X and the server using language Y, both having IDL mappings. The first step is to define the interface in IDL. This interface is written directly in IDL or mapped from an interface definition in UML. The IDL interface consists of operations (methods) with parameters expressed in various data types.

The IDL interface is compiled into language X on the client side as a set of headers and the stub implementation using an IDL compiler that embodies the language mappings for the target programming languages. The *stub* implementation contains the logic for marshalling the data types from the language-specific form to a standard format for transfer to the server. An X-language programmer writes the

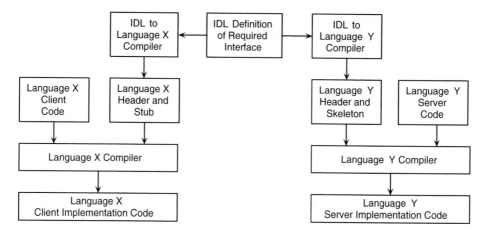

Figure 3.19. Processes for achieving interoperability of differently implemented objects based on IDL.

implementation of the client object. The header and stub are compiled together with the client implementation by an X-language compiler into client-side executable code.

The server object is to be implemented using programming language Y. A corresponding set of operations generates the server header and *skeleton*, which is compiled together with the server implementation into executable code. The skeleton is responsible for de-marshalling the received data into the form required by the server-side code.

The stub and skeleton require the support of a mechanism such as CORBA to transfer requests and return responses.

An example of an IDL interface definition in the TINA Service Architecture described in Chapter 6 follows.

```
interface i_Initial {
void requestNamedAccess (
  in TINACommonTypes::t_UserId userId,
  in TINACommonTypes::t_UserProperties userProperties,
  out Object PAaccessIR,
  out TINAAccessCommonTypes::t_AccessSessionId asId
) raises (
  TINAProviderInitial::e_AccessNotPossible,
  TINAAccessCommonTypes::e_UserPropertiesError
);
}; /* interface iInitial */
```

3.5.4 CLASSICAL TELECOMMUNICATIONS PROTOCOLS

Several important telecommunications protocols were defined before object-orientation became the dominant paradigm for protocol definition, for example Q.931, ISUP and MAP. Each protocols targets a specific environment. For example Q.931 is intended for signalling between ISDN terminals and the end exchange. The specification of such protocols generally has four parts.

First, protocol operations and their purposes are defined, for example Setup, Setup Acknowledge and Alerting in Q.931. Second, individual parameters are defined. Third, mandatory and optional parameters are specified for each operation. For example, the Q.931 Setup operation requires four mandatory parameters: protocol discriminator, message type, call reference and bearer capability. Up to 17 optional parameters are permitted. For example Setup can have a called party number, calling party number and transit network selection parameters present. Fourth, procedures are defined using message sequence charts and SDL diagrams.

3.5.5   METHOD CALLS AS APPLICATION PROTOCOLS

The software system analysis and design methodology results in components representing co-located objects that are accessible through defined interfaces. Each interface contains a number of method calls. Each method call specifies in general the operation to be performed, the parameters, the return value and exceptions. These method calls represent the application-to-application interaction. The method calls play the same role as application layer protocol operations.

3.5.6   IETF TEXT-BASED PROTOCOLS

A number of IETF protocols that are important in the convergence process use a common approach to defining the protocol messages. These protocols include the Simple Mail Transfer Protocol (SMTP), the Hypertext Transfer Protocol (HTTP), the Session Initiation Protocol (SIP) for controlling multimedia sessions and two protocols (PINT and SPIRITS) for facilitating interworking between Internet clients and IN Service Control Points. All protocols are defined in a human-readable, common text format and are known as *text-based* protocols.

The protocol messages fall into one of two types: *request* and *response*. Each message has a *start line, headers* and, optionally, a *message body*. The start line of the message differs depending on whether it is a request or a response. A request identifies a method that is requested on a resource. A response contains a status code, informing the requestor of the outcome of the method call. All messages identify the protocol and version. For example, to download a Web page, an HTTP request would start with the line [95]:

```
GET http://www.ee.wits.ac.za/pub/HypotheticalProject.html HTTP/1.1
```

A response starts with a *status line* consisting of the protocol and version, a numeric *status code* and a phrase that explains the code. For example an HTTP server that does not wish to be contacted directly by the requesting party could respond to the GET method as follows:

```
HTTP/1.1  305 Use proxy
```

The start line of a request or response is followed by one or more headers carrying the message parameters. A header is rigidly formatted according to rules for text messages in the Internet. The format of headers and a number of commonly used headers are defined in [42]. Further headers may be defined in individual standards

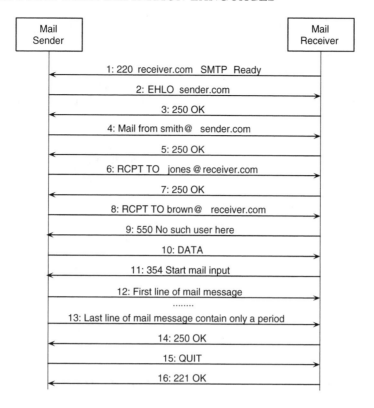

Figure 3.20. Example of an IETF text-based protocol: transfer of an e-mail message using SMTP.

based on RFC822. A header consists of a field name and a field body separated by a colon, for example headers from an e-mail message [146].

```
From: "Joe Teacher" <j.teacher@ee.wits.ac.za>
To: "Kathy Boss" <k.boss@ee.wits.ac.za>
Subject: About graduate study
Date: Wed, 2 Feb 2005 11:40:01 +0200
Message-ID: <003101c5090b$2b2b5ec0$b2108d92@ee.wits.ac.za>
```

The type of header is always identified, for example To and From. The field body may contain several items of information according to definitions in either RFC822 or the protocol standard. For example a Received header may have up to seven items including from, by and via fields.

The message may have a body that follows the headers. The body is in general a number of lines of ASCII text whose format and method of delimitation depend on the application protocol. For example, in SMTP the body of the e-mail message is transmitted as lines of text where the last line contains only a period. In HTML, discussed below, tags delimit the message.

The use of request and response messages is illustrated in Figure 3.20 for the transfer of an e-mail message using the Simple Mail Transfer Protocol [146]. Requests contain method names shown in capitals: `EHLO`, `MAIL`, `RCPT`, `DATA` and `QUIT`. Responses contain numbers and a text elaboration, for example 250 `OK`.

Hypertext Transfer Protocol

Hypertext Transfer Protocol, an Internet protocol conforming to RFC822, is intended to support the transfer of hypermedia between applications [95]. The prime application is therefore the support of Web browsers. While its name refers to hypertext file transfer, HTTP is a generic protocol in the sense that it supports diverse applications. HTTP requires a reliable network transport and normally runs over TCP.

HTTP is a request/response protocol supporting a client server model. Clients may make direct requests to the server or requests may be forwarded via a proxy server. Intermediate hosts such as proxy servers may modify message headers. HTTP defines a number of request methods, for example GET and POST.

The protocol is extendable, allowing further request methods and headers to be defined.

An HTTP request message illustrates the use of headers. This request has no message body.

```
GET / HTTP/1.1
Accept: image/gif, image/x-xbitmap, image/jpeg, image/pjpeg, */*
Accept-Language: en-us
Accept-Encoding: gzip, deflate
User-Agent: Mozilla/4.0 (compatible; MSIE 5.01; Windows NT)
Host: hypothetical.ora.com
Connection: Keep-Alive
```

A response to a successfully executed `GET` method follows. The body of this message contains the Web page encoded in HTML that is to be downloaded.

```
HTTP/1.1 200 OK
Date: Mon, 06 Dec 1999 20:54:26 GMT
Server: Apache/1.3.6 (Unix)
Last-Modified: Fri, 04 Oct 1996 14:06:11 GMT
ETag: "2f5cd-964-381e1bd6"
Accept-Ranges: bytes
Content-length: 327
Connection: close
Content-type: text/html
<!DOCTYPE HTML PUBLIC "-//W3C//DTD HTML 4.0 Transitional//EN">
<HTML>
<HEAD>
    <META HTTP-EQUIV="Content-Type" CONTENT="text/html; charset=iso-8859-1">
    <META NAME="Author" CONTENT="Joe Teacher">
    <META NAME="GENERATOR" CONTENT="Mozilla/4.08 [en] (Win95; I) [Netscape]">
    <TITLE>What is a Mark-up Language</TITLE>
</HEAD>
```

```
<BODY>
<CENTER> <H2>
<FONT COLOR="#CC0000">What is A Mark-Up Language?</FONT></H2></CENTER>
<FONT SIZE=+1>We define a <B>Markup Language </B>as a method of defining the
content of a document as well as instructions to format the document. A
markup language contains only printable characters. Formatting instructions
are distinguished from content by using special characters to show the start
and finish. For example, consider a bulleted list:</FONT>
<UL>
<LI> <FONT SIZE=+1><B>SGML:</B> an ISO standard markup language</FONT></LI>
<LI> <FONT SIZE=+1><B>HTML:</B> used for defining Web  pages</FONT></LI>
<LI><FONT SIZE=+1><B>XML: </B>Extensible Markup Language</FONT></LI>
</UL>
</BODY>
</HTML>
```

3.5.7   HYPERTEXT MARKUP LANGUAGE

The message body in the previous example is formatted in Hypertext Markup Language (HTML), the lingua franca for publishing hypertext on the World Wide Web. HTML is a non-proprietary format for describing the structure of hypermedia documents.   As the example above shows, HTML uses codes embedded in the content for logical markup. Tags are used to demarcate content having particular significance. The section between the `<HEAD>` `</HEAD>` tags contains information about the document. The `<BODY>` and `</BODY>` tags demarcate the page definition. Other tags specify formatting within the document. For example, tags `<H2>` and `</H2>` make a second level heading and `<B>` and `</B>` make bold text. Other tags create bulleted or numbered lists, structure text into tables, and make hypertext links, interactive forms, headings, paragraphs, lists, and produce many more effects. A Web browser knows how to process the HTML tags to produce the correct appearance when the page is displayed on the screen. HTML documents can be created and processed in a wide range of tools from simple plain text editors to easy-to-use authoring tools.

3.5.8   EXTENSIBLE MARKUP LANGUAGE

The Extensible Markup Language (XML) is a pervasive method for defining the structure of information for specific applications or application domains. XML is often introduced as an outgrowth of HTML that allows the information and the formatting of displays to be separated. Rather, XML is a generalised way of defining data elements for various application contexts that contains both the information and its significance in the same document. XML finds application at many levels. The familiar use is definition of data for specific applications, for example a user profile in a telephone system shown in Figure 3.22. A less familiar example is the use of XML to define the basic structure of UML through meta-modelling. Figure 3.21 shows how XML relates to a structured analysis and design software project. The project could deliver XML documents that encode requests or data structures or both. The XML standards provide two means of defining the elements that make up a document for a particular application or application domain. The Document Type Definition (DTD) shown in

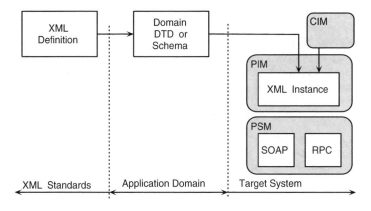

Figure 3.21. Relationship between XML standard, application domain data type definition and data instance specification.

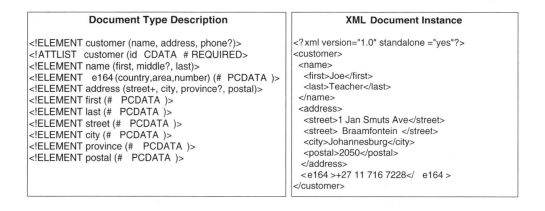

Figure 3.22. XML Document Type Definition (left) and sample document (right).

Figure 3.22 defines the tags and structure of the document. Alternatively, an XML Schema defines the document.

XML is essentially a method of exchanging information between applications in documents that simultaneously identifies the data fields and contains the data in those fields. XML documents have found widespread acceptance due to the ability to define documents or schemas for application domains. The easy readability of XML documents by humans has also aided acceptance.

The number of XML applications is substantial. Four examples typify the application of XML in specific contexts:

- *Internet Protocol Data Record* (IPDR): definition of document formats for accounting and billing information for a variety of ICT services [109].

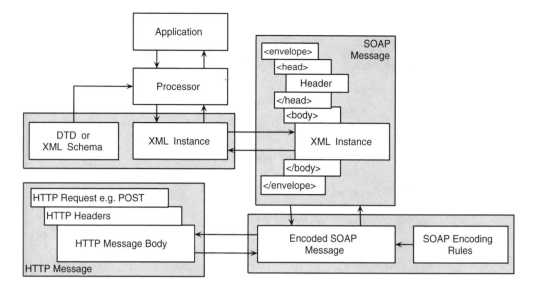

Figure 3.23. Relationship between XML, SOAP and HTTP.

- *VoiceXML*: definition of data and scripts for control of interactive voice response units [154].

- *Web Services Definition Language* (WSDL): a way of describing Web services that allows potential users to establish the suitability of the services for their needs [36].

- The OMG *Meta Object Facility* (MOF): a basic definition the constructs in UML, as shown in Figure 3.15.

Use of XML documents to transfer data requires several supporting protocols and operations, typically those shown in Figure 3.23. The Application requires the support of a processor that uses the DTD or Schema to extract data from and insert into the XML instance document exchanged with another application. A common practice is to encapsulate the XML document in a Simple Object Access Protocol (SOAP) message.

The SOAP message uses XML tags. The XML message is contained in a SOAP message between <envelope> and </envelope> tags. A number of SOAP headers (<head> </head>) may follow to control the transfer of the message. In a request/response type of application, the SOAP message is transported by a HTTP request or response in the body section.

The resulting SOAP message has the overhead of the tags and may require encoding for efficient transmission. An XML definition, for example in a DTD, has a tree structure with compulsory and option elements at each level of the tree. XML definitions therefore express containment relationships. The document contains compulsory or optional elements which in turn may contain elements. An XML

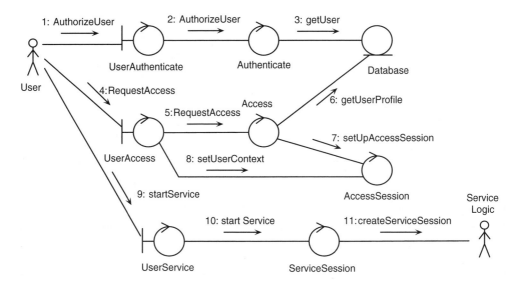

Figure 3.24. Communication (robustness) diagram for the User Access Control System.

document may therefore be represented as a directed graph. Embedding the XML document in a SOAP message simply extends the tree. This structural property is used in serialisation of XML data for transmission in a SOAP envelope. The SOAP layer must therefore be aware of the application-specific type DTD or schema.

The encoded SOAP message is then transferred as the message body in an HTTP request with a POST method.

3.6   DYNAMIC MODELLING NOTATIONS

3.6.1   ACTIVITY, COMMUNICATION AND ROBUSTNESS DIAGRAMS

Dynamic system models depict the interaction of the objects that constitute the system. Several notations known as activity, collaboration, robustness and communication diagrams are in use. We review the communications diagram that was introduced in the ICONIX process as the robustness diagram but is now part of the UML 2 specification. These diagrams serve to expand the use case definition into a more detailed view of scenarios involving interacting objects that perform the actions needed to implement the use cases. The communication diagram shows the objects that participate in the use case and identifies the relationships between pairs of objects. The objects are identified by other analysis steps, for example in the development of the Information View.

Figure 3.24 shows the communication diagram notation. Objects are shown using three symbols: an edge entity, a processing entity and an entity such as a database that does not invoke further entities to service requests. Operations defined for each object are shown on the communication diagram as numbered actions.

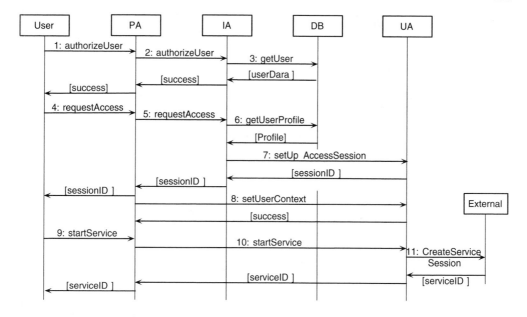

Figure 3.25. Message sequence chart for system shown in Figure 3.8.

We illustrate the use of the communication diagram in the context of example 1 in Figure 3.24. The processes are those that make up the User Access Control System. Interactions numbered 1–8 allow a user to gain access and set up an access session in the Authenticate User use case in Figure 3.14. Interactions 9–11 belong to the Start Service use case. The communication diagram therefore builds on the use case diagram by identifying the processes and how they interact to implement individual use cases.

3.6.2 MESSAGE SEQUENCE CHARTS

A message sequence chart (MSC), or simply a sequence chart, is a construction showing the entities involved in processing and the messages exchanged between entities. The MSC shows the ordering of messages in time but not the exact times of emission. Figure 3.25 shows a message sequence chart for the User Access Control System introduced in Figure 3.8. The four entities within the system (PA, IA, DB and UA) are shown together with the user and an external service provider. The processes shown in Figure 3.24 have been implemented within the four entities. The messages and responses in a successful sequence from the user first requesting authorisation to the starting of a service by the external service logic provider are shown.

Figure 3.25 is one of several message sequence chart formats that are defined in various standards or have been formed by common usage. Sequence charts may be simple or elaborate. In an enterprise view of a system, a sequence chart may show interactions with natural language identifiers rather than formally defined information flows. For formal descriptions, the sequence charts uses actual method calls or protocol operations and may be elaborated by showing, for example, the lifetime of objects,

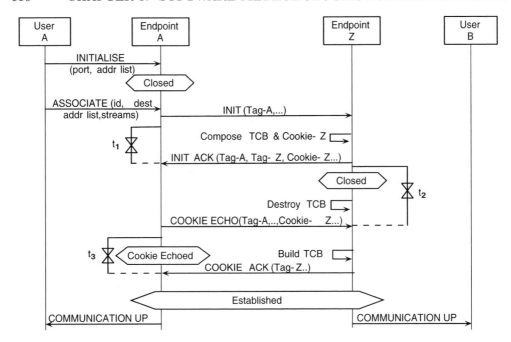

Figure 3.26. Message sequence chart for setup logic for the SCTP protocol expressed in ITU-T convention.

timers, conditions encountered and references to descriptions of internal processes. Message sequence charts may not show returned result fields. One convention for showing returned results is shown in Figure 3.25 using square brackets. Alternatively, a dotted arrow may be used. We review three standards for sequence charts, each emphasising different additional information to the request messages.

ITU-T Message Sequence Charts

While the MSC gives a view of the external behaviour of functional entities, some internal states and actions can be reflected. The ITU-T standard MSC defined in Recommendation Z.103 is possibly the most comprehensive form of MSC. Figure 3.26 uses the familiar notation for messages and primitives invoked on one entity by another. In addition, the notation allows *conditions* to be indicated. A condition may correspond to an SDL state such as Closed or the outcome of a decision, for example a tag is valid, or the occurrence an exception. In the last two cases different sequences would ensue from different outcomes of the decision and possible exceptions.

The ITU-T MSC notation shows the setting of timers and the arrival of the expected event. Not shown is the expiry of a timer before the arrival of an awaited event. As in other notations, the ITU-T MSC allows the indication of internal actions by message arrows folded back to the entity's timeline.

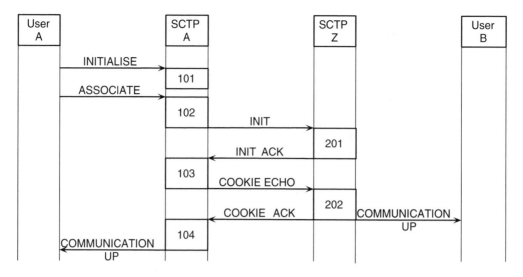

Figure 3.27. UFM information sequence chart for setup logic for the SCTP protocol.

Figure 3.26 summarises key aspects of the ITU-T MSC notation to give an enhanced external view by applying it to the SCTP protocol association initiation introduced in Example 1.

Example 2 continued.

Figure 3.26 uses the familiar notation for messages and primitives invoked on one entity by another. In addition, the notation allows *conditions* to be indicated. A condition may correspond to an SDL state such as Closed or the outcome of a decision, for example, a Verification Tag is valid, or the occurrence of an exception. In the last two cases different sequences would ensue from different outcomes of the decision and possible exceptions.

The ITU-T MSC notation shows the setting of timers, the arrival of the expected event before the timeout, for example INIT ACK before t_1 expires in Figure 3.26. Not shown is the expiry of a timer before the arrival of an awaited event. In such a case, the alternative sequence must also be shown. As in other notations, the ITU-T MSC allows the indication of internal actions by message arrows folded back in the entity's timeline, for example Destroy TCP.

Information Sequence Diagrams

The Unified Functional Method Recommendation defines a form of sequence chart that proves useful in correlating SDL and MSC diagrams. This form is shown in Figure 3.27 in the context of Example 2.

Transitions are shown by blocks on the entity's timeline, staring with the receipt of a message and ending with the emission of one or more message. This method is used in the ITU-T Intelligent Network Distributed Functional Plane definition of

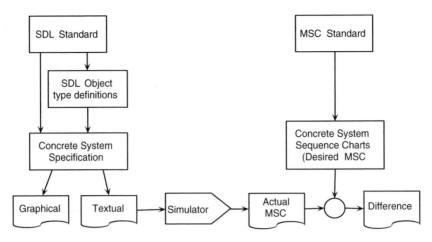

Figure 3.28. System definition process using SDL and MSCs.

functionality to be incorporated in reusable software element. Because of its usage in the UFM to show information flows between functional entities, this type of diagram is known as an *information sequence diagram.*

Example 2 continued.

The blocks marked 101–202 in the information sequence diagram in Figure 3.27 correspond to the transitions similarly labelled in Figure 3.9. For example, transition 102 applies to SCTP-A and commences in the Closed state and ends with the emission of an INIT message to SCTP-Z with the process in the Cookie Wait state. Transition 201 occurs in SCTP-Z, starting in the Closed state with the receipt of the INIT message and ending with the return of an INIT ACK message in the Closed state.

Joint Use of SDL and MSC

Specification and Description Language provides a rigorous description of the internal behaviour of an object or functional entity. The way that this entity interacts with others is depicted using a message sequence chart. In critical designs, SDL and MSC modelling is used to ensure consistency between the internal detail and external behaviour. Often, MSC descriptions show only the main, successful operation sequence. Proper use of SDL compels the analyst and designer to define alternative execution sequences, for example a party unavailable, timer expiry, or exceptional conditions, for example, illegal parameter values. The MSC traces for the alternative and exceptional cases are as important as the main sequence in critical system design.

Figure 3.28 shows the joint use of SDL and MSCs in the analysis and design. The system is designed using SDL for the internal detail and MSC for external interactions. SDL tools provide the ability to simulate the system and produce the actual MSC

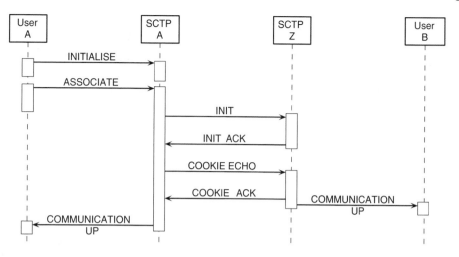

Figure 3.29. UML sequence diagram for setup logic for the SCTP protocol.

corresponding to the internal logic. Differences are then fed back into modification of the design.

UML Sequence Charts

Possibly the most commonly used sequence diagram for new designs and standards is the UML sequence diagram. This notation is applied in the object-oriented context whereas the ITU-T and UFM versions were defined prior to object orientation becoming the de-facto approach. The UML sequence diagram is illustrated in Figure 3.29, again using Example 2.

The diagram is constructed around object lifelines, usually shown as dotted lines. The description at the head of a lifeline may be an object name, an interface, an actor, or a boundary, control or entity element from a communication (robustness) diagram. Synchronous messages shown with solid arrow heads may have implicit return values that are not shown. Asynchronous messages have line arrowheads and, if shown, the return is drawn dotted. The lifeline is thickened to a rectangle to show the object activation. Self-messages folded back in the lifeline show a local activity in the object.

The sequence diagram may be further elaborated by showing the instantiation and destruction of objects. Alternative sequences are seldom shown but are indicated by annotations.

3.6.3 STATE DIAGRAMS

Processes in ICT systems generally involve two or more communicating state machines. Several notations are available for use in describing dynamic system behaviour. State machines are defined by states and transitions between states. In general a state is an identified condition of the process. A process may simply be waiting for an event or could be performing specified processing that could be affected by the occurrence of

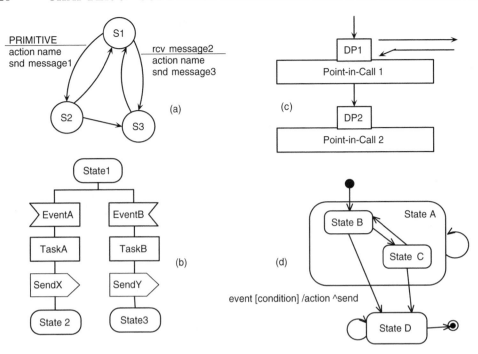

Figure 3.30. Four methods of describing state-transition processes.

an event. For example, in a hypothetical communications protocol, the process is in a state called Idle until a request to open a connection occurs. This type of state is conveniently distinguished as a *waiting state*. Once the connection is open, the process is in the Open state and can process transmission and reception of packets, but could be taken out of the Open state by a request to close the connection. We refer to this type of state as a *processing state*.

Figure 3.30(a) shows a form of *state transition diagram* used in IETF protocol standards, for example [173]. The states, S1, S2, and S3, may be waiting or processing states. The notations:

PRIMITIVE	rcv EVENT
action name	action name
snd MESSAGE	snd Message

indicate that a transition is triggered by receiving a command or primitive or an event occurring. An action specified by name happens, for example Delete TCB and a message is emitted. The system then enters the next state shown by the arrow.

States in SDL, illustrated in Figure 3.30(b), are strictly of the waiting type. All processing is reflected in the transitions between states, for example by task and decision boxes. Events shown in parallel start transitions representing alternative execution paths. The events or messages that can trigger a state transition are shown as inputs following the state. Before entering a state, one or more outputs occur or a timer is set.

The notation shown in Figure 3.30(c) is used in the IN Basic Call State Models (BCSM) [120]. The *Point-in-Call* represents a set of processing steps that can include the receipt and emission of messages to other entities and SDL-type states. A *Detection Point*, such as DP1, represents a particular state in process at which external logic may be invoked. Prior to entering the state, the process notifies the Service Switching Function (SSF) that it has reached the detection point and waits for a response from the SSF. Processing resumes at a specified point-in-call.

Figure 3.30(d) shows the UML *statechart* notation. A state is defined in UML as a situation in which some condition is fulfilled, specified processing takes place or the process waits for an event. Thus, a UML state can be equivalent to an SDL state or could be an abstraction for part of the process. A state can contain substates which may be of the waiting or processing types. For example, states B and C are substates of state A. State transitions occur due to different events. UML provides a notation with up to four fields of the form:

event received [guard condition] /action ^ send message

The event received element must be present. If all four elements are present, the notation means: when the specified event occurs, if the guard condition is true, perform the specified action and send the specified message.

3.6.4 CONNECTION MODELS

The actions of logic in functional entities in ICT systems may control connections and media processing elements in an underlying network. Figure 3.2 depicts a physical entity containing both functional entities and a physical processing element. Connections in multiparty, multimedia systems are often complex, as are those made through media gateways. Connection specifications include the parties to be connected, the media flows, the required quality of service and the connection topology. Several methods of visualizing the function performed and state of the PPE are in use.

The Connection View State (CVS) is used in the IN Capability Set 2 Recommendations is one such notation [125]. The CVS diagram describes the various bearer connection topologies and transitions from one topology to another, together with the events or signals that cause the change. The notation allows multiparty connections as well as connections to Specialised Resource Peripherals to be described.

Figure 3.31 shows an number of pre-defined permitted states with possible transitions. A *call segment* is represented by an unshaded rounded box with a circle indicating the connection. A party connected is shown by a solid line to the circle while a party being set up or held is shown dashed. One or more call segments are contained in a *call segment association* shown as a shaded rounded box. The controlling party is designated by c and other parties are identified by p1 and p2.

The possible views are shown by an number of named stereotype patterns. The *Null* view exists before the call is initiated, an *Originating Setup* state exists from the time of the originating attempt to the authorisation of the call. A *Stable-2-Party* state correspond to an active call.

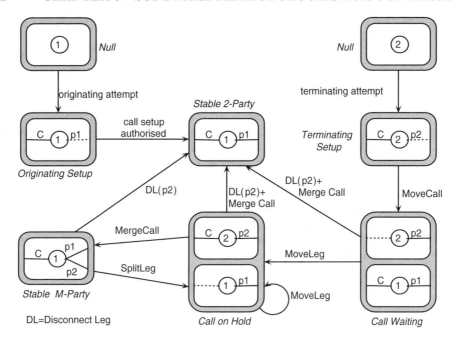

Figure 3.31. Connection View State notation for describing connections in IN Capability Set 2. State names are shown in *italics*.

The CVS construction is useful in depicting call party handling in complex services. Assume that call 1 has been set up. A second call is initiated but with one of the active parties of call 1 as the called party. Call waiting logic is initiated. The waiting call is placed in a call segment of the *Call Waiting* type with the existing call segment. No connection is made so that p2 is associated with a dotted line showing a virtual controlling party. The party on hold is connected and an existing party p1 is put on hold by an INAP Move Leg operation.

The three parties could then be reconnected as a conference call, as shown in the *Stable M-party* state, by merging the two call segments. The conference could revert to the call waiting state by splitting a leg into a second segment. Similarly, the call could revert to a two party call by disconnecting any one leg.

Transitions form one view to another occur when the call process passes a specified detection point or an INAP operation is sent to the switch.

3.7   COMPONENT AND INTERFACE NOTATIONS

The PIM–CV reflects the logical packaging of the software system that emerges from the PIM–IV. Functionality is gathered in *components*. Requests or method calls are gathered in *interfaces*. A component is a unit that must be deployed on a

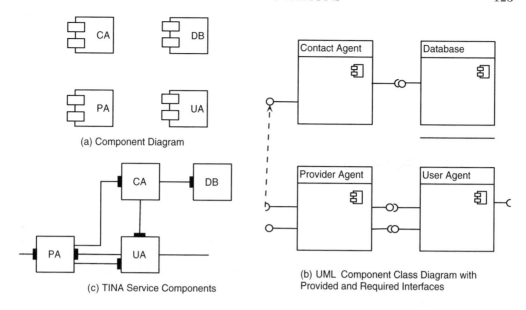

Figure 3.32. Three notations for depicting components and interfaces.

single computing node. Reference points may be created at which conformance with definitions can be verified. The component-level description of the system is, in the absence of constraints, platform independent.

Several notations are available for depiction of the PIM–CV. Three such notations are shown in Figure 3.32. The UML component diagram shown in (a) identifies the components and possibly the interfaces depicted by the small boxes. The UML notation in (b) provides more detail. The component is depicted as a stereotype object. Each component has a set of interfaces that it provides. Similarly, each component indicates the interfaces that it requires for use on other components. Relationships between required and provided interfaces are shown.

The third notation in Figure 3.32(c) is that used in the TINA standards. Here we show the components required in Example 1. The location of the interface is shown by the dark rectangle and the line indicates the component that requires the interface. An example follows of how the interfaces on an object are defined and how this object depends on interfaces belonging to other objects:

```
object ObjPA {
requires
    i_Access,
    i_AccessInitialise,
    i_ProviderInitial,
    i_ProviderAuthenticate,
    i_ProviderNamedAccess,
```

```
supports
    i_Init,
    i_Initial,
    i_Access,
    i_ProviderAuthenticate,
    i_AccountingPull,
};
```

3.8   DISTRIBUTED SYSTEMS

The distributed nature of software processes to support services and applications is a pervading consideration in emerging networks. The early emphasis fell on separate data and telecommunications networks, for example for connecting terminals to mainframe computers. With convergence and high dependence on logic and data, we are concerned with truly distributed systems. As these distributed systems become increasingly complex, we resort to technologies that hide the complexity of distribution.

Computer networking introduced the need to deal with the distributed nature of computing through the Open Systems Interconnection (OSI) approach. While this initiative was concerned initially with managing the complexity of communications protocols through a layered reference model, there was also a strong emphasis on *openness*. An open system presents a standard interface that allows different implementations of the communicating systems to work together. For example, an Ethernet adapter from any vendor should be capable of being inserted into a personal computer and support network connections. Its physical interface is compatible with the Ethernet medium and other interfaces connected to the medium. Its software interface is compatible with the network layer software provided in the computer's operating system. In the OSI reference model, standard interfaces occur at each layer, enabling protocol stacks to be constructed to support particular applications or classes of application.

The concept of openness has been extended to software as well. Definition of applications programming interfaces is a means of partitioning a software system into one part that provides services to another part. The standard way of invoking services is defined as software interfaces with defined method calls, relying on accompanying data definitions. The APIs may be defined in an implementation language independent form, for example UML, IDL or XML, or in a standardised programming language such as Java or C++.

While software applications are distributed across computing nodes, topologies have developed from a simple client–server model, where both parties are on nodes identified by transport addresses, to fully distributed systems. In a fully distributed system, the components that interact may be distributed over several computing nodes, but, more significantly, are unaware of the implementation detail or absolute location of components that they invoke. An open distributed system has public interface definitions for all its constituent modules, hides the implementation details of components and has a standard way of dealing with the arbitrary, and possibly dynamic, location of components on computing nodes.

3.8.1 NETWORK PROTOCOL-BASED DISTRIBUTED SYSTEMS

At the simplest level, the required communication platform for distributed systems could be provided using networking technologies only, for example a TCP/IP network. Routing of packets in an Internet Protocol network and delivery to the correct recipient process on the target host rely on the host IP address and application port number. While port numbers are generally agreed values, the client and server in an Internet Protocol network seldom know each other's IP addresses in advance. Servers may have permanently assigned IP addresses, but end-user hosts usually have per session address allocation. A client usually knows the identity of the server in the form of a URL. How then, are the IP addresses found to allow communication? Two protocols help solve this problem: the Dynamic Host Configuration Protocol (DHCP) and the Domain Name Service (DNS).

A host, on first connecting to the network, invokes a DHCP server that returns an IP address and other configuration information. The IP address is valid for a defined time, called a lease. We now follow the process at the client side, leading to sending packets toward the server.

The client knows the server URL, for example `www.ietf.org`. and the standard port number (80) for Web servers. The application requests the local DNS resolver to return the IP address corresponding to the URL. The resolver queries a name server, a host that is part of a distributed set of name servers holding information for translating between names such as URLs and network addresses, including Internet addresses. The host's IP address is returned.

Having received the server IP address, and having cached it for future use, the host can send application-related packets. The application invokes the TCP layer, supplying the source IP address, source port number, destination IP address and destination port number. The two port numbers are inserted into fields in every TCP frame header. The two IP addresses are passed to the IP layer. The IP layer inserts the two IP addresses into every IP header and also identifies the upper layer user protocol as TCP in the protocol field.

When the IP datagram is received at the server, the protocol field is used to direct the payload, a TCP frame, to the TCP layer. The TCP layer delivers data to the process identified by the destination port.

Network protocols such as TCP/IP offer little distribution transparency, for example against the failure or relocation of a server. We now examine methods of achieving distribution transparency to differing extents.

3.8.2 REMOTE METHOD INVOCATION

We first look at the process of invoking a method on a remote computer when the object implementation and platforms are in the same language but not necessarily running on the same platforms on client and server sides. The most common instance is the Java language with its Java Remote Method Invocation (RMI) technique. Before examining RMI, we review the reasons for Java's ability to run on different platforms. The Java compiler produces an intermediate form of the programme called bytecode that is identical irrespective of the target computer platform. Each platform has a bytecode to native machine instruction interpreter, called the Java Virtual Machine. The Java bytecode provides a compact representation of Java classes that is processed by the virtual machine. The Java Virtual Machine (JVM) provides platform independence, hiding heterogeneity in operating systems and networks.

The RMI system architecture has three components [29]. First, the client side application communicates with a *stub* while the corresponding element on the server side is called a *skeleton*. The server's bytecode is to be invoked by the remote client. The server bytecode is incorporated into a proxy class representing remote objects. Data to be transferred to a remote server in general consists of objects that have defined structures. The objects must be converted (marshalled) into serial bytestreams for transmission. When a remote invocation is made, the client stub initiates the call, marshalls data and demarshalls returned values. At the server side, the received data is demarshalled, the object implementation is invoked and the return values are marshalled.

Second, the *remote reference layer* is a component that allows invocation modes other than a simple unicast client–server type and supports replication of objects and dealing with failure. The presence of replication layer functions keep the stubs and skeletons simple.

Third, the transport layer is based on TCP and is responsible for setting up connection, passing data to the remote node and listening for incoming data. In addition, a table of remote objects containing object identifiers and endpoints is maintained.

3.8.3 WEB SERVICES MODEL

Web services are conceived as a way of achieving application-to-application interaction intended primarily for electronic business (e-business) but finds widespread use. The emphasis is on sharing functionality and data across computing platforms in application-to-application interactions. Web services represents a form of distributed computing in which functionality that already exists can be made available for use by other programmes [210].

A Web service is a software system with a number of characteristics. The interfaces are defined and described using XML. The Web service is identified by a Uniform Resource Identifier (URI). The service description is registered in a repository and can be discovered by would-be users. The service is invoked by passing XML-formatted messages conveyed by Internet protocols, such as SOAP and HTTP, as described in Figure 3.21.

The three parties to Web service provision and usage are shown in the triangular relationship in Figure 3.33. The service provider possesses a software programme, the *service*, with an interface that accepts requests and returns results contained in XML documents. Beyond this requirement, the implementation details are not material. The service is described in an XML format called Web Services Description Language (WSDL). A service description in WSDL has two parts. First, an abstract interface identifies the operations supported, parameters and defined data types. Second, the implementation description binds the abstract interface definition to concrete implementation details: a network address (port and URL), protocol used (for example SOAP) and data structures.

The service description is published in a service registry. The Universal Description, Discovery and Integration (UDDI) standard supports both requests for service

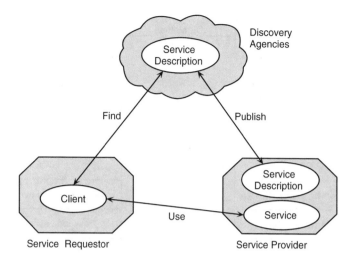

Figure 3.33. Representation of Web services as a triangular relationship.

information as well as posting service information. The UDDI specifies a data model for registry information, again in the form of XML schemas.

The Web services model represents a form of distributed computing. Deployment of complex applications relying on Web services requires further support for reliable message passing between applications and services. One such approach is the enterprise service bus described in Section 5.2.4.

3.8.4 CORBA-BASED SYSTEMS

CORBA is the part of the OMG's Object Management Architecture that supports client and server objects to communicate in a distributed environment. Figure 3.34 shows the essential elements involved in supporting distribution transparency. The object is to support the interaction of a client and server object, irrespective of location and implementation detail. The role of the Interface Definition Language (IDL) in supporting interoperability of differently-implemented objects is discussed in Section 3.5.3. The client–server interactions at application level are those contained in the IDL interface definition.

The method of generating executable code, together with the stub on the client side and skeleton on the server side, is described in Figure 3.19. Two further considerations are the essential ORB services and the transport mechanism.

ORB services provide essential functions to support distribution transparency. Location transparency is enabled by the use of object references. An *object reference* is a name or address. The name and other details are registered by the server in a repository. Only the name need be used to invoke the server object. When different ORBs interoperate, the object reference takes the form of an Interoperable Object Reference (IOR).

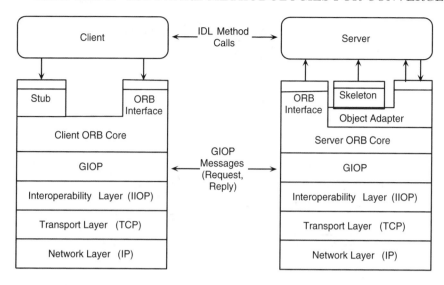

Figure 3.34. Essential elements in CORBA support for object interactions. Examples of protocols and messages are shown in brackets.

The client uses a *naming service* provided by the ORB to locate the server object and then binds to that object (that is, becomes logically associated with it). The CORBA naming service allows objects that are registered with a repository to be found by name. The IOR is returned and requests can be made on the server.

The CORBA *trading service* allows a server object to be located from a description of the services it offers. The trading service allows an application to invoke methods such as query() to find a serving object from a description of the required service. Consider as an example trying to find a database that maintains the user profile for subscribers with a certain range of E.164 numbers. A description of this service is registered as a *service offer* with a trading service named T3. The client seeking the service has access to a trader called T1. The trading service may have links to traders. Rules limit the number of linked traders in a single query and avoid loops. In this example, the application queries trader T1. T1 cannot find the attributes of the requested service and, from its list of links, forwards the query to trader T2. T2 is similarly unable to match a service offer to the description and forwards the query to T3, where a successful match is found. The IOR to the target object is returned to the requestor.

Figure 3.34 shows the client and server side architecture. The ORB interface gives access to ORB services such as naming and trading. The client–server relationship between the communicating objects takes place through the stub and skeleton.

The *object adapter* on the server side supports server objects by performing a number of functions: it registers server classes with the implementation repository; instantiates new objects; generates and manage object references; advertises their availability; handles incoming object calls; and routes the object call to the correct method via the skeleton. The object adapter allows the actual ORB to be minimised. Rather than have an ORB which is designed to meet all possible requirements, the object adapter is specialised for the actual server.

Communication between client and server requires the ORB to be adapted to the underlying communications protocol stack. The CORBA interoperability architecture defines the *General Inter-ORB Protocol* (GIOP) as the interoperability protocol for heterogeneous ORB communication. The GIOP protocol provides an abstract specification that can be mapped onto conventional connection-oriented transport protocols. The GIOP specification defines the Common Data Representation (CDR), a formal mapping of the data types used in a CORBA invocation. The Interface Definition Language types are mapped from their native host formats to a low-level bi-canonical representation. CDR-encoded messages transmit CORBA requests and server responses across a network

The General Inter-Orb Protocol has eight messages that support client–server interactions. Request messages carry all the information necessary to invoke a remote object located on the server. Reply messages are sent in response to Request messages from the server. The LocateRequest and LocateReply messages are used to query the current location of an object. Requests may be multiplexed requests on connections: a client can issue new Request or LocateRequest messages on a given connection before replies to previously issued requests have been received.

The GIOP messages must be adapted to the transport protocol. The OMG defines the adaptation to run over the TCP transport layer. The adapter for this case is called the Internet Inter-ORB Protocol (IIOP). Adapters to other protocols may be defined as required. The underlying transport protocol must have the following characteristics:

- It must be connection-oriented, supporting full duplex, symmetric connections.

- The transport protocol must be reliable, guaranteeing that messages are delivered once only and in order. Positive acknowledgement of delivery is available.

- Information to be transported is viewed as a byte stream: no arbitrary message size limitations, fragmentation or alignments are enforced.

- Notification of disorderly connection loss is provided: both ends of the connection receive an error indication of such a breakdown.

3.9 CREATING A UNIFIED FRAMEWORK

The point of departure of this chapter is the identification of functional entities with interfaces and reference points in the NGN Framework. Functional entities are software elements that require systematic analysis, design and implementation. Frameworks, architecture and formal notations for software systems are therefore important. Two main frameworks for complex systems are the RM-ODP and the MDA. The number of notations is large: we have touched on UML, SDL, MSC, ASN.1 and XML.

In the same way that we developed a single framework, the *NGN Framework*, for analysing and designing ICT systems in Chapter 2, this chapter develops an integrated approach to the analysis, design and implementation of complex ICT software systems. The integrated approach to software systems involves two considerations:

- As an organising principle, the RM-ODP viewpoints and the Model Driven Architecture provide a process with up to five stages, as necessary in each project. We have identified these phases by combinations of RM-ODM and MDA viewpoints as: CIM–EV, PIM–IV, PIM–CV, PSM–EV and PSM–TV.

- The most appropriate notations drawn from the many available should be used to support analysis and design.

Chapter 4

An NGN: the Managed Voice over IP Network

Telephony was traditionally the preserve of switched circuit networks. The Internet uses the TCP/IP protocols and is a best effort network, basically unsuited to real-time transport of signals such as voice. This long-standing division of functions between the Internet and traditional telecommunications networks started to break down in the 1990s. The Internet had progressed from its initial role of providing remote login to mainframe computers, and e-mail and the World Wide Web had become major applications. With the ubiquity of the Internet and its culture of innovation, it was natural that attention would turn to Internet services involving real-time signals: telephony, videoconferencing and media streaming. Agendas driving packet telephony varied. One one hand, the H.323 initiative sought to exploit packet connectivity in local area networks to provide voice in addition to the conventional data communications. Others saw the ability to communicate by voice over the public Internet as an opportunity and were not deterred by the best-effort nature of the network.

Simultaneously, development of the PSTN/ISDN was limited by the basic 64 kbit/s bearer networks and circuit switching from developing into a multimedia network. The Broadband ISDN concept for a multimedia network had not been realised. Telcos looked to the emerging packet voice technology as a cost-effective means of expanding existing telephony services and deploying new networks. Issues of interworking between packet- and circuit-switched voice networks became a major focus of standardisation.

By the late 1990s, the first widespread usage of the term next generation network occurred to describe packet voice networks capable of interworking with existing switched circuit telephone networks. While the distinction was not always made clearly, there are two cases of real-time packet networking: best-effort voice services on the Internet and managed voice services provided by a telco. The latter draw heavily on the standards of the former, but must, in addition, deal with the telco concerns such as customer care, billing and ensured quality of service. The telecommunications standards bodies are therefore active in packet-based multimedia services.

This chapter focusses mainly on telephony and multimedia services in a managed private or public network. We commence with a historical review in Section 4.1, leading to a description of how this chapter is organised. We conclude the chapter with a brief review of voice services in the best-effort public Internet in Section 4.11.

Network Convergence: Services, Applications, Transport, and Operations Support
Hu Hanrahan © 2007 John Wiley & Sons, Ltd

4.1 DEVELOPMENT OF PACKET MULTIMEDIA STANDARDS

A *multimedia service* involves the possibility of communication using different media: audio, video, image, text, hypertext or other forms of data under co-ordinated control. The control mechanism allows changes to parties participating in the service and the media used during service usage. A packet voice service is a special case of multimedia service.

Recognition that digitised voice and video could be encoded in packets led to two initial multimedia standards initiatives: first, an ITU-T initiative to support conferencing on local area networks that would interwork with the PSTN and ISDN led to the H.323 suite of standards; second, IETF initiatives developed the Internet conferencing suite of standards. The IETF standards are currently identified mainly with the Session Initiation Protocol (SIP). As both these sets of standards generally rely on the Internet Protocol for layer 3 routing, services are referred to as Voice over the Internet Protocol (VoIP).

The H.323 suite of standards, or simply H.323, has undergone several stages of development from its original local area network and switched circuit network interworking orientation. The suite now specifies a full multimedia communication standard supporting conferences with voice, video and data exchange.

The IETF conferencing standards rely on SIP for setup, modification and termination of connections. In turn, SIP uses the Session Description Protocol (SDP) to specify the characteristics of the media. The Media Gateway Control Protocol (MGCP) was introduced to support interworking between a SIP-controlled VoIP network and a legacy telephone network. The architectures of SIP-based systems and multimedia session control based on SIP are examined in Section 4.8.

Over time, significant efforts occurred to harmonise the different approached to packet-based multimedia. In version 3 of the H.323 standards, the switched-circuit gateway architecture was changed to allow harmonisation with IETF standards and to expose the signalling transfer between the two types of networks. We examine the harmonised form of H.323 in Section 4.6. The constituent protocols of H.323 are examined: registration admission and status, Q.931, H.245 and media gateway control through H.248 (Megaco). The H.323 system architecture and message sequence charts are presented.

Both standards, H.323 and SIP, concentrate on setting up control associations between the call parties and negotiating the characteristics of the media streams to be transferred between endpoints. Both standards are based on the softswitch principle of separating call control from the transport of user media streams. Except for using the Real-time Transport Protocol RTP for monitoring the delivery of packets containing encoded signals, neither standard deals with the actual transport of packets. The IETF standards include a number of approaches to specifying and, in some cases, ensuring quality of service in IP transport networks.

Both H.323 and SIP inherently support number translation type services and call party handling. Both, however, lack a call state-related triggering method for invoking service features similar to that in IN. This chapter examines calls and sessions. The provision of value-added services in IP networks is left for Chapters 7 and 8.

Section 4.10 reviews the result of the Bearer-independent Call Control (BICC) initiative. This protocol sought to ease interworking between circuit- and packet-switched networks through adapting the ISUP call control protocol for packet bearer networks by separating the call association control from the bearer control.

The subject of multimedia communications in a packet network environment consists of three standards thrusts, H.323, SIP and BICC, as well as two harmonisation models. To build understanding in this complex subject, we first examine the properties and requirements on real-time signals in Section 4.3. We then establish general patterns and principles for multimedia service control in Section 4.4.

The SIP and H.323 standards are essentially protocol definitions with identified entities or agents. The NGN Framework established in Chapter 2 is used to classify functions and aid understanding.

This chapter is concerned mainly with the softswitch control of calls and services. Aspects of media transfer are also considered.

4.2 REQUIREMENTS ON A MANAGED VOICE NETWORK

A *managed voice service* has several attributes. Roles are well-defined: an identified provider exists and users have a subscription with the provider. Service and quality of service levels are agreed. Usage is measured and services can be billed. Such services are also referred to as *carrier-grade* services.

The principal requirement to be satisfied by a managed voice service provided over an IP network is summed up by the phrase 'parity with the PSTN'. Voice services must be comparable with respect to a number of metrics:

- *Ease of use*: a phone on an IP network has a range of implementations: at one extreme a VoIP phone looks like a PSTN plain phone while at another telephony is implemented in a computer, along with other services and applications. There is high availability of dial tone: the probability of not getting dial tone should be equal to or lower than in the PSTN.

- *Low blocking or good grade of service* (GoS): the probability of a call setup failing due to overloaded infrastructure must be comparable to the PSTN. In the PSTN, a call setup fails due to all circuits on a trunk being busy, which is a hard limit. In a packet-switched network, overload has a softer onset. As the demands for packet transport increase, delays and packet loss increase. Deciding when a call is to be blocked is therefore more complex than just running out of circuits between exchanges. Rather, call attempts should be blocked once a performance metric such as delay exceeds a threshold.

- *Post-dial delay*: the speed of call setup should be comparable with that in the PSTN.

- *Acceptable voice quality*: the public is accustomed to the voice quality of the PSTN and the slightly lower value in mobile networks. Low bit rate coding is attractive in IP networks and the codecs used must give quality in an acceptable range.

- *Access control*: as the service is billed, users require control of access either on a per-terminal (telephone like) or per-user (user-id + PIN) basis.

- *Accounting for usage*: the network must be capable of recording the essential billing ticket information so that detailed bills can be produced.

- *Value added (IN-type) services*: the NGN should not be poorer than the PSTN with respect to voice-related value added services such as freephone, call forwarding, caller identification and telephony VPN.

In addition, types of services not readily available in the PSTN may be provided. For example:

- *User mobility*: it should be possible for users to logon to a VoIP terminal and have the calls charged to the user's own account.

- *Always-on*: the terminal equipment must be ready to receive incoming calls and requires minimal initialisation for outgoing traffic. The always-on status incurs no time-based charges.

The requirements on a multimedia service such as videoconferencing follow similar principles. Details such as quality of service for video transport must be defined.

4.3   PROPERTIES OF PACKETISED VOICE

4.3.1   KEY PROPERTIES OF SPEECH

Speech manifests itself as an acoustic wave in air, excited by the human vocal mechanism and detected by the ear. In a communication system, the emitted wave is converted to an analogue electrical signal by a microphone; the analogue signal is reconverted to an acoustic wave by means of a loudspeaker. Between microphone and loudspeaker, the signal is represented in digital form, that is a sequence of numerical values.

User satisfaction in a communication system is measured by three questions:

- Is the received speech *intelligible*?

- Can the talker be *recognised*?

- Subjectively, how close does the speech sound relative to the original: what is its *fidelity*?

The primary characteristic of a communications channel is the highest frequency that is transmitted without attenuation. This value is, in terms of one use of the word, the *bandwidth*. The communications channel bandwidth can be limited by means of a lowpass filter. With a lowpass filter that cuts out frequency components over 7 kHz, the fidelity of speech is considered good and could be described as FM Radio quality. Increasing the bandwidth to 16 kHz makes little difference to the fidelity of speech – music fidelity would, however, be enhanced to CD-quality. Reducing the channel bandwidth from 7 to 3.4 kHz, as is done in telephone systems, lowers the fidelity, but leaves speech intelligibility and the ability to recognise the talker little changed as experienced on telephone circuits.

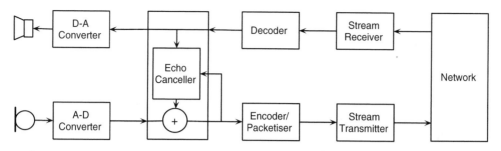

Figure 4.1. Processes involved in encoding and decoding of speech for transmission over a packet network.

4.3.2 DIGITISING SPEECH

The key to convergence is the ability to represent all communications media in digital form, that is a set of numbers that allow the original signal to be reconstructed. Numerous methods of analogue-to-digital signal conversion exist. All start with what is known as waveform coding: representing the time-varying speech signal as numerical values taken at regularly spaced instants. This process relies on *Nyquist's Sampling Theorem*: an analogue signal can be reconstructed from the values of the signal recorded every T seconds where the spacing T is no greater than $1/2B$ where B is the bandwidth of the signal. The sampling frequency is $f_s = 1/T$.

Having sampled the signal, each sample is represented by a number. The number is derived by dividing the range of signal values into N intervals. The intervals are identified by integers k, $0 \le k \le N - 1$. The number corresponding to a particular sample is just the sample in which it lies. The number of intervals is chosen to be $N = 2^b$ where b, an integer, is the number of bits required to encode the signal. The two operations, sampling and quantising, constitute analogue-to-digital conversion, shown in Figure 4.1 digitising the microphone output. The rate at which the conversion process produces data is just $f_s b$.

4.3.3 REDUCING THE BIT-RATE

Since the start of pulse code modulation applied to telephone speech transmission, many standards have been developed for reducing the average bit rate. Table 4.1 summarises the properties of several codecs in common use. The speech coders are identified by the ITU-T recommendation number or other identifier such as the GSM coder.

Several principles are used that exploit the properties of speech and the way that humans perceive speech to reduce the average bit rate.

Waveform coders with instantaneous compression (PCM): the A-law and μ-law PCM codecs fall into this category. These codecs are equivalent to a 14-bit linear analogue-to-digital converter followed by a reduction to 8 bits, which compresses the instantaneous amplitudes according to an approximate logarithmic law.

Adaptive differential pulse code modulation (ADPCM): these coders exploit two ideas. First, a signal formed from the difference of successive samples of a band-limited

Table 4.1. Characteristics of speech codecs

Codec	Principle	Audio BW, kHz	MOS	Bit Rate, kbit/s	Delay, ms
G.711	A/μ-law	3.4	4+	64	0.125
G.722	ADPCM	7.0	4++	64	40
G.722.2/AMR	ACELP	3.4	4+	12.2–4.75	
G.723.1		7.0	4++	5.3/6.3	
G.726	ADPCM	3.4	4–	40–16	5–10+
G.727	ADPCM	3.4		40–16	
G.728	LD-CELP	3.4		16	
G.729	CS-CELP	3.4	4+	8	35
GSM	RELP	3.4	4––	13	60
AMR-WB		7.0	4+	23.6–12.6	25

signal has a smaller amplitude range than the original signal, requiring fewer bits to encode. Second, the signal tends to be confined to a range for a period and the analogue to digital converter can adapt. If the signal is generally low the range of the analogue to digital converter is adjusted down, and when the signal rises the range is adjusted up. This method requires that the signal be processed over an interval and therefore introduces a delay.

Linear predictive coders (LP): this class of coder achieves a significant reduction in the required bit rate by analysing blocks of the signal to create a model of the speech production method. The parameters of the model are transmitted and the speech is reconstructed from the model. Since the algorithm needs to analyse a block of data, there is an inherent delay.

4.3.4 SPEECH QUALITY METRICS

The quality of speech when transmitted over a given channel is determined by a sequence of factors. The first and fundamental determinant of quality is the bandwidth of the speech signal before processing. Telephone quality speech occupies the voice band, 200–3400 Hz. High-quality speech comparable with the quality of FM radio requires a 7 kHz bandwidth.

The second factor is encoding for digital transmission. This step introduces possible subjective degradation of the signal as well as a delay due to processing finite blocks of data and packetisation. The delay is determined to a large part by the frame size. For example, G.726 at 32 kbit/s has a frame size of 5 ms. The minimum delay is twice this value, once for encoding and once for decoding.

The third stage arises from the actual transmission. In a TDM system, bit errors result in added noise. In packet-based systems frames are lost due to transmission errors. Also, the average delay is increased by the transmission time and the delay has variation jitter caused by differing queue lengths in routers along the transmission path. Other factors such as the ambient noise at the source and receiver also affect the perceived quality.

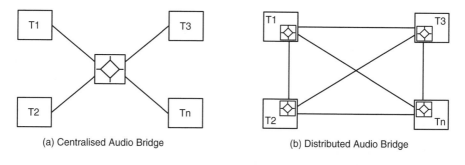

(a) Centralised Audio Bridge   (b) Distributed Audio Bridge

Figure 4.2. Audio bridge configurations. (a) Centralised, and (b) distributed.

Table 4.1 provides approximate measures of the speech quality that users typically experience with particular codecs. The *mean opinion score* (MOS) is one of several ways of expressing the user's satisfaction with voice transmission quality [142]. A MOS value of four corresponds to a judgement of 'Good'. This rating captures the effect of the speech coding method mainly in the absence of other transmission impairments, for example an excessive bit-error rate on the channel. Rather than give exact numerical values, the table ranks codecs in terms of improvements above and below MOS = 4.

Taking the familiar example of GSM mobile telephony, the best speech quality attainable is that determined by the voice bandwidth and coding method used. Together, these result in a MOS below that for fixed line telephony. In an actual call, errors on the air interface give further degradation leading to the phenomenon of 'breaking up' in severe cases.

4.3.5 AUDIO PROCESSING FOR CONFERENCING

When the number of parties to a voice call exceeds two, we term the call a *conference*. Conference calls require voice processing. In general, the received signal of a particular party must be made up of the sum of the transmit signals of all other parties plus an attenuated form of the parties own transmitted signal. The latter component is similar to the sidetone in an analogue phone: voice feedback that allows a talker to control the voice level. For example, for parties 1, 2 and 3 of an N-party conference, the received signal must be:

$$v_{1R} = av_{1T} + v_{2T} + v_{3T} + \ldots + v_{NT}$$
$$v_{2R} = v_{1T} + av_{2T} + v_{3T} + \ldots + v_{NT}$$
$$v_{3R} = av_{1T} + v_{2T} + av_{3T} + \ldots + v_{NT}$$
$$\ldots$$

$$(4.1)$$

where $v_{1T}, v_{2T}, v_{3T}, \ldots, v_{NT}$ are the transmitted signals of the N parties. The constant a reflects the attenuation of the party's own signal to give an acceptable sidetone. The N equations define the total audio bridging function that is required.

Two forms of audio bridge implementations found in packet telephony network are shown in Figure 4.2.

- *Single, central audio bridge*: this type of bridge is a common equipment and, as it is commonly implemented using digital signal processing, can in general support a number of conference calls simultaneously. Each phone in a call is connected to the bridge. The bridge computes the N-receive signals from the set of equations (4.1). The signals are then transmitted to their respective recipients. A variant on the central bridge is to co-locate the bridge with one of the parties, say T1. Only $N - 1$ flows occur in the network but must traverse the access network to which T1 is connected.

- *Distributed audio bridging*: the parties to the call are connected in a full mesh, that is each terminal has a streamed connection to every other terminal. Multi-stream receivers feed into a local bridge. Each terminal must implement a local calculation of its received signal: terminal T_k computes equation k, $k = 1, \ldots, N$ in equations (4.1). The number of bidirectional streams flows in the network is $N(N - 1)/2$.

In general, addition of digitised signals requires decoding and decompression of the signal into a linear quantised form. After addition, encoding and compression must be performed. Even in the simple case of A-law coding, signals must be expanded before being added and the sum converted to A-law.

The type of bridging determines the type of stream connections. For a central bridge, each endpoint makes a unicast connection to the central bridge that supports bidirectional transmission. The bridge performs the matrix calculation according to equation (4.1) and directs each output to the appropriate interface for the destination.

For the case of a distributed bridge, each endpoint makes a multicast bi-directional connection to the remaining conference participants. The N multicast connections are equivalent to a full mesh of unicast connections.

4.4   GENERAL CONCEPTS OF MULTIMEDIA COMMUNICATIONS

Figure 4.3 provides a reference architecture for introducing packet voice networks. Components are identified in several domains. The customer premises and access networks are described in Section 4.4.1. Packet traffic is concentrated in one or more edge network elements that may perform packet-level admission control. Two types of servers are shown: the call agent and the media gateway controller. Dotted lines show signalling relationships. The packet network interworks with switched circuit network via two elements. The media gateway (MG), described in Section 4.6.3, performs the conversion between TDM and packet voice signal formats. The signalling gateway (SG) transfers signalling messages between the packet and switched circuit networks by mechanisms described in Section 4.5.

The term *softswitch* is a generic description for an architecture in which service control is separated from the voice packet transport mechanism in a telecommunications system. A *call agent* is a computing node that performs some or all of the following functions:

- receives and processes access signalling messages;

- performs admission control;

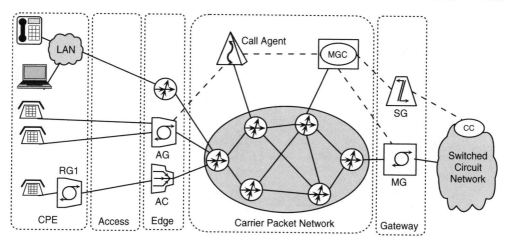

Figure 4.3. Generalised architecture of a packet multimedia network and relationship to switched circuit networks.

- executes call control logic, maintains call state;

- signals to other call agents or control nodes, including those in other networks;

- translates addresses;

- performs call routing in the sense of determining the gateway to be used to a transit or terminating network;

- hands-off control to other servers;

- controls connections with switched circuit networks at a media gateway;

- generates call detail records (CDR) and transfers these to a billing system;

- implements or invokes service features (optional);

- initiates calls in response to requests by application servers (optional).

The terms *call manager* and *call controller* are synonyms for call agent. The call agent is often referred to by terms that are specific to particular standards. A *SIP server* is structured according to roles defined in the SIP standard and uses the SIP and SDP protocols for access and inter-server signalling. A call agent may implement *H.323 gatekeeper* functions and it may then be loosely called a *gatekeeper*. A node controlling stream connections between switched circuit and packet networks, as shown in Figure 4.3, is called a *media gateway controller*. This term is frequently applied to a call agent implementing additional functions.

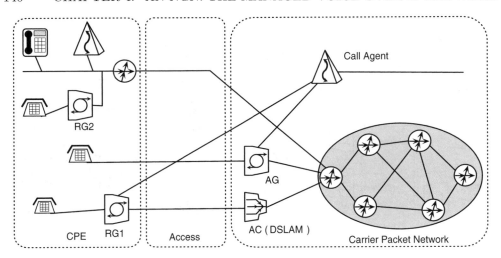

Figure 4.4. Three methods of providing voice access to a carrier's packet network. (a) Access gateway in provider domain. (b) Residential gateway with WAN connection to access concentrator (e.g. DSLAM). (c) Via CPE LAN/WAN network.

4.4.1 ACCESS CONFIGURATIONS

Figure 4.4 shows several access mechanisms for voice packet networks. Several types of gateways form common physical entities in packet voice networks.

The term gateway signifies an element between two systems working according to different standards, particularly application layer protocols. A media gateway provides adaptation and processing of signals between two networks using different encoding principles, for example A-law PCM in a switched circuit network and packet voice in the carrier packet network. Several specific types of gateways are found at the edge of packet multimedia networks.

- *Access gateway:* terminates analogue telephone lines and their signalling, encodes and packetises the speech; is located in the service provider domain.

- *Access concentrator:* a gateway element that receives already packetised traffic via a wide area network from several packet based access devices or gateways and concentrates the traffic toward an edge router.

- *Signalling gateway:* a gateway that terminates SS7 or ISDN D-channel signalling transport on the SCN side and adapts the application layer signalling to a packet transport protocol for transmission to a call agent or similar call control entity. The application layer protocol is relayed across the gateway, not converted.

- *Residential gateway:* located in the customer premises and terminates a limited number of analogue telephones and their signalling.

- *Subscriber gateway:* synonymous with residential gateway.

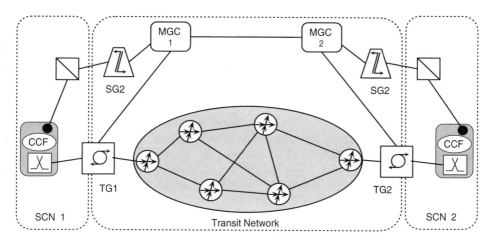

Figure 4.5. Use of a packet voice network for trunking between two switched circuit networks.

Table 4.2. Functions normally associated with types of media processing node

Function	Type of Element				
	IN-SRP	MS	IVR	MCU	MG
On-off hook detection					X
Digit reception	X	X	X		X
Tone generation					X
Announcement replay	X	X	X		
Text-to-speech conversion			X		
Speech recognition			X		
Transcoding				X	X
Signal bridging				X	

- *Trunking gateway:* terminates multiplexed bearer traffic on the SCN side, converting to packet format on the packet network side. Signalling relating to the connections is carried separately, usually through a signalling gateway.

An application of trunking gateways is shown in Figure 4.5. A packet network acts as a transit network between two switched circuit networks. Multiplexed TDM trunks connect the circuit switch to the media gateway. Call control signalling is transferred to the media gateway controllers via signalling gateways.

A packet telephony network contains *media processing nodes* that transmit and receive streams and perform various processes on the signals. Table 4.2 lists a number of media-processing nodes and typical functions performed. For reference, we include the Intelligent Network *Specialised Resource Peripheral* that plays announcements and receives digits dialled by a user. The counterpart in a packet network is called a *media server* (MS). A more powerful physical entity embodying text-to-speech conversion and speech recognition is called an *interactive voice response* (IVR) Unit. A Multipoint Control Unit (MCU) has a specific significance in H.323 systems. Several additional functions shown in Table 4.2 are often incorporated in a media gateway.

4.4.2 TERMINOLOGY

The concepts of call and connection have simple meanings in the PSTN context. In packet-based communication, we also have connections and calls, but the detailed meaning of the terms differs. The concept of a session, well known in data communications, is also used. The idea of a call involving two parties is extended to the idea of a conference in which the number of parties is three or more.

- *Connection*: meaning is dependent on context. In a switched circuit network a connection is a set of reserved resources such as time a slots on a number of trunks. In a packet-switched network, a connection is an association between end stations at a specified level in the protocol stack, for example at the transport layer. In a packet network, the route to be taken by the packets may be determined for a particular connection as a virtual circuit.

- *Call*: a point-to-point communication between two endpoints involving call set-up, a connection during which information is streamed and a termination process.

- *Session:* in general, a session is a period during which a relationship exists among participating entities to serve a defined purpose. Session has specific meanings within specific protocols. A session provides a context for related activities or objects. In packet voice networks, session usually applies to relationships among the media flows.

- *Conference*: a communication between three or more endpoints involving a set-up, a connection during which information is streamed between the parties, addition and removal of parties and a termination process.

- *Media*: the end user information to be transferred and in particular the codec used, bit rates and other parameters such as quality of service.

- *Streaming:* the process of transferring the end-user information or media.

4.4.3 GENERALISED SOFTSWITCHED SYSTEM CONFIGURATIONS

Figure 2.16 identifies the main constituents of a real-time packet-based network and the separation of call control from media streaming. One call manager is active in each administrative domain.

Figure 4.6 illustrates aspects of a call in a softswitch-based system in a number of different configurations. In general, a signalling configuration involves up to four entities: endpoint or terminal A, an associated call manager CM1 contacted by TA, terminal B and the associated call manager CMn. The number of call managers involved in the call may vary; call managers are therefore shown as belonging to a cloud with n members. The four elements TA, TB, CM1 and CMn are visualised as

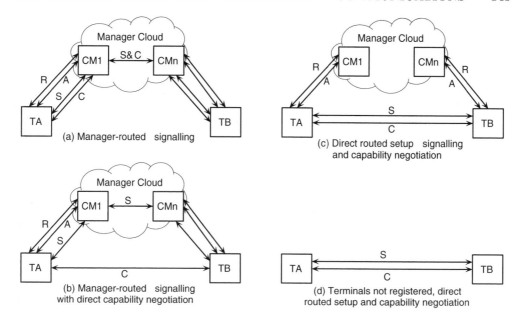

Figure 4.6. Signalling patterns in a softswitch-based network.

the four corners of a trapezoid. Depending on the configuration, signals flow along various edges of the trapezoid, never diagonally.

Four signalling operations are shown. *Registration* (R) allows a terminal to be bound to a call manager. A registered terminal performs *access control* (A) signalling before attempting to set up a call. *Call setup* (S) signalling establishes the association between the call parties. *Capability negotiation* (C) signalling exchanges information on media streams and other call or session parameters and supports negotiation of these parameters.

Four signalling configurations are shown. Case (a) is typical of a managed carrier-grade VoIP telephony. The simplest case (d) would only be encountered in unmanaged or local area network services. A call typically requires up to five sets of actions:

Call setup: establishing the association between call parties;

Capability exchange: transfer of information between endpoints on communication capabilities—may occur simultaneously with call setup actions;

Stream flow: establishment of audio visual communications;

Call modification: addition, suspension or termination of parties, change of bandwidth or media during the call;

Call termination: ending media flows and association between parties.

Scenario (a) shown in Figure 4.6 considers two endpoints TA and TB, which are already registered to two different call managers (CM1 and CMn). A registered terminal must be admitted by the call manager in order to participate in a call. Call signalling is routed through the call managers.

1. TA initiates the call by making an admission request to its call manager CM1. The called party number and media details are supplied with this request.

2. CM1 may not know the identity of TB's call manager and adopts a measure such as consulting a Domain Name Service (DNS) or broadcasting a request to all servers. In the latter case, CMn responds with its own transport address for call signalling. CM1 can now signal to CMn.

3. CM1 acknowledges admission of TA.

4. TA issues a call request to CM1.

5. CM1 relays the call request to CMn. A call progress response with a trying indicator may be returned.

6. CMn sends a call request message to TB.

7. When TB receives the call request and is willing to accept a call, it must request admission.

8. On confirmation of admission, TB sends a call progress response with a ringing indication in the backward direction.

9. When the B-party answers, a call answer response is returned.

The setup phase is complete. Further media negotiation may follow.

In cases (b) and (c) in Figure 4.6, one or both of the call setup and capability negotiation are performed directly between the endpoints once the terminals are admitted by their respective managers.

The generalised cases (a)–(c) have concrete forms in various H.323 and SIP configurations, discussed later in this chapter.

4.4.4 GENERALISED MULTIMEDIA CALL SIGNALLING

In this chapter we consider several signalling protocols for setting up calls and multimedia sessions. While details of the messages and procedures differ, a general message sequence pattern exists. This pattern is also observed in ISUP and Q.931 signalling in switched-circuit networks. Figure 4.7 generalises the call setup message sequence. For generality, three call managers are assumed, one serving each of the A and B parties with a third (CM 2) in a transit network. Three generalised messages are shown. The Call Request message conveys the request to set up the call, with parameters including the B-party identifier and a description of the required media flows. Call Progress sends indications toward the initiating party of the result of processing at each node. For example an indicator Trying gives confidence that the request is successful thus far and has been passed to the next entity. Similarly Ringing indicates that the called party is being alerted. When the B-party answers, a Call Answer message is returned.

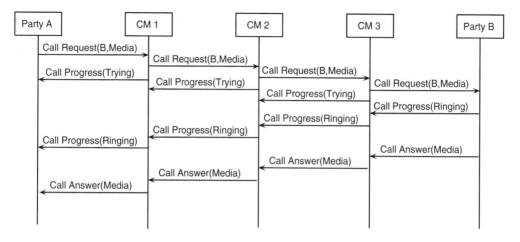

Figure 4.7. Generalised signalling sequence for call set up in voice networks. See Table 4.3 for mapping to specific protocols.

Table 4.3. Mapping from generalised call signalling to specific protocols

Generalised Protocol	ISUP	H.225.0/Q.931	SIP
Call Request	IAM	Setup	INVITE
Call Progress(Trying)		Call Proceeding	100 Trying
Call Progress(Ringing)	ACM/CPG	Alerting	180 Ringing
Call Answer	ANM	Connect	200 OK
			ACK
Call Clear	REL	End Session	BYE
Call Cleared	RLC	Release Complete	200 OK

Not shown are call clearing messages (Call Clear) or those indicating failure of the call, modification and release, with their respective acknowledgements.

Table 4.3 maps the generalised signals into the messages of the ISUP, H.323 family and SIP protocols.

4.5 SIGNALLING PLANE FOR PACKET MULTIMEDIA

Following the convention established in Section 2.3.3, the signalling plane is concerned with the protocols that support application-to-application interactions. Two types of transport are needed in packet multimedia networks: transfer of real-time streams and signalling messages. The conventional protocol stack in an Internet Protocol network uses the TCP protocol at the transport layer. TCP provides reliable transport in the sense of ensuring receipt and correct ordering of all packets in a stream. TCP has features that render it unacceptable for both real-time and critical signalling message transfer applications.

TCP exhibits behaviour known as *head-of-line blocking*. When a TCP entity fails to receive an acknowledgement of receipt of a segment, it must retransmit from the lost segment onward. A delay in transmission ensues. TCP also adjusts the acceptable data rate downward and this rate is slowly readjusted upward until further loss occurs. This behaviour introduces unacceptable variable latency in signalling message transfer. In the case of a real-time signal, TCP cannot ensure that the mean delay and jitter are kept within limits. Recovery from lost frames is of lesser importance in VoIP. TCP's characteristics are therefore unsuited to real-time signal transmission.

The *byte-stream orientation* of TCP means that it treats the data to be transferred as a sequence of bytes and has no notion of a message as an entity. Other mechanisms, for example in the application or application layer, must deal with the delineation of messages. While TCP delivers segments in order, it does not support in-sequence delivery of messages to the upper layer protocol.

The *simple connection model* of TCP is unsuited to high availability computer and network interface configurations that are necessary in carrier-grade signalling applications. Multi-homing of a host computer to give redundant connections and failover is an essential enabler of high availability. The connection in TCP is essentially point-to-point and multi-homing is not achieved efficiently in TCP systems. Mechanisms for recovery from failed connections are better suited to non-critical applications.

TCP is *vulnerable to attack*, for example a SYN attack when a malicious host sends requests for synchronisation that interfere with normal transmission.

Two protocols have been developed for the IP multimedia environment to avoid the use of TCP in real-time and critical situations. The Real-time Transport Protocol (RTP), described in Section 4.5.1 supports real-time stream transfer. The Signalling Control Transport Protocol (SCTP) described in Section 4.5.2 is suited to reliable delivery of critical messages and the configuration of high-availability systems.

4.5.1   REAL-TIME TRANSPORT PROTOCOL

The Real-Time Transport Protocol (RTP) is intended to support end-to-end network transport of streams such as audio and video that have real-time delivery requirements, for example as required in packet telephony or conferencing [190]. RTP may be applied to multicast or unicast services. RTP provides support for sequencing received packets, monitoring delay and jitter and for identifying and specifying different types of payloads. RTP is extensible in that profiles for payload types and formats may be defined. An RTP implementation must follow not only RFC1889 but two further specifications: first, a set of payload codes and the media encoding to which they map, and second, a payload format specification.

RTP has a number of limitations. Having no port identification field, RTP must rely on a protocol such as the User Datagram Protocol (UDP). Unlike TCP, RTP does not attempt to provide reliable transport over an unreliable network. RTP does not address resource reservation; neither does it guarantee quality-of-service for real-time services. RTP provides information to other mechanisms outside the standard for ensuring acceptable delivery of real-time streams. These mechanisms could be as simple as an endpoint adjusting its transmitted bit-rate or as complex as a central bandwidth management system.

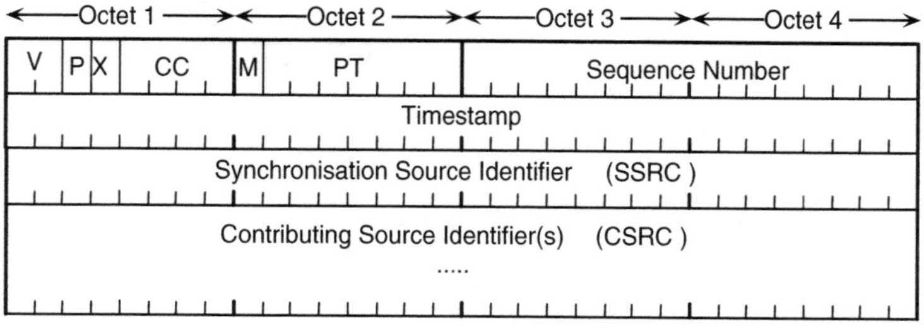

Key:

Field	Definition
V	RTP version
P	Padding indicator
X	Header extension indicator
CC	Number of CSRC identifiers to follow SSRC
M	Marker with significance dependent on profile
PT	Payload format identifier

Figure 4.8. Real-time Protocol header and field definition.

The RTP standard includes the Real-time Control Protocol (RTCP) that supports reporting on the quality of real-time data delivery. RTP and RTCP are designed to be independent of the underlying transport and network layers.

RTP Concepts and Data

The RTP protocol centres on the RTP header and the data that it carries as well as the services offered by the RTCP. Processing of RTP information and reporting by RTP is usually an application layer function rather than a classical OSI layer service.

The fields in the RTP packet header shown in Figure 4.8 indicate the capabilities and limitations of RTP. The first 12 octets are always present. Three important fields are:

- Sequence Number (16 bits): increments by one for each packet sent and is therefore of use in detecting missing packets and resequencing received packets at the receiver.

- Timestamp (32 bits): sampling instant of the first octet in the RTP data packet derived from a sampling clock with resolution defined for the payload format, that is defined in the profile.

- Synchronisation Source (SSRC) *identifier* (32 bits): a number that identifies the source of a real-time packet stream. Every source is identified by a randomly chosen value that is checked to ensure that all SSRCs are unique within a network.

The header may contain Contributing Source (CSRC) identifiers. These are the SSRC values of sources that contribute streams to a media processing element, typically a

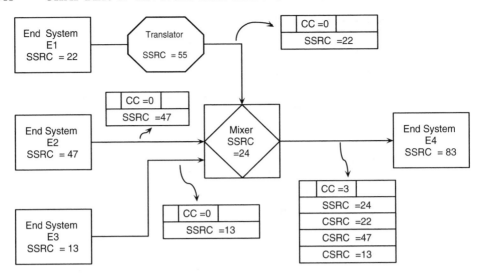

Figure 4.9. Identification of synchronisation and contributing sources in RTP.

mixer or audio bridge. The Contributing Source Count in the header indicates the number of contributing sources, including zero, that follow the fixed header.

A number of entities participate in an *RTP Network*. Entities are classified into four types:

- *End System*: a source or sink of RTP streams that acts as a synchronisation source.

- *Mixer*: combines streams and may change the stream format but does not alter the SSRC. The mixer makes timing adjustments among streams and generates new timing for the outgoing stream. The SSRC identifier of the outgoing stream is that of the mixer.

- *Translator*: changes the format of a stream but keeps the synchronisation source identifier. Translators can be transcoders or replicators from unicast to multicast transmission or devices used to traverse firewalls. While the SSRC value is unchanged across a translator, the payload type may change.

- *Monitor*: an entity that receives RTCP packets and estimates QoS parameters. Monitor functions are not part of the RTP standard.

Figure 4.9 shows an illustrative RTP network. Four End Systems, E1–E4, a Translator and a Mixer make up the network. Every source of a real-time packet stream is identified by a Synchronisation Source identifier. For example, the End Systems on the left have SSRC values 22, 47 and 13, respectively. The RTP packets emitted by a source carry the SSRC value in their headers. The SSRC is an alias for an end system or mixer and could change if the entity has to restart transmission during a session. Each source therefore has a canonical name, CNAME, of the form `user@host.domain`.

At a Translator, the payload format is changed. For example, packetised G.711 64 kbit/s A-law stream from end system E1 is translated to 8 kbit/s G.729 packets. The translated signal carries the original SSRC value. Some information identifying the payload and the packet count changes to describe the new payload.

At a Mixer, a number of streams already in or converted to compatible payload format are combined to form a single signal that is transmitted onward. In the example, streams from sources E2 and E3 as well as the output of the Translator are combined for transmission to end system E4. The mixer has a unique SSRC value, 24 in the figure. Packet headers have CC = 3, indicating that three CSRC values $(22, 47, 13)$ follow for the three contributing sources. The output data format may be changed. The RTP headers thus preserve information about the origin of each stream and changes in synchronisation source for every translation and mixing operation.

Real-time Control Protocol

Information gathered from SSRC identification, sequence numbers and time stamps on each packet allows each entity, other than a translator, to measure the real-time performance of each link. The Real-time Protocol Control Protocol (RTCP) allows entities to exchange performance information via Sender Reports (SR) and Receiver Reports (RR). Generally, reports are exchanged between two entities in unicast communication or between one entity and a number participating in multicast distribution.

The RTCP has three mandatory functions. First, RTP provides feedback on the quality of real-time data distribution. Second, the identity of the original sources is maintained through the use of the canonical name (CNAME) identifier in a packet of the Source Description (SDES) type. This measure is necessary because of the possible change of SSRC and to allow receivers to keep track of all participants and to associate all the data streams for a participant. Third, all participants must send RTCP packets to other members of the RTP network that exchange real-time streams. RTCP may also be used to carry simple session control information.

RTCP provides the mechanism for exchanging data on the performance of the call or conference. Congestion and flow control are the responsibility of other protocols. For example, a third party monitor may collect information from RTCP packets which could be used by a call manager to determine whether new calls can be admitted to the network.

RTCP measures and exchanges statistics such as the total number of RTP data packets and octets transmitted, the fraction and cumulative number of RTP data packets lost and the inter-arrival jitter. Each endpoint is either a sender or receiver of RTP packets or both. A sender periodically reports the information listed in Table 4.4, called sender information, to receivers. Each receiver accumulates the information shown in Table 4.4 in a structure called a report block over the reporting interval on streams received from a particular SSRC.

The RTCP report packets are divided into two types: Sender Reports and Receiver Reports. A Sender Report contains a header and sender information. An entity, other than a translator, generates a sender report if it is a source of RTP packets.

Table 4.4. RTCP Sender and Receiver Report fields

Sender Report	
NTP time stamp:	An absolute wallclock time
RTP timestamp:	Value of the clock used in RTP timestamps corresponding to NTP time.
Sender's packet count:	Cumulative number of packets sent since start of transmission.
Sender's octet count:	Cumulative number of payload octets sent
Receiver Report	
fraction lost:	Since the last report
cumulative packets lost:	Since beginning of reception
highest sequence number:	Received
interarrival jitter:	Estimate of variance of arrival timestamps
last SR:	NTP timestamp of last sender report received
delay since last SR:	Between last sender report packet and this reply.

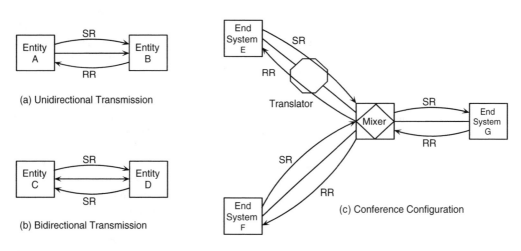

Figure 4.10. Exchange of Sender and Receiver Report blocks using RTCP.

If the sender also receives data, a **report block** follows for each source, identified by its SSRC. A receiver report consists of the RTP header and one or more **report blocks**. RTCP requires that report and Source Description (SDES) packets be transmitted as a compound packet within a single UDP datagram. Thus, the CNAME is available with every report.

Figure 4.10 illustrates the exchange of sender and receiver report packets. End systems and mixers perform reporting. Translators forward reports but do not generate their own data and reports. In the simple unidirectional transmission case in Figure 4.10(a), Entity A is the sole source and generates periodic SRs. Entity B emits only RRs. In the bidirectional transmission case shown in (b) both ends send and receive SRs. Case (c) shows a conferencing configuration, considering only the transmission of signals from Endpoints E and F to G. Sender and Receiver Reports between endpoint E and the Mixer are forwarded by the Translator. Otherwise, SR and RR are exchanged between adjacent entities as shown.

The RTP standard does not specify how the data reported in SRs and RRs is to be processed: that is the concern of the system designer and depends on the system configuration. In a conference situation, shown in Figure 4.10(c), the audio bridge is a potential site for gathering information for use in bandwidth management. All receivers measure the variance of jitter. Let σ_{ij}^2 be the variance measured at node j for the stream coming from node i. In the example of Figure 4.10, denote the endpoints by $i = 1, \ldots, 3$ and the mixer by $i = m$. The mixer generates jitter measures σ_{1m}^2, σ_{2m}^2 and σ_{3m}^2. The mixer receives metrics σ_{m1}^2 from end system 1, σ_{m2}^2 from end system 2, and σ_{m3}^2 from end system 3. Thus, values of all jitter variance metrics are available at the mixer. Similarly, the mixer has access to six measures of packet loss.

4.5.2 RELIABLE SIGNALLING TRANSPORT IN PACKET NETWORKS: SCTP

Signalling in the PSTN is carried between entities such as switches and service control points by the robust, high-performance Signalling System No. 7. In IP telephony architectures, signalling messages must be transported by the packet network between elements such as signalling gateways, call managers and media gateway controllers. Two principal requirements arise. First, for *interworking* between circuit-switched and packet telephony networks, call control signalling must be relayed between the gateway circuit switch and the call manager or media gateway controller. Second, signalling arrangements in the IP network must have *dependability* and *availability* characteristics similar to those of SS7.

The Stream Control Transfer Protocol (SCTP) is designed to be a reliable transport protocol for signalling messages operating on a transport network that does not provide mechanisms to ensure reliable delivery [168, 197]. This approach is in contrast to SS7 where the mechanism to ensure reliable delivery is located at the data link layer. Reliability is achieved in SCTP through the following services:

- establishment of an association between endpoints;

- acknowledged error-free non-duplicated transfer of user data;

- supporting multiple streams within an association that do not block each other;

- sequenced delivery of user messages within multiple streams;

- optional unsequenced (order-of-arrival) delivery of messages;

- bundling of multiple user messages into a single SCTP packet;

- data fragmentation to conform to MTU size requirements;

- multi-homing at either or both ends of an association for fault-tolerance;

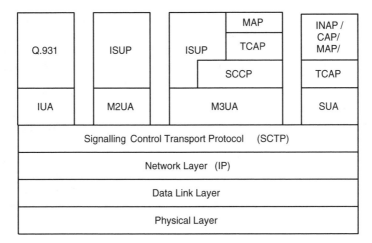

Figure 4.11. Protocol stack with SCTP as the transport layer. Adapters give interfaces corresponding to ISDN D-channel and various levels in the SS7 stack.

- congestion avoidance behavior; and

- resistance to security attacks.

Protocol Stacks Involving SCTP

In general, the SCTP is a stream-oriented transport layer protocol that operates over a connectionless IP network layer. SCTP is intended to transport messages belonging to PSTN, ISDN, IN and mobile network protocols over an IP network. Protocols include Q.931, ISUP, MAP, CAP and INAP. These message may originate and terminate in the switched circuit network or in the IP network.

Each of the SCN-oriented protocols assumes the services of lower layers in the SS7 protocol model. The ISUP circuit-oriented protocol requires SCCP and MTP-3. Transaction-oriented protocols such as MAP, CAP and INAP require the support of TCAP and SCCP. Application layer protocols such as INAP and MAP, rely on Global Title Translation provided by the SCCP layer. With the possibility of signalling points being located in the IP network, the significance of global title goes beyond E.164 address translation. To accommodate the range of application protocols and to support TCAP and SCCP, the SCTP standard defines a single set of services provided by the SCTP layer and an adaptation layer for each of the upper layer protocols.

Figure 4.11 shows the basic protocol stack with the generic SCTP transport layer. For each user protocol, a User Adaptation (UA) module is defined outside the basic SCTP standard and is located as shown in Figure 4.11.

- IUA: this UA adapts Q.931 signalling to the SCTP layer, for example for delivery from a signalling gateway that receives basic or primary rate D-channel signalling to the media gateway controller.

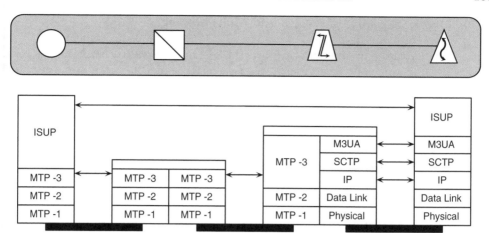

Figure 4.12. Signalling plane definition for interworking between switched circuit networks and packet voice networks.

- M2UA: this UA allows the transport of MTP-3 messages, for example to a media gateway controller. M2UA provides an interface equivalent to the MTP-3/MTP-2 boundary. Within the packet network, M2UA gives an effective point-to-point link: layer 3 IP routing routes packets between the points of the connection.

- M3UA: this UA allows transport of signalling messages belonging to protocols that use MTP-3 for message routing in the SS7 network, for example ISUP and SCCP.

- SUA: this UA allows the TCAP layer supporting transaction-oriented protocols such as INAP, CAP and MAP to access the SCTP layer while utilising SCCP-type services.

Figure 4.12 shows the signalling plane supporting ISUP, one of the possible application protocols, signalling between a PSTN switch and a call manager in an IP network. The protocol stacks are shown at significant stages: the originating Signalling Point (SP), a Signal Transfer Point (STP) in the SS7 network, the signalling gateway (SG) and the call manager. Protocol stacks are shown for each of the entities. The stacks at the SP and STP are conventional SS7 types used for transporting ISUP messages over MTP-3. At the signalling gateway, the SCN-side stack must match the STP stack, namely MTP-3, MTP-2 and MTP-1. On the IP network side, the M3UA layer presents an MTP-3 interface at the upper side, allowing the user messages (ISUP in this example) to be relayed [193]. The remainder of the stack is SCTP, IP and the prevailing data link and physical layers. The end-system stack at the call manager is a peer-to-peer mirror of the SG stack, with an ISUP application layer.

The signalling gateway does not process the application layer messages: the SG function is rather to change the supporting protocols from a SS7 stack to an SCTP/IP stack with the appropriate user adapter.

Associations, Streams, Messages and Multi-homing

Before two entities can communicate over SCTP, an *association* must be set up at the request of an upper layer user. An association exists between two endpoints, conveniently called A and Z. The association involves sets of permissible network addresses, a verification mechanism and the maintenance of state information.

When an association is established, a number of *streams* can be specified. A stream is a logical channel identified by a Stream Identifier that is used for transferring user messages. Each stream has a Stream Sequence Number for use in ensuring in-order delivery. A stream may become blocked while waiting for the next message in the sequence but does not block other streams between the same pair of endpoints.

A *message* in the SCTP context is the unit of data submitted by the sending ULP for transport across the network.

The SCTP packet format is shown in Figure 4.13. Every frame contains the 12-octet general header. Like TCP and UDP, the header contains the source and destination port numbers. The second field is the Verification Tag. On the establishment of an association, each endpoint generates a 32-bit random number that is exchanged during the association setup. Subsequently, every packet must carry the Verification Tag. Each endpoint verifies the tag value received and discards the packet if an invalid value is detected. A checksum is computed over the entire SCTP packet.

The remainder of the SCTP packet consists of one or more *chunks*. A chunk has the general structure shown in Figure 4.13. The Chunk Type field identifies whether the chunk contains user data or a protocol operation and associated data. For example, the DATA type of chunk is formatted as shown in Figure 4.13. Three chunk flags are defined. The U-flag indicates whether sequenced delivery is waived. The B-flag indicates whether the beginning of the message is in the present chunk. Similarly, the E-flag indicates that the end of the message is in the chunk. Fragments carry the same Stream Sequence Number. The B and E flags aid in the reconstruction of a message that must be fragmented over different chunks. For example a chunk with B = 1 and E = 1 carries a non-fragmented message.

A DATA chunk carries the Stream Identifier (S). Every chunk carries a Transmission Sequence Number (TSN), a sequence number that increments with each chunk that is used for acknowledgement of receipt. The DATA chunk carries the Stream Sequence Number that is significant only within the stream and is used for reconstruction of the message sequence.

Other types of chunks are defined for each of the SCTP protocol operations and acknowledgements. For example, a DATA chunk is acknowledged by a Selective Acknowledgement (SACK) chunk.

The ability to carry more than one DATA chunk in a packet is called *chunk bundling*.

Association setup at the SCTP level is accomplished by a four-way handshake, described in Example 2 in Chapter 3. The initiating SCTP entity sends an INIT chunk that specifies the number of outbound streams to be used and the acceptable number of inbound streams that the peer may establish. The initiator gives its own host address or host name. The latter would be resolved by a mechanism such as DNS. The Initiation Acknowledge (INIT ACK) chunk defines the Z-endpoint's required number of outbound streams and may cap the number of streams requested by the initiator.

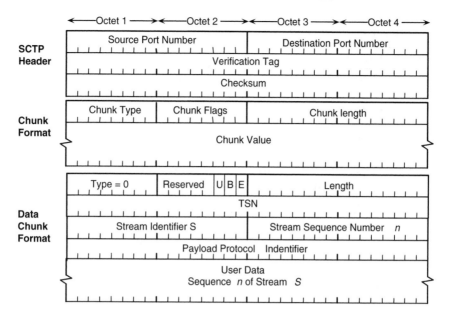

Figure 4.13. Frame format used in SCTP showing the general header, the chunk structure and the structure of a DATA chunk.

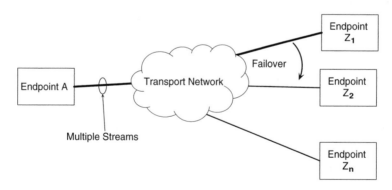

Figure 4.14. Illustrating the concept of an association between one source and several destinations in SCTP.

This chunk contains a State Cookie, a structure containing all state information that must be exchanged at this stage

Figure 4.14 shows the principle of multiple homing in SCTP communication. Endpoint Z is a single entity that is reachable via more than one network interface, $Z_1, Z_2, \ldots$. The ability to reach a single entity via different network interfaces, and hence transport addresses, is called *multi-homing*. Endpoint A normally sends packets to interface Z_1. If this interface is unavailable, transmission would be redirected by

Table 4.5. Principal ITU-T Recommendations in the H.323 series

Number	Coverage
H.225.0	Call signalling protocols and media stream packetization
H.245	Control protocol for multimedia communication
H.248.1	Gateway control protocol: version 2
H.248.2–35	Packages for H.248
H.323	Packet-based multimedia communications systems
H.350.1	Directory services architecture for H.323
H.450.1	Generic functional protocol for the support of supplementary services
H.450.2–12	Specific supplementary services
H.460.1–16	Guidelines for the use of the generic extensible framework

the sending endpoint A to an alternate interface, say, Z_2. The process of changing from an inactive interface to a working interface is called *failover*.

Multi-homing is achieved by each endpoint being reachable by more than one transport address. The multiple transport addresses are specified by a port number, usually a common value, and a list of network addresses. The address list is exchanged during association setup and one address is selected as the primary address.

The SCTP standard gives broad guidance on failover procedures. When a transport address becomes inactive, for example due to failure, or an upper layer user explicitly requests transmission to a transport address that is inactive, the endpoint attempts to send the data to another transport address on the list. Raising an error notification is a last resort. The strategy for choosing the alternative transport address is regarded as an implementation decision.

4.6   THE H.323 SUITE OF MULTIMEDIA COMMUNICATIONS STANDARDS

4.6.1   DEVELOPMENT AND STRUCTURE OF H.323.

The designation *H.323* is both an ITU-T Recommendation number and a common name used for one of the main approaches to or architectures for packet-based telephony and multimedia communications. In the latter sense, H.323 is used to denote a suite of ITU-T Recommendations. The principal Recommendations in the H.323 suite are listed in Table 4.5. We concentrate on H.323 itself, H.225.0, H.245 and H.248.1.

The initial intended application of H.323 was in a local area network (LAN) environment where, provided the LAN is not overloaded, the quality of service would be adequate. H.323 draws heavily in existing ITU-T recommendations, particularly ISDN and Broadband ISDN. The focus of H.323 has shifted as indicated by a name change. Originally, H.323 was entitled "Visual telephone systems and equipment for local area networks which provide a non-guaranteed quality of service". After 1996, the title became simply "Packet-based multimedia communication systems" [137]. From early versions, H.323 addressed interworking with switched circuit networks. The early emphasis fell on interworking between packet-mode and PSTN and ISDN terminals.

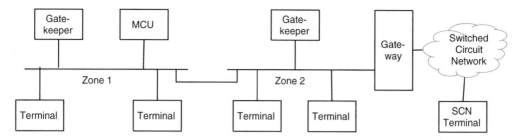

Figure 4.15. Principal entities defined in the H.323 Recommendations version 3.

The current standards support switching of calls between circuit-switched and packet-switched networks. H.323 is a true multimedia standard covering conferences with voice, video and data communication. H.323 is of special interest for voice over IP networks.

From its first incarnation as a LAN-based audio-visual system, H.323 evolved to its present version 5. Version 2 was approved in 1998 and improved the security and solved a problem in version 1 that call setup could take a long time. Also, ISDN-style supplementary services were introduced in the H.450 series of recommendations, for example call forwarding (unconditional, on busy and on no reply). Version 2 improved the setup of data services by requiring endpoints to support data services as well. Version 3 represented a set of enhancements to version 2. Version 4, approved in 2000, made a major architectural change, namely the decomposition of the H.323 Gateway, discussed in Section 4.6.3, and other enhancements to make H.323 better suited to large-scale carrier-grade applications. A second significant addition is the incorporation of an extensibility framework that will allow the already comprehensive base protocol to remain unchanged when new features are added. The emphasis in the current version 5 is maintaining stability. With the move toward large scale VoIP networks H.323 is a viable voice telephony standard.

4.6.2   THE H.323 ARCHITECTURE

The entities and general architecture of a H.323-based system are shown in Figure 4.15. The underlying packet network is not specified and is typically based on IP, Ethernet or ATM protocols. Figure 4.15 indicates interworking with the PSTN through a H.323 Gateway entity. We do not consider the earlier emphasis in H.323 on interworking with compatible terminals on switched circuit networks.

H.323 uses the notion of a *call* as point-to-point multimedia communication using a collection of channels between endpoints, possibly involving intermediate elements such as an audio bridge. An *endpoint* is an entity that can call or be called, that is, it can act as a source or sink of information streams.

The H.323 Recommendation recognises a number of entities:

- *H.323 Terminal*: an endpoint on a network which provides real-time, two-way communication with another H.323 terminal, a Gateway or a Multipoint Control

Unit. In its simplest implementation, an H.323 terminal must provide voice communications.

- *H.323 Gateway*: allows communication between H.323 packet-network based terminals and other networks, principally switched circuit networks. Before H.323 version 4, the gateway is a combination of media and signalling gateway with control functionality. The gateway is decomposed into these three elements in version 4. The Gateway is required only if SCN interworking is supported.

- *H.323 Gatekeeper*: as its name indicates is an entity on the packet network which controls access to the network by H.323 terminals, Gateways and MCUs. The Gatekeeper is a form of call manager providing services such as address translation, locating Gateways and bandwidth management. The Gatekeeper while essential in carrier-grade applications, is not required in all networks.

- *Multipoint Control Unit* (MCU): an endpoint in the H.323 context which controls three or more terminals participating in a multipoint conference. H.323 has two support functional entities (a concept not used in the H.323 specifications):

  - *Multipoint Controller* (MC): the set of functions required to control a multiparty conference.
  - *Multipoint Processor* (MP): provides a range of stream processing operations including switching and mixing.

Three elements, the H.323 terminal, the Gateway and the Multipoint Control Unit are sources and sinks of stream traffic. These three elements are therefore described as endpoints. The Gatekeeper is essentially a control function and does not handle media streams. For every Gatekeeper there is a *zone*. A number of terminals, gateways and MCUs registered with a Gatekeeper occupy the zone. A zone is not necessarily defined geographically or by network topology but is simply the set of registered endpoints.

H.323 systems can be implemented on various scales. For example, a minimal H.323 system implemented in a LAN with no access control or SCN interworking requires only terminals. A carrier-scale implementation requires Gatekeepers and Gateways as well.

4.6.3 GATEWAY DECOMPOSITION

The H.323 Gateway is a complex, monolithic entity with three sets of important functions shown in Figure 4.16(a). First, at the connection/bearer level, the Gateway terminates streams and performs format conversions. On the packet network side it must function as an endpoint for information streams. On the SCN side, the gateway must terminate trunks. Second, the Gateway is also involved in signalling interworking. In early versions, the concern was with conversion of signalling formats between H.225 and those used by devices in the switched circuit networks. In large-scale networks, signalling conversion is not performed. Third, call control functions may be performed.

The complex, monolithic gateway reduces flexibility for implementing H.323 networks and limits the scalability of the architecture. Development of multimedia and packet telephony standards in the IETF favoured separate entities for media

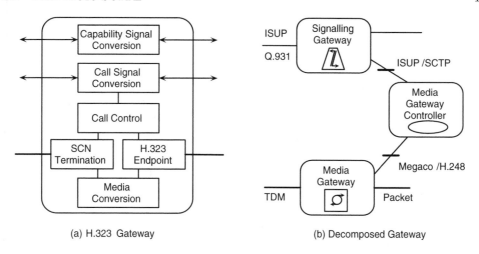

(a) H.323 Gateway   (b) Decomposed Gateway

Figure 4.16. Decomposition of the H.323 Gateway (a) into three elements shown in (b): Signalling Gateway, Media Gateway Controller and Media Gateway.

connection and conversion, call control and signalling conversion as shown in Figure 4.16(b). The H.323 Gateway was therefore decomposed in version 4 of the Recommendation. Three resulting building blocks for packet multimedia systems are the Media Gateway (MG) and Media Gateway Controller (MGC) discussed in Section 4.7 and the Signalling Gateway (SG) described in Section 4.5.2.

The decomposed gateway architecture requires the support of two protocols. First, the Signalling Gateway must carry signalling messages between the Media Gateway Controller and the switched circuit network. The SCTP protocol described in Section 4.5.2 allows the transport of PSTN signalling messages across the IP network.

Second, the interface between the MGC and the MG is exposed and had to be standardised. Two protocols respond to this need. The Megaco Protocol is a comprehensive media gateway control protocol and was developed by the Megaco Working Group of the IETF and was adopted as ITU-T Recommendation H.248.1, simply named the Gateway Control Protocol [136]. Megaco is reviewed in Section 4.7.2. The Media Gateway Control Protocol (MGCP) reviewed in Section 4.7.3 is a lighter-weight protocol that provides a set of messages which allows the Media Gateway Controller to control actions in the Media Gateway. The Media Gateway Controller is a functional entity that establishes and tears down connections and controls media conversion processes in the gateway.

With the decomposed gateway, the possible mappings of functional entities in H.323 onto the NGN Framework are shown in Figure 4.17. The two physical processing elements, the Media Gateway and the Multipoint Processor are allocated to the SCMF layer. The Multipoint Controller and the strictly defined Media Gateway Controller perform resource control functions and are located in the RCF layer. The Gatekeeper is predominantly a service control function, as are the endpoint functions of the H.323 Terminal.

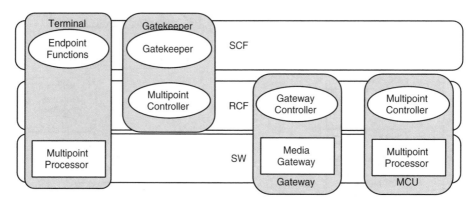

Figure 4.17. Functional entities defined in the H.323 standards mapped onto framework.

4.6.4   ADDRESSING IN H.323 SYSTEMS

Addressing in a H.323-based communication system is complex and the following types of address are recognised:

- *Alias*: a VoIP telephony user must have an E.164 number to be reachable from any telephone. A PSTN call to be terminated on a H.323 terminal will carry a E.164 called party number. The E.164 number has no significance in the packet network. The E.164 number is therefore an alias when viewed from within the network.

- *Network Address*: the address in the packet network of an endpoint or Gatekeeper. In an IP network, addresses may be allocated temporarily.

- *Transport Address*: the combination of Network Address and the identifier of the upper layer destination, called the *Transport Service Access Point* (TSAP). In a TCP/IP network, Transport Address and socket (IP address + Port Number) are synonymous. Multiple upper layer entities share a network address but have different port numbers.

The Gatekeeper translates the Alias to the Network Address. H.323 uses the terms addressable and callable to classify entities. An *addressable* entity has a Transport Address and can receive and generate signalling. In H.323, a call is the multimedia communication between endpoints, lasting from the call setup procedure to call cleardown. All entities except the MC and MP are addressable: these entities must be co-located with another addressable element. A *callable* entity is capable of being identified by a user to take part in some service involving information streams. For example, a Gatekeeper is not callable while Terminals are.

4.6.5   H.323 PROTOCOL STACKS

The protocol stack for H.323 is shown in Figure 4.18. The low layer protocols are assumed to be IP at the networking layer supported by the data link and physical

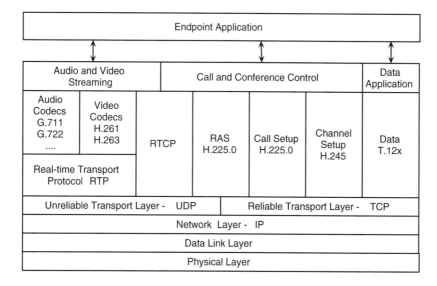

Figure 4.18. Protocol model for H.323 Endpoints.

Table 4.6. Call and Registration, Admission and Status messages

H.225/Q.931 Messages	H.225/Q.932 Messages
Alerting	Facility
Call Proceeding	Notify
Connect	
User Information	**H.225/RAS Messages**
Progress	GatekeeperRequest (GRQ)
Release Complete	RegistrationRequest (RRQ)
Setup	UnregistrationRequest (URQ)
Setup Acknowledge	BandwidthRequest (BRQ)
Status	LocationRequest (LRQ)
Status Inquiry	DisengageRequest (DRQ)
	InfoRequest (IRQ)

layers in use. Some upper layer protocols require reliable transport and use TCP or
SCTP as the transport layer. For others, such as stream delivery using RTP, the User
Datagram Protocol (UDP) suffices. All that is needed is to identify the upper layer
entity via a port number, the PDU length and the user layer protocol.

Each upper layer must have a unique Transport Service Access Point identifier.
The application executing on the endpoint uses up to three types of interface shown
in Figure 4.18: call and conference control, media stream and data application.

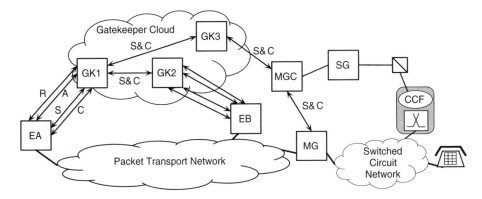

Figure 4.19.   H.323 trapezoid formed by two terminals and their associated gatekeepers, with extended trapezoids to media gateway and switched circuit network terminal.

The *call and conference control* part uses three protocols. First, the Registration, Admission and Status (RAS) part of H.225.0 is used by terminals to register with the Gatekeeper and to be admitted to a call, either as the calling or called party [127]. Second, call setup signalling uses the Q.931-based part of H.225.0. Third, a logical channel control protocol defined in Recommendation H.245 is used to negotiate channel usage and capabilities. Logical channels are established for audio, video and data as required and for control information. A logical channel is identified by an integer (0–65535) and is unidirectional. Logical channel numbers are not preassigned except for the H.245 control channel (0). Table 4.6 lists the messages defined in Recommendation H.225.0. Only those Q.931 messages listed in the table are permitted. Other Q.931 messages such as Connect Acknowledge and Disconnect are not of use in the packet-switched environment.

The *audio and video streaming* interfaces are the actual signal connections to the codecs. A H.323 system must have at least a standard 64 kbit/s G.711 codec. A number of ITU-T lower bit rate codecs are allowed. For example, G.729 is a 8 kbit/s codec that gives speech quality close to the G.711 type. Video codecs, if present, follow the ITU-T H.261 and H.263 Recommendations for compressed video codecs for conferencing with bit rates at multiples of 64 kbit/s.

The Real-time Transport Protocol (RTP) described in Section 4.5.1 supports reconstruction of the real-time signals within monitored delay and latency variation constraints. The RTCP protocol reports metrics gathered from the information streams.

The last part of the H.323 protocol set, *the data applications part*, supports the exchange of data in a conferencing environment, that is, once a call is set up, data channels may be opened between endpoints.

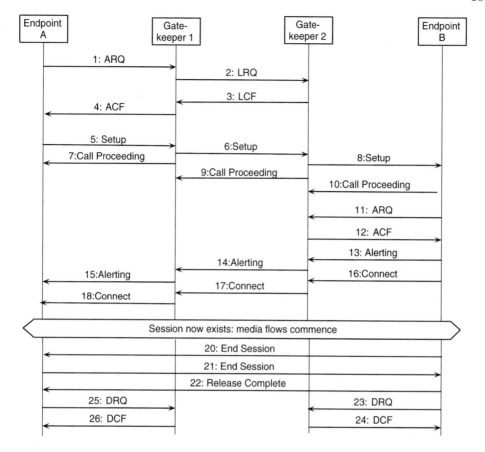

Figure 4.20.   Call signalling with two H.323 Endpoints, each registered to a Gatekeeper.

Call Setup and Media Negotiation

Call control in H.323 is illustrated by three examples based on the reference model in Figure 4.19. First, a two-party call in a carrier network environment is described in this section using the message sequence chart given in Figure 4.20. Second, we examine the sequence of messages needed to create and add parties to a multiparty conference. After examining the Megaco protocol in Section 4.7.2, an example of a PSTN to packet network call is given.

Several modes of call control signalling are accommodated in the H.323 Recommendation similar to the softswitch signalling patterns shown in Figure 4.6. The most complex situation has each endpoint registered to a gatekeeper. The call setup and teardown signalling sequence is shown in Figure 4.20.

1. Endpoint A, registered to Gatekeeper 1 (GK1), sends a H225.0 Admission Request (ARQ), containing among other information an identifier for the called party.

2. The called party, B is not registered to GK1 but another gatekeeper. Some mechanism, for example broadcasting a Location Request (LRQ) to other Gatekeepers, is necessary to identify the Gatekeeper to which B is registered.

3. Gatekeeper 2 (GK2), having the identified terminal registered to it, replies with a LocationConfirm (LCF) message. Additional address information on the destination may be returned.

4. GK1 confirms the admission request to EA using AdmissionConfirm (ACF).

5, 7. EA initiates the call by issuing a Setup message. Assuming that the fast start option is used, this message contains one or more H.245 OpenLogicalChannel data structures, defining the required logical channels and media parameters. A Call Proceeding message provides a local acknowledgement.

6, 9. The Setup message, with the capability information, is forwarded to GK2, with local progress information using Call Proceeding.

8, 10. GK2 sends a Setup message with the capability information to the called endpoint. The endpoint provides a local acknowledgement.

11, 12. Endpoint B requests admission by GK2 using ARQ. Admission is confirmed using ACF.

13–15. Endpoint B then enters a ringing state, indicated by the Alerting message, that is relayed via the Gatekeepers to EA. The Alerting messages carry replies to the OpenLogicalChannel structures, confirming or refusing the proposed media parameters.

16–18. When the called party answers, the Connect message is returned and relayed by the gatekeepers to EA.

Media flows are now in progress and the call is active. The call termination sequence is shown messages 20–26.

20. The endpoint wishing to terminate the call, B in this case, stops transmitting video, data and audio and then issues an EndSessionCommand to the opposite endpoint.

21. Endpoint B waits for an EndSessionCommand from A and closes the media control channel.

22. Endpoint B sends a Release Complete message, terminating call signalling.

23, 24. The Gatekeepers must be informed that the call has been completed. Endpoint B sends a Disengage Request (DRQ) that is confirmed by a Disengage Confirm (DCF).

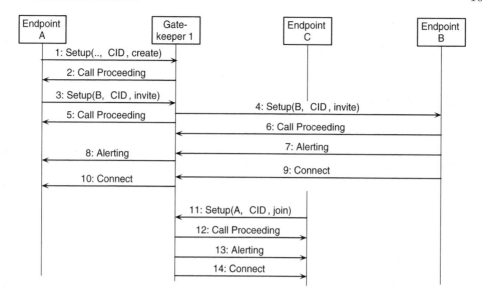

Figure 4.21. Adding parties to a H.323 conference via a Gateway with an MC.

24, 25. Similarly, Endpoint A disengages using the DRQ and DCF sequence.

This example illustrates the distinction between call and session in H.323. The call lasts from the first **Setup** message to the **Release Complete** message. The session starts with the establishment of a H.245 control channel to the **EndSession** command.

Fast start signalling can be carried by messages from **Setup** to **Connect** but not after the latter. If the fast start procedure is not used, additional end-to-end signalling is needed after the **Connect** message establishes the end-to-end connection.

Conference Management

The H.323 suite supports multiparty conferencing. Parties may become attached to the conference in various ways. The **Setup** message allows a **Conference Goal** parameter to be set to indicate whether the party issuing the **Setup** message wants to *create* a new conference, *invite* a party to an existing conference or *join* an existing conference. The sequence in Figure 4.21 illustrates the process. The Gatekeeper performs the Multipoint Controller function in this example. Admission requests are not shown.

1. Endpoint A (EA) initiates an new conference by issuing **Setup** with a conference identifier (CID) and the goal to create a conference.

2. The **Call Proceeding** message reassures EA that the request is acceptable.

3. EA now issues a **Setup** request to the Gatekeeper with Endpoint B's (EB) number, the same CID and an invite goal.

4. The Gatekeeper issues a **Setup** to EB, relaying the invite goal.

5. The Call Proceeding message reassures EA that the request is acceptable.

6. EB issues a Call Proceeding.

7. EB issues an Alerting message to the GK.

8. The GK relays the Alerting message to EA.

9. EB answers and the terminal issues a Connect message to the GK.

10. The GK relays the Connect message to EA. The media streams may now flow. A and B are now members of the conference.

11. Endpoint C wishes to join the conference and knows A's number and the CID. C issues a Setup to the GK, indicating that the goal is to join the conference.

12. The Gatekeeper reassures EC with a Call Proceeding message.

13. The Gatekeeper issues an Alerting message

14. The Gatekeeper issues a Connect to EC. Endpoint C is now connected to the conference.

4.7   MEDIA GATEWAY FUNCTIONS AND CONTROL

4.7.1   MEDIA GATEWAY ENTITIES

The *media gateway* is a physical entity that performs a number of functions at the signal, encoding, multiplexing and packetisation levels at the interface between dissimilar networks. At the signal level, functions include audio bridging, DTMF digit reception and recognition, voice recognition and tone and announcement generation.

At the encoding level, the media gateway must format signals in the A-law or μ-law format on the SCN side and in the desired encoding on the packet network side, for example G.729 encoding. The MG must perform transcoding: conversion of a signal from one encoding format to another. Similarly, fax signals must be demodulated and converted to a form suitable for transmission in the packet network.

The MG must terminate switched circuit network (SCN) connections in the form of TDM trunks and, in small gateways, subscriber loops. The MG must therefore multiplex and demultiplex TDM signals. At the packet level, the MG must function as an RTP endpoint for each voice connection.

The key technology enabling the implementation of media gateways capable of handling large numbers of simultaneous connections is digital signal processing (DSP). Tone and announcement generation is readily implemented in a digital signal processor. A media gateway must terminate voiceband modems on the switched circuit network side. A single digital signal processor can implement the functionality of a number of modems. Transcoding, for example from G.711 at 64 kbit/s, A-Law PCM to G.729 at 8 kbit/s, is largely a signal processing operation and is readily implemented in DSP apparatus.

The media gateway is located in the SCMF layer of the NGN Framework defined in Chapter 2 as its functions centre on media streams passing between a SCN and a packet network. The media gateway is controlled by a media gateway controller in a master–slave relationship. The media gateway controller is placed in the RCMF layer as shown in Figure 4.17. The physical entity representation abstraction of the MGC in Figure 4.16 shows two protocol interfaces. First, an interface to the Signalling Gateway carrying PSTN signalling (ISUP or Q.931) uses SCTP as the transport layer. The MGC sends and receives PSTN call control signalling via this interface. Second, the MGC to Media Gateway interface has a specialised protocol to support the master–slave control relationship. Two standardised protocols are the Megaco or H.248 Gateway Control Protocol and the Media Gateway Control Protocol (MGCP). We review these protocols in subsequent sections. In addition, MGC-to-MGC signalling may be performed as shown in Figure 4.5 and an additional interface is required.

4.7.2 MEDIA GATEWAY CONTROL PROTOCOL: MEGACO

The Megaco or H.248 Gateway Control Protocol is intended for use in the physically decomposed gateway with various packet multimedia standards. The Megaco Protocol supports the control of a Media Gateway by a Media Gateway Controller (MGC) [39]. The Megaco standard sees the MGC as controlling 'the parts of the of the call state that pertain to connection of media channels in a MG' [136].

Megaco is based on a *connection model* embodying abstractions for the terminations on the SCN and packet network sides of the gateway and the manner in which these are grouped and manipulated. Connection models are similar to the IN CS-2 Connection View States described in Section 3.6.4 but embody circuit- and packet-mode terminations and a wide range of media. The abstractions in the connection model are:

- *Terminations*: represent the sources and sinks of single or multi-media streams, described by stream parameters, bearer characteristics and the properties of the terminating unit, for example a modem.

- *Contexts*: a context represents an association between terminations. For example an RTP termination is associated with an SCN bearer channel termination to make a connection between an SCN and a packet network. A context describes the connection topology, multiplexing, the connection state and media processing such as mixing and switching.

An example of a connection model is shown in Figure 4.22.

A connection model is manipulated by Megaco *commands*. The commands are described in the form of an application programming interface. Each API command specified the command parameters known as *descriptors* in Megaco. Descriptors are defined for common constituents of a termination such as multiplexer types, modem types, the media and the state of the termination. Eight commands are supported:

- Add: adds a termination to a context, creating the context if not already specified.

- Modify: changes or supplies missing values of connection properties, events to be notified and signals for a termination.

- Subtract: removes a termination from a context, reports usage statistics to the MGC and deletes the context in the case of the last termination.

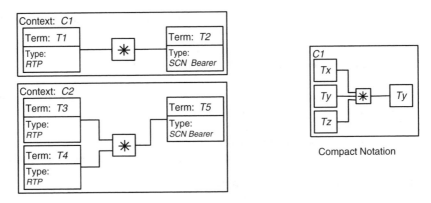

Figure 4.22. Connection model used in the Megaco/H.248 Gateway Protocol.

- Move: moves a termination from one context to another.

- AuditValue: returns the current state of specific terminations.

- AuditCapabilities: returns all the possible values for termination properties.

- Notify: allows the Media Gateway to report specified events to the Media Gateway Controller.

- ServiceChange: supports taking terminations into and out of service.

The media gateway must support a variety of signals and media at terminations. A termination of a given type is described by a *package*. A package is a definition of the properties of terminations, events that may be detected and reported, signal formats and statistics that may be gathered. Basic packages include Time Division Multiplex, RTP, Data Modems and DTMF Receivers. Megaco has an extensibility mechanism based on the definition of new *packages*. Packages define additional properties, events, signals, statistics and additional error codes for particular terminations.

A media gateway typically supports many simultaneous connections. The interactions of the MGC in controlling the MG therefore relate to many transactions. Megaco therefore has a transaction co-ordination mechanism. The Transaction API allows commands to be grouped at two levels: an *action* is a set of *commands* relating to a specific context and whose sequence must be preserved. A *transaction* is a set of actions. Common parameters include the transaction and context identifiers.

The Media Gateway Controller is often viewed as a call control entity, comparable to the H.323 gatekeeper. Megaco, however, is not designed for interacting with intelligent terminals since it has no rich methods corresponding to the Q.931 Setup (or INVITE in SIP). In transit networks, however, the MGC is able to set up the connections at the trunking gateways on ingress and egress to the packet network. Figure 4.5 illustrates this case. The media gateway controllers must exchange ISUP signalling messages with the switched circuit networks. Signalling between Media Gateway Controllers relates to co-ordinating the connections across their respective

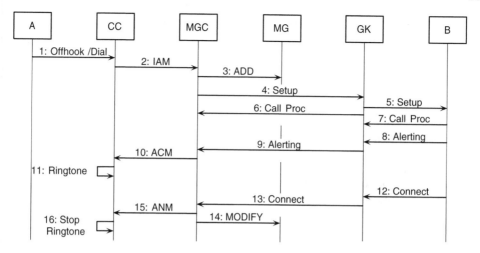

Figure 4.23. Message sequence for a call originating in a SCN and terminating in a packet network using H.323 and Megaco.

media gateways. Megaco is therefore able to transfer the required information at bearer and media level between MGCs.

Figure 4.23 shows signalling in setting up a call originating in a circuit-switched network and terminating in an H.323 network.

1. The user dials the B-party's E.164 number.

2. The end exchange (with possible transit exchanges not shown) send an IAM request to the Media Gateway Controller. The IAM is relayed transparently across a Signalling Gateway.

3. The MGC uses the ADD command to instruct the MG to create a context for the call and initialise the PSTN-side termination values.

4–8. A Setup message is sent to the Gatekeeper to which the called party is registered. The familiar sequence of H.323 messages follows.

9. The Gatekeeper relays the Alerting message to the MGC.

10, 11. The MGC sends the ACM message toward the originating exchange.

12, 13. When the B-party answers, the Connect message is relayed to the MGC.

14. The MODIFY command supplies missing values for the termination, completing the connection.

15, 16. The end exchange is notified by the ANM that media streams can now flow.

4.7.3   MEDIA GATEWAY CONTROL PROTOCOL

The Media Gateway Control Protocol is concerned with the control of connections in a media gateway by an intelligent entity outside the gateway [23]. This entity is called the Call Agent (CA) in the MGCP standard. The Call Agent is assumed to take care of signalling to other entities, for example circuit-switched networks via ISUP messages. MGCP is concerned solely with the relationship between the CA and the MG. Gateway control is assumed to reside in the Call Agent.

MGCP is designed to control a number of types of gateways ranging from residential gateways where analogue phones plug into the gateway, to a gateway that terminates trunks carrying a large number of voice circuits. MGCP describes sources and sinks to be connected as *endpoints*, for example an interface on the gateway that terminates a trunk carrying individual circuits. Similarly a single analogue telephone interface is an endpoint. Endpoints may be virtual, for example ports on an audio content server. A *connection* is an association of endpoints. The connection may have different topologies, for example point-to-point or point-to-multipoint. The specification for how the media is handled across the connection is specified using the Session Description Protocol, described in Section 4.8.4. For example, a circuit on a trunk carrying 64 kbit/s A-law PCM must be converted to G.729 packet speech to be transmitted over the packet network.

MCGP is simpler than Megaco in many respects. Protocol operations defined in MGP are summarised below.

- CreateConnection: (CA to MG) creates a connection between two endpoints; uses SDP to define the receive capabilities of the participating endpoints.

- ModifyConnection: (CA to MG) modifies the properties of a connection.

- DeleteConnection: (Both ways) terminates a connection and collects statistics on the execution of the connection.

- NotificationRequest: (CA to MG) requests the media gateway to send notifications on the occurrence of specified events in an endpoint.

- Notify: (MG to CA) informs the media gateway controller when observed events occur.

- AuditEndpoint: (CA to MG) determines the status of an endpoint.

- AuditConnection: (CA to MG) retrieves the parameters related to a connection.

- RestartInProgress: (MG to CA) signals that an endpoint or group of endpoints is taken into or out of service.

4.8   MULTIMEDIA COMMUNICATIONS BASED ON SIP

4.8.1   IETF MULTIMEDIA CONFERENCING PROTOCOLS

The second major suite of protocols for voice over IP networks and multimedia communications arises from the Internet Multimedia Conferencing Architecture [104]. This architecture was based on the premise that, while the Internet had not been

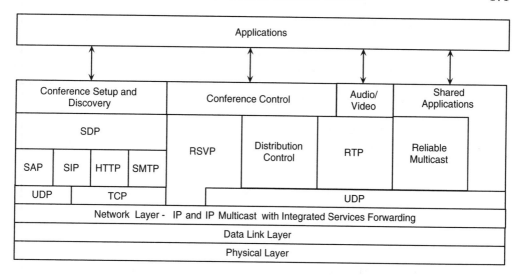

Figure 4.24. The Internet Multimedia Conferencing Architecture showing the protocols used.

designed for real-time signal transport, there was an expectation that improvements would allow the Internet to carry acceptable quality audio and video. It was also felt that it would be possible to go beyond the tightly controlled PSTN and ISDN conferences for a limited number of participants. Internet conferences would range from small to very large, described as having 'television-sized' audiences. A form of Internet–broadcasting convergence was envisaged. The enabler of this notion is the ability in the Internet to *multicast*, that is efficiently broadcast packets from a source to multiple recipients by a router-based mechanism. IP multicast distributes packets to multiple destinations via data distribution trees. Routers at the nodes of the tree replicate packets for the routes that spread out from the node [55]. The sender addresses the packets to a single multicast address and port number and requires no knowledge of the recipients' individual IP addresses to transmit packets.

The Internet Multimedia Conferencing Architecture is essentially a suite of protocols required to overcome the limitations of the best-effort network, to set up and control multimedia conferences and to control the flow of audio-visual streams. Figure 4.24 shows the constituent protocols and their relationships. The transport network is assumed to be an IP network with multicast ability. The IP network is supported by QoS-enhanced forwarding based on Integrated Services (RSVP) or Differentiated Services. As in the case of H.323, protocols requiring reliable transport use TCP at the network layer while those that do not use UDP.

Audio and video streams are transported using RTP and RTCP as described in Section 4.5.1. Reliable multicast attempts to provide ordered, duplicate-free delivery of packets [195].

The conference setup and discovery group consists of a number of protocols. The Session Description Protocol provides a convention for describing the details of a conference. The Session Announcement Protocol informs potential participants that the conference will take place, as a more efficient variant to using e-mail. The Session Initiation Protocol (SIP), which has now become the basis for control of VoIP calls and multimedia calls in 3G networks, is available for issuing specific invitations to multimedia conferences.

While the multicast protocol standards underpinning the Internet Multimedia Conferencing Architecture remain experimental, transport (RTP/RTCP) and session control protocols (SIP and SDP) underpin an important approach to telephony and tightly-coupled multimedia conference in IP networks. The remainder of Section 4.8 describes the use of SIP and SDP as the basis for multimedia conferencing, including telephony in packet networks.

4.8.2 ARCHITECTURE OF SIP MULTIMEDIA SYSTEMS

The purpose of the SIP protocol is to support the creation, modification and termination of multimedia, multiparty sessions. The session refers to the multimedia streams: SIP has no notion of a session at the service control level. The application may be a voice call, a video conference or media distribution. The multimedia attribute relates SIP's ability to transfer comprehensive descriptions of multimedia stream requirements using the Session Description Protocol. The SIP protocol has no inherent limit on the number of session participants. Other mechanisms must support multiparty issues such as audio bridging and keeping track of the parties to a conference. SIP does not provide services but rather enables transfer of service control information as well as other information such as the media to be used for communication.

Multimedia and packet telephony systems using SIP have end stations and servers with particular roles. The general configurations shown in Figure 4.6 are applicable to SIP-based systems. At its simplest, two terminals could set up a session without the aid of a server. In a managed service environment, terminals are likely to be registered to servers, resulting in a signalling configuration as in Figure 4.6(a–c). Registration in SIP is concerned with recording an address that will find the end user rather than admission control. Other mechanisms are required for authentication and admission control, for example the Radius protocol [181].

SIP service control is based on the notion of a session as composed of a number of sources and receivers of media streams and the streams used in the communication [105]. The term call has no formal meaning in the SIP context but, if used, has a similar meaning to that in H.323, namely the period from the message initiating the association between parties to the message finally ending that association. The emphasis of SIP is to establish agreement on media that are compatible for all parties to the session.

The servers in a SIP-based system have several functional roles. Two SIP addressing concepts are essential to defining the types of server. A *contact address* is an address that can be reached by normal routing mechanisms. The *address of record* is a published address that identifies a user or resource that may need translation or additional processing to determine the corresponding contact address.

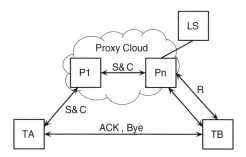

Figure 4.25. The *SIP trapezoid* formed by two terminals and associated proxy servers.

The various types of server, or more correctly, service are:

- *Proxy server*: makes requests to other entities on behalf of a client and returns responses to the client. While processing the request or response, the proxy server may alter parts of the message. Thus, the proxy server's function is to route requests and responses but could also be the location for enforcement of policies. The proxy server first contacted by a user to initiate a session is called the *outbound proxy*.

- *Location service*: assists other types of servers to determine the actual location of a called party by holding bindings between address of record and contact address. These mappings are registered by users. The protocol for accessing the location server is not specified.

- *Registrar:* a server that receives requests to register an address mapping and places this information in a location service.

- *Redirect server*: responds to requests from users with information on the current locations of a party. A redirect server uses a location service. Unlike the proxy server, the redirect server does not process subsequent call processing requests and responses.

- *Back-to-back user agent* (B2BUA): an entity that initiates a multimedia session without a request from any party that will be involved in the session, unlike the proxy or redirect servers that respond to requests from call parties.

The relationship between various SIP server types in a managed environment is illustrated in Figure 4.25. A similar trapezoidal relationship exists to the general pattern established in the case of H.323. Terminal TA contacts outbound proxy server P1. Server P1 may make use of a location service to identify proxy server P2 that is 'closer' to the destination. Proxy server P2 uses a location service to find the actual URI for the called party, terminal B. Setup and configuration signalling are exchanged via the proxy servers. Proxy servers do not supervise the session. Confirmation of session establishment (ACK) and termination (BYE) may be exchanged directly between terminals.

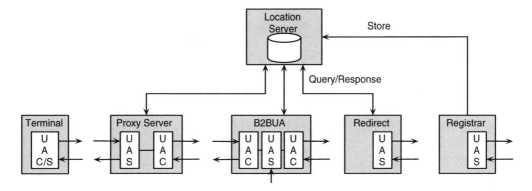

Figure 4.26. Illustrating the entities identified in the SIP standards.

The SIP standard uses the concept of a *user agent* to denote an entity taking part in a transaction. A user agent may have client or server roles. If an user agent has initiated a request, it is involved in a transaction until the appropriate response is received. In this mode, the user agent is called a *user-agent client* (UAC). A user agent that has received a request and is processing the transaction is called a *user-agent server* (UAS) until an appropriate response is issued. An entity can change from client to server mode. Figure 4.26 shows the various SIP servers and terminal and their possible roles as user-agent clients and user-agent servers.

4.8.3 SIP METHODS AND MODES OF OPERATION

Messages in SIP conform to the Internet text-based protocols described in Section 3.5.6. SIP is therefore often described as an extension of the Hypertext Transfer Protocol, but this is more in format than semantics. Messages fall into two types: requests and responses. Requests and responses have two or three parts: a *start line*, a *message header* and an optional *message body*.

The start line has different fields in requests and responses. In the case of a request, the start line consists of the SIP method to be invoked, the user or service to which the message is addressed (the Request-URI) and the protocol version, for example INVITE sip:joe@flybynight.com SIP/2.0. Table 4.7 lists the six basic methods defined in the SIP standard [188]. A number of SIP extensions defined subsequently are also listed.

The response format contains the protocol version and a status code (number plus descriptive part). An example of a response start line is SIP/2.0 407 Proxy Authorization Required. The number of status codes is large and a selection is listed in Table 4.7.

Headers follow the start line and are formatted according to RFC822. Commonly used headers are listed in Table 4.8. Headers may be altered by intermediate nodes. For example, a proxy server passing on a request may insert contact addresses or add its URI to the Via header to ensure that responses are routed via an identical path.

Table 4.7. Summary of methods defined in SIP

Method	Definition	Purpose
INVITE	RFC3261	Used to initiate a session, relay initiation requests or modify session
ACK	RFC3261	Acknowledgement for final responses
OPTIONS	RFC3261	Allows a SIP UA to query the capabilities of another UA or proxy server
BYE	RFC3261	Terminates a session and dialogue
CANCEL	RFC3261	Client asks for processing of previous request that it sent to be halted
REGISTER	RFC3261	Add a new binding between an address-of-record and one or more contact addresses
SIP Extensions		
INFO	RFC2976	Carries mid-session control information, such as ISUP and Q.931 messages for telephony call control [46]
PRACK	RFC3262	Acknowledgement for provisional responses
PUBLISH	RFC3903	Publishes event state for distribution through the SIP Events Framework [163]
REFER	RFC3515	Requests the recipient to refer to a specified resource and inform the sender of the outcome [194]
SUBSCRIBE	RFC3265	Requests asynchronous notification of events from a co-operating remote node [182]
NOTIFY	RFC3265	Informs the SUBSCRIBEd party on the state of a resource
UPDATE	RFC3311	Allows a client to update session parameters without impacting on the state of a dialogue [145]
COMET		Indicates fulfillment of preconditions between user agents
MESSAGE	RFC3428	Transfers an instant message in the body part without starting a dialogue [32]
Response Codes		
100 Trying		Request has been received and unspecified action taken
180 Ringing		Recipient of INVITE is alerting the user
181 Call being Forwarded		call is being forwarded to a different destination(s)
182 Queued		User is unavailable and server will try again
183 Session Progress		Progress as indicated in headers or message body
200 OK		The request has been succesful
300 Multiple Choices		Resolved address is not unique: user must decide
301 Moved Permanently		User can not be found:use new address
302 Moved Temporarily		Retry the request with address supplied
305 Use Proxy		Request must be directed to proxy server
4xx Request Failure		Various failure causes
5xx Server Failure		Various failure causes
6xx Global Failure		Various failure causes

Table 4.8. Selected SIP Headers

Header	Purpose
Call-ID	Unique identifier: phone identifier plus random string
Contact	Direct route to called party: username and a fully qualified domain name
Content-Type	Describes the type of message body
Content-Length	Length of the message body in octets
CSeq	Sequence number: integer and method name
From	Display name and SIP/SIPS Uniform Resource Identifier of requestor
Proxy-Authenticate	Authentication challenge information
Proxy-Authentication	Client identification to proxy requiring authentication
Route	Required routing of requests to proxy servers
To	Display name and SIP/SIPS Uniform Resource Identifier of recipient
Via	Address at which sender expects to receive responses to the request

The message body allows numerous types of information to be transported. A common form of the body is the session description described in Section 4.8.4. The body could be a Web page formatted using HTML or a document encoded using XML. Equally, the body could be an instant text message. The message body is optional and its use depends on the request or response purpose and type. Transport of information in the message body is transparent: intermediate nodes such as proxy servers do not process the message body.

The SIP protocol is independent of the type of session being initiated. The methods, the selection of headers and the independence of the message body allow different types of sessions to be set up, ranging form a packet voice call to a large multimedia conference. The protocol makes no assumptions about the underlying network or the domains the messages traverse. Sessions may therefore be set up involving different types of transport networks, for example IP and ATM and using different transport layers such as UDP and TCP.

SIP Methods and Responses

SIP follows a request–response model. A session setup always starts with an INVITE message that carries the full state information. Intermediate responses such as 100 Trying may be returned before a response such as 200 OK is emitted. The ACK method is used to complete an invitation as a three-way handshake by acknowledging the final response. A client also uses ACK to acknowledge a response indication failure, for example 404 Not Found in response to an INVITE. The BYE method terminates a session.

The CANCEL method allows an operation in progress to be halted. For example, if a user aborts a call before the called party answers, a CANCEL would be issued.

The `REGISTER` method allows a user to register contact addresses with a location service. The contact addresses are bound to the user's address of record. Figure 4.27 shows the registration sequence, in messages 1–5. To protect the registration server from attack, transport layer security (TLS) is used to transfer the requests and responses. The registrar server requires the client to be authenticated, by returning a `401 Unauthorized` response. The `REGISTER` request in message 3 carries authentication information.

The registration may be updated or cancelled by the user. Registration information may be queried by servers involved in session setup and updated addresses returned to calling parties. SIP thus supports user mobility.

Addressing in SIP

SIP uses Uniform Resource Identifiers (URI) for users and services.[4] A URI is an identifier in the form of a character string in a defined syntax that identifies a resource, such as a document, service or person. The URI convention allows alternative types of identifiers, name or location or both. In SIP three types of URI are possible:

- *Normal SIP URI*: an identifier for a user or resource such as a user agent, proxy server or network gateway of the form:
  User agent: `sip:joe@comms.flybynight.com`
  Proxy server: `sip:sps.supertel.com`
  Network gateway: `sip:mgw.supertel.com`

- *Secure SIP URI*: similar to the SIP URI but secure transfer of requests and responses is required. For example:
  User agent: `sips:joe@comms.flybynight.com`

- *Telephone URI*: allows the use of local or global E.164 telephone numbers to address entities in SIP. Examples of global and local tel URI addresses are:
  `tel:+27117165470`
  `tel:5470;phone-context=+2711716`

A SIP user is identified by two types of address. The *address of record* (AOR) is an address in the form of a SIP URI that would be given out or published, analogously to a telephone number. The actual location of the user is specified by an IP address or an address consisting of a username and a fully qualified domain name that enables messages to be routed directly to the user. This form of address is called a *contact address*. Translation from address of record to contact address is effected by means of a *location service*.

Several types of addresses are used in a SIP message. The `Request-URI` in a request start line identifies the user or service to which the request is directed. The `To:` header contains the logical address of the recipient of the message. Similarly, the `From:` header contains the logical address of the sender. The `Via:` address is the address to which the sender wishes replies to be directed.

---

[4]The familiar URL is a special case of a URI where the identifier provides an access mechanism or network location.

SIP Extensions

In 1998, SIP was described as a simple protocol with a specification of only 99 pages [191]. This had grown to 269 pages with the publication of RFC3261 in 2002 [188]. Over 30 companion specifications to SIP have been published. Several of these specifications introduce new SIP methods listed in Table 4.7. For example, the PRACK method allows a provisional acknowledgement while complex sessions are being setup. With the basic SIP methods, there is no notion of a provisional response.

In the basic SIP specification, servers are not required to maintain state information. Rather, the various user-agent clients and servers have state machines that can be used to ensure that transactions are managed properly. The basic SIP protocol deals with users being in different autonomous domains by normal Internet techniques. For example, the number of hops a message may take is may be limited, loop detection is applied to finding proxy servers and DNS mechanisms are used. With the application of SIP to more complex services possibly spanning different administrative domains, the need arises to maintain state information and to notify events to call parties. The SUBSCRIBE and NOTIFY methods allow an entity to register for notifications to receive these notifications.

The MESSAGE method exploits the ability of the message body to carry any kind of information to carry (Instant) messages.

4.8.4   SESSION DESCRIPTION PROTOCOL

The Session Description Protocol (SDP) allows a user to define the details of a multimedia session, for example the creator of the session, information about the session and its timing. A specification for the media to be transferred and codecs to be used [105] may also be transferred to allow the parties to agree on these details.

The session description normally follows the headers of the INVITE message but is also carried on response messages. In the case of multicast sessions, with one sender and multiple recipients, the sender determines the session parameters. In the case of unicast sessions, the two parties may need to agree on the session parameters, for example the audio and video codecs. When a two-party session is set up by a third party call controller (B2BUA), the controller may not know the parameters acceptable to the parties and must assist in arriving at a mutually acceptable session description. The SDP does not itself support negotiation of media parameters but provides data so that other mechanisms can reach agreement. The offer/answer mechanism proposed in [186] provides a way of using the SDP to arrive at a session description that is mutually acceptable to the parties.

Table 4.9 summarises the SDP fields. Only the r, o, s, t and m fields are mandatory. The two principal fields that enable the parties to agree on the media to be used are the media announcement (m=) and attributes (a=). The former specifies the media (for example audio, video, application or data), the port to be used, the transport protocol and the format (profile), for example m= 40442 RTP/AVP 0. A medium may have several attributes, for example a unidirectional or bidirectional stream is required. A unidirectional stream would have a=recvonly. The attribute field when RTP is used gives the RTP map, that is the binding of the codec to the payload type. For example, a stream using standard payload profile 0 and μ-law PCM encoding with 8000 samples/s is described by a=rtpmap:0 PCMU/8000.

Table 4.9. Session Description Protocol fields (mandatory lines shown in bold)

Field	Description	Example
Session-level descriptions		
v	**protocol version**	`v=0`
o	**owner/creator, session Id**	
s	**session name**	`s=Lunchtime Conference`
i	session information	`i=Briefing on new acquisition`
u	URI of description	`u=www.nightfly.net/brief.html`
e	email address	`e=joe@flybynight.net`
p	phone number	`p=+27 11 716 5470`
c	connection information	`c=IN IP4 141.146.16.1/127`
b	bandwidth information	`b=CT:10000`
z	time zone adjustments	`z=2898723442 2h`
k	encryption key	`k=uri:ss.flybynight.com`
a	session attribute	
Time-level description		
t	**time the session is active**	`t=0 0`
r	zero or more repeat times	`r=7d`
Media-level description		
m	**media name, transport address**	`m=audio 49172 RTP/AVP 0`
i	media title	
c	connection information	`c=IN IP4 ws99.flybynight.net`
b	bandwidth information	
k	encryption key	
a	more media attribute lines	`a=rtpmap:0 PCMU/8000`

4.8.5 USE CASES FOR SIP

Many service scenarios using SIP and SDP are possible. These range from a peer-to-peer relationship between the two parties, to the use of a proxy servers and to application-initiated sessions. Five telephony-oriented use cases for SIP and SDP are illustrated in Figures 4.27–4.31. These five examples are illustrative but represent only a sample of call sequences possible with SIP. Further cases are described in the best current practice documents [143, 144, 185], where full details of headers used and session descriptions are given.

The richness of possible call scenarios with SIP is attributable in part to the various ways that proxy servers may be used. The outbound proxy may not know the contact address for the called party but has various ways of finding a server that does [187]. One mechanism is forking the **INVITE** request to all servers on a list and relying on a response from one sever that knows how to find the user.

Two-party Call Using a Proxy Server

The registration process, messages 1–5 in Figure 4.27, is described above. The remaining sequence describes the session setup using a single proxy server.

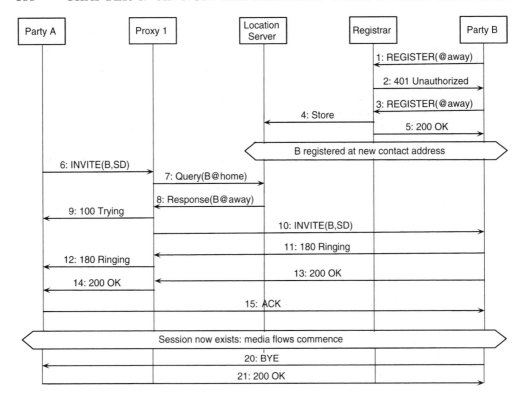

Figure 4.27. Session setup and clearing showing registration and user of proxy server.

6. Party A wishes to set up a session involving party B. A sends a request containing an INVITE method, B's known URI, say sip:B@home.net, together with the session description (SD) to the proxy server.

7, 8. The proxy server queries the location service by a mechanism that is not part of the SIP standard and receives the address to use, for example B@away.net.

9, 10. After sending a 100 Trying reassurance message to inform A that the request is being processed, the proxy server forwards the INVITE request to party B.

11–12. If B is available, the user is alerted and the 180 Ringing message is returned via the proxy server to A.

13, 14. When B answers, the 200 OK response is returned to A via the proxy server.

15. A completes the process by sending an ACK method to party B.

Note that a successful setup involves a three-way handshake: the INVITE, 200 OK and ACK messages.

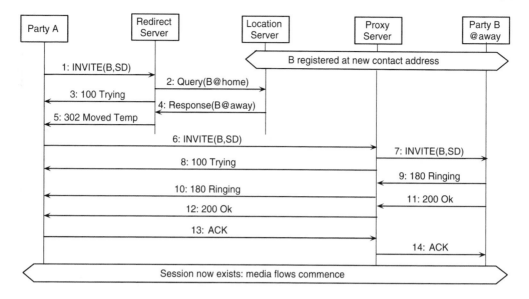

Figure 4.28. Two-party call using SIP illustrating the use of a redirect server.

Once the ACK has been sent, both parties stream media signals using encodings, protocols and addresses conveyed in the session description.

Clearing a SIP call is a simple process shown in messages 20 and 21 in Figure 4.27. One party, B in this example, issues a request with a BYE method to the other party. The other party responds with a 200 OK message. Both parties stop streaming signals and revert to an idle state.

Two-party Call Using a Redirect Server

This example illustrates support for user mobility in a SIP environment. The sequence shown in Figure 4.28 starts with the assumption that the called party (B) has registered an address to which calls should be directed temporarily, for example as shown in Figure 4.27.

Messages 1–5 show the typical sequence for consulting a redirect server. The redirect server receives the initial INVITE message from the calling party and consults a location service to obtain the called party (B) address information by sending the INVITE with B's known address. The redirect server consults the location service and establishes that party B has registered a temporary location. The redirect server sends a 302 Moved Temporarily response containing the proxy server to use rather than a contact address. In this example, the B party is then contacted using the proxy server identified in message 5. The redirect server does not forward the INVITE message toward the called party. Having received the users address or proxy to use, the called party signals in the normal way in messages 6–14.

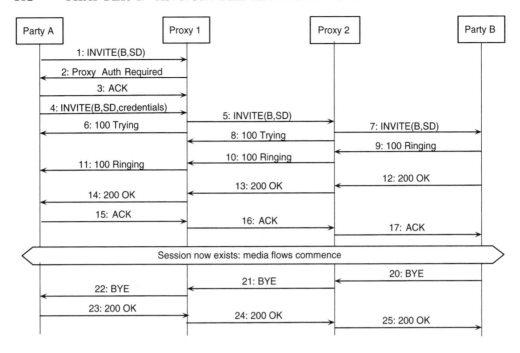

Figure 4.29. Two-party call using SIP, with calling party authentication and using two Proxy Servers to locate the called party.

Use of Multiple Proxy Servers

The message sequence in Figure 4.29 illustrates the use of an outbound proxy (Proxy 1) that verifies the credentials of the caller. The outbound proxy determines the proxy server (Proxy 2) that can locate the user B. Messages 1–4 show a mechanism for authentication.

1. The caller issues an INVITE with no credentials.

2. The outbound proxy server issues a 407 Proxy Authorization Required response.

3. This ACK acknowledges the 407 response.

4. The caller then re-issues the INVITE with credentials.

5, 6. The outbound proxy accepts the credentials and forwards the INVITE to Proxy 2. The proxy's address is inserted in a Record-Route header to ensure that all responses are returned via itself. A 100 Trying reassurance response is returned to caller A.

7, 8. Proxy 2 establishes the contact address for the called party B, reassuring Proxy 1 with the 100 Trying message.

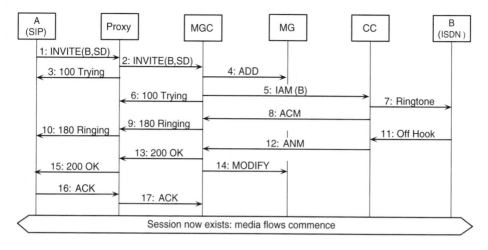

Figure 4.30. SIP-initiated, PSTN-terminated call setup signalling.

9–11. The 180 Ringing response is returned to A via both proxies.

12–15. When B answers, the 200 OK response is returned to A via both proxies.

15–17. The ACK method completes the session setup and media flows take place.

Messages 20–25 shows the termination of the session with the B-party requesting the termination.

Call Involving a Party in the PSTN

Figure 4.30 shows a sequence of messages that set up a call between a SIP user and a PSTN subscriber. In this use case the A-party is in an IP network and the called party is a PSTN subscriber. The SIP specifications do not describe the gateway to the PSTN in detail. We use the H.323-like media gateway controller (MGC) as the network gateway element, similar to that in Figure 4.23. The MGC is assumed to be able to process both SIP and ISUP messages.

Caller A uses an outbound proxy that identifies the media gateway controller as the gateway to use.

1, 2. The INVITE message is forwarded by the outbound proxy to the MGC.

4. The media gateway controller uses the ADD method to create a connection across the media gateway.

5. The MGC forms an ISUP IAM message and sends it to a PSTN switch call controller (CC). (Only one shown.)

8–10. The ACM received at the MGC results in the emission of a 180 Ringing message, relayed by the proxy to caller A.

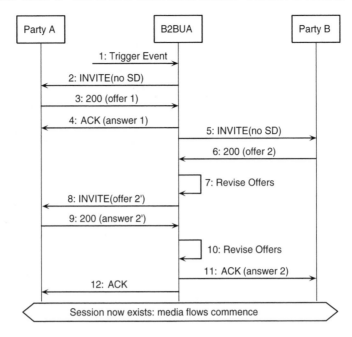

Figure 4.31. Third party session initiation by a SIP back-to-back user agent.

11, 12. When the B-party answers, the ISUP ANM is sent to the MGC.

13, 15. A 200 OK message is relayed by the proxy to the caller.

14. The termination parameters are updated at the gateway by the MODIFY command, completing the connection.

16, 17. The ACK method is local to the SIP network.

The session description conveyed by messages 1 and 2 is used by the media gateway to ADD the IP termination at the media gateway. The MGC must MODIFY the TDM termination to complete the connection.

Third Party Initiated Call

A third party call in the softswitch context is a call that is initiated by a party other than one that will take part in the media session. Examples of services involving third party calls are click-to-dial, where a Web browser user initiates a call between two parties, and managed conferences.

As an Internet protocol, SIP has a strong end-to-end orientation, that is application logic should be located in endpoints. With increasing interest in using SIP in a converged environment, the possibility arises of using SIP messages to initiate sessions by a party not involved in the session [185]. Figure 4.31 illustrates the message sequence in such a scenario.

The server issues invitations to both parties to the call. The server is therefore called a back-to-back user agent (B2BUA). In general, the preferred session description of the two participants is unknown to the server. On receiving the event that triggers the call set-up, the B2BUA interacts with both parties to agree on a session description using an Offer/Answer process [186]. The sequence in Figure 4.31 starts with an event represented by message 1.

2. The B2BUA sends an INVITE request with no session description present to party A.

3. If A is willing to participate in the call, a 200 OK message is returned with user A's preferred session description (offer 1). The offer specifies the media that are proposed in SDP a-fields and the desired codecs in m-fields.

4. The server acknowledges the offer.

5, 6. The server INVITEs party B, again with no session description in the message. Party B responds with its session description offer, media and codecs, in the body of a 200 message. This is offer 2.

7. The server compares and revises the offers to attempt to create compatible descriptions.

8, 9. The server put the revised session description to party A in an INVITE message and receives a possibly altered answer in the 200 response. An answer may reject a medium (by specifying port 0) but must not change the number of media in the description.

10–12. The server again revises the offers and completes the process by communicating the final session description to party B in message 11. Message 12 confirms the session with party A.

4.9 SUPPLEMENTARY SERVICES IN PACKET TELEPHONY

Both H.323 and SIP provide signalling protocols for basic call and multimedia session control. Additional services are provided according to frameworks defined in the respective standards. The H.323 Recommendation identifies several types of services and mode of provision.

4.9.1 SUPPLEMENTARY SERVICES IN H.323

Network services include authentication, call admission control, address translation, mobility management and call distribution. The Gatekeeper is the most appropriate location for network service logic. The Gatekeeper is also the appropriate locus for accounting and production of call data records (CDR) for time-based billing.

Supplementary services in H.323 are similar to those defined in the ISDN, for example call transfer and call waiting. Unlike ISDN, where supplementary services are switch-based, implementation in H.323 occurs in the endpoints. The H.323 Recommendation seeks to allow vendors to provide a range of supplementary services and, in view of the implementation in endpoints, ensure a high degree of interoperability.

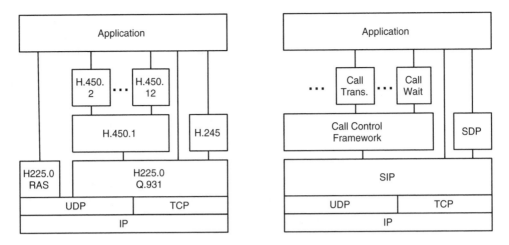

Figure 4.32. Methods of providing supplementary services in H.323 and SIP.

The H.450 series of Recommendations therefore defines supplementary services at the protocol and state machine procedural levels [128].

In addition to the peer-to-peer type of supplementary services, the H.323 suite envisages a class of master–slave services, for example implemented in a media gateway. The H.323 supplementary service model is shown in Figure 4.32 in relation to the call control protocol stack. The basic call control is supported by the H.225.0/Q.931 protocol operations and state machine. The H.323 Recommendation does not define a state machine but, by implication, relies on the SDL diagrams in the original Q.931 Recommendation [110]. Recommendation H.450.1 defines the generic functions for supporting supplementary services [128]. Individual recommendations define particular supplementary services, for example H.450.6 defines a call waiting service [130].

Generic functions provide for the transport of supplementary service application protocol data units (APDU) on the Setup, Call Proceeding, Alerting and Progress messages as well as the Facility message. Supplementary service APDUs are constructed according to the *remote operation service* (ROS) model. Under the ROS model, a supplementary service is initiated using an *invoke* message. The serving peer replies with a *return result* message or a *return error* or *reject* message. Invocations are uniquely identified by an Invoke Id field. A response is linked to its invocation by carrying the same Invoke Id value. A further invocation may and may be linked to a currently active invocation by a Linked ID field.

The logic for specific supplementary services is defined in terms of the APDUs particular to the service and the state machine and message sequences. Figure 4.33 illustrates the operation of a call waiting service. An H.323 call is in progress between parties A and B. Party C wishes to call B. This example assumes that call waiting logic is available at endpoint B. Caller C receives an Alerting message with a call waiting indication. If user B accepts the waiting call, the call hold logic, assumed to be on

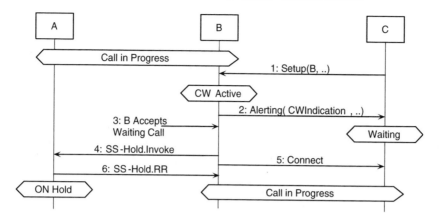

Figure 4.33. Call waiting as a supplementary service in H.323.

endpoint A, is invoked. User C's media streams are then connected to B. The ROS invocation and return result messages are 4 and 6 in Figure 4.33.

The model in Figure 4.32 shows the relationship between basic call services, generic support for supplementary services and individual services defined in the H.450 series. More than one supplementary service may be active and feature interaction rules are necessary to prevent unwanted and faulty conditions. All supplementary services interact with the basic call function through the H.450.1 generic layer. The location for implementation of feature interaction rules is therefore the generic supplementary services layer.

4.9.2 CALL SERVICES IN SIP TELEPHONY

The SIP protocol is a set of methods and headers that, together with the information in the message body allows sessions to be set up and controlled. Procedures are defined for transactions initiated by a user agent rather than for an entire service scenario. SIP is therefore a much less regimented protocol than H.323 and several approaches to enhancing the protocol and adding service features are possible and are in a state of evolution. While the notion of a call is informal in the SIP context, SIP's increasing use in call setup, control and clear-down, a considerable standardisation effort concentrates on call scenarios and call features.

Standardisation of call control enhancements and multiparty sessions is in a developing stage.[5] We therefore review only the broad approaches to enhancing SIP and services in SIP-based systems.

SIP is an extensible protocol in that new methods, headers and parameters may be defined. Also new body types can be accommodated. SIP has been extended by the addition of a number of methods listed in Table 4.7. The IETF has instituted a change control process to ensure orderly extension of SIP [152]. This process requires

---

[5]At the time of writing this work is an Internet Draft: R. Mahy *et al.*   A Call Control and Multi-party usage framework for the Session Initiation Protocol (SIP), February 2005, http://www.ietf.org/internet-drafts/draft-ietf-sipping-cc-framework-04.txt.

that added features have general utility, that is do not support niche applications, and are simple and robust.

The Call Processing Language supports end user applications communicating with servers to request particular treatments of calls [148] on specific conditions, for example user busy or no answer.

The emphasis of SIP falls on simple call services with features provided end-to-end; server-based services are not contemplated. Figure 4.32 shows an approach to provision of supplementary service in SIP [206] similar to that of H.323.

With SIP well-established as a two-party call control protocol, interest naturally shifts to multiparty, tightly coupled conferences and multimedia services. With the move to multiparty services, it becomes necessary to consider matters such as audio bridging topologies and how the protocol is used with different bridges. The goals of SIP conferencing are consistent with those of SIP itself, namely to use the simple protocol in flexible ways to perform the required functions, in contrast with H.323 with its emphasis on standardised services. SIP conferencing supports both centralised (third party) and peer-to-peer control.

Envisaged call features are simple and about of the complexity of IN CS-1 services, for example various types of call transfer, call waiting and in- and outbound call screening. For simplicity, service features should affect one party only, the invoker, and other parties should not be aware that a feature has been of activated. As calls get more complex, new architectural elements come into play, for example announcement and digit reception units. SIP serves to specify the session parameters and make media connections to such units. Other protocols such as VoiceXML may be required to specify the operations performed by these devices.

4.10   ITU-T EVOLUTIONARY PROTOCOLS: BICC

The use of ISUP for setting up connections in a switched circuit network is well established in the telecommunications world. Close coupling exists in switched circuit networks between the Call Control Function (CCF) and the TDM-based bearer switching. Inter-CCF signalling messages using the ISUP protocol carry parameters such as the Circuit Identification Code (CIC) that are specific to the TDM environment. Bearer control is thus intimately linked to call control. The ISUP protocol is transported in switched circuit networks over the SS7 network.

With the move to packet-based bearer networks, the ITU-T recognised the need to decouple service support from the underlying bearer and signalling network technology. The Bearer Independent Call Control Protocol (BICC) was developed to support narrowband services in packet networks. The objective is to remove detailed bearer consideration from the call setup and cleardown signalling and to provide simple interworking with existing ISDNs that use ISUP for call control [135]. The Call Control Function of the SCN switch shown in Figure 4.34 is decomposed into a Call Service Function (CSF) concerned with the association of the parties and a Bearer Control Function (BCF), concerned with the transport of user streams. With this decomposition, two types of nodes are identified. A *serving node* has both CSF and BCF functionality while a *switching node* has only bearer control functions.

Figure 4.35 compares ISUP and BICC signalling by considering the first part in the call setup sequence. In ISUP, an IAM message is sent between adjacent pairs of

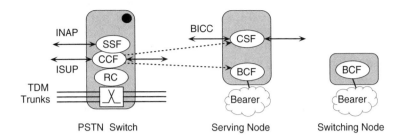

Figure 4.34. Decomposition of control functions into Call Service Function (CSF) using the BICC protocol and Bearer Control Function (BCF).

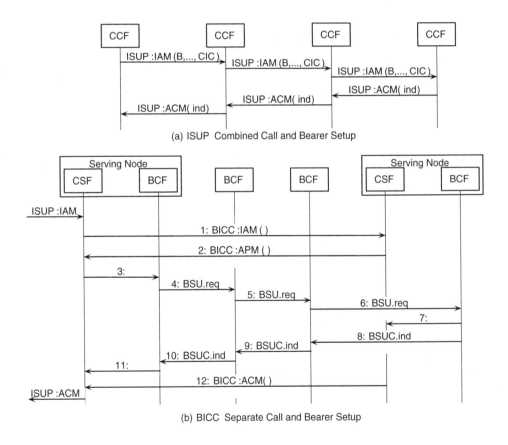

Figure 4.35. Comparison of the initial phase of call setup using (a) ISUP, and (b) BICC. The bearer control protocol (BSU) is hypothetical.

exchanges along the route to be followed by the switched circuit bearer connection. The IAM message parameters contain association-related parameters including the called party number as well as bearer-related parameters. If there are free circuits and the called party is free, an ACM message is returned via the same route. In BICC, the IAM message and its acknowledging ACM are sent between CSF entities in the serving nodes but not to the switching nodes. Bearer setup messages are sent to all involved Bearer Control Functions. BICC does not define the bearer control protocol.

The similarity between ISUP and BICC messages is intended to make signal transfer between circuit and packet networks simple. For example, association parameters are similar in ISUP and BICC messages. Bearer parameters change at the transition between switched circuit and packet network. For example the CIC in the switched circuit network termination identifies a dedicated 64 kbit/s TDM channel. In the packet network, the CIC identifies a bearer connection, a logical association between points of ingress and egress to the packet network. Bearer control ideas in BICC resonate with setting up virtual circuit connections in ATM networks.

4.11   VOICE ON THE INTERNET

The H.323 standards emerged from an initiative to support multimedia communications on local area networks, interworking with PSTN terminals. The SIP/SDP protocols evolved from multimedia session control in the Internet to adoption in managed networks including in 3G mobile networks. Protocols to support voice communications in the Internet are well established. A strong movement has understandably emerged to promote voice communications in the public Internet, conveniently referred to as *Voice on the Net* (VoN), accompanied by the emergence of new types of providers using standards-based and proprietary technology.

Voice on the Net is seen as holding the promise of benefits to both users and service providers [207]. Benefits to users are seen as cost savings, particularly in long distance call charges, and innovative services enabled by the use of multifunction terminals such as personal computers and personal digital assistants that already support Internet services such as e-mail and Web browsing. Benefits to providers are said to be lower capital and operating expenses, leveraging an increased number of broadband access connections and the ability to profit from innovative services. Arguments are advanced that mandatory services such as emergency response can be better provided on the Internet. Similarly, provision of access to services for the disabled is seen as greater via the Internet than the PSTN. In short, Voice on the Net is seen as a valuable manifestation of convergence.

A number of contested and unresolved areas surround the widespread adoption of Voice on the Net. Many countries are trying to resolve the regulatory regime and in particular interconnection with traditional telco networks. Shortcomings of VoN precluding telco quality service include low call completion rates, dropped calls, quality not guaranteed, inability to route past firewalls and when Network Address Translation (NAT) is used, and complex configuration of the end user terminal. While improvements are being made, the approach remains a best effort, without the service guarantees provided by traditional telcos. While add-ons to basic Internet protocols have been and are still being developed to improve the quality of Internet telephony, guaranteed voice quality and call completion still elude the user. Services that are

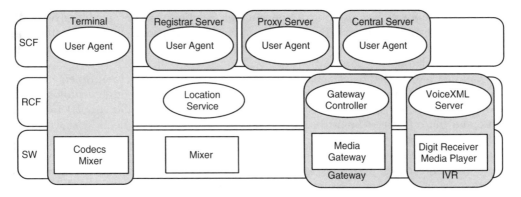

Figure 4.36. Mapping of entities occurring in the SIP context to the NGN Framework.

mandatory in telco networks, including emergency response and legal intercept, are difficult to provide in a VoN environment.

Thus, Voice on the Net remains an area where there is significant interest from many parties but with many unresolved issues.

4.12 CONCLUSION

This chapter reviews the principles of packet multimedia networks. Two main call and session control protocol families are reviewed, namely H.323 and SIP. Their architectures and protocols for basic and supplementary services are examined. Voice over Internet Protocol is a special case of packet multimedia. A number of important issues in packet multimedia communications are deferred to later chapters. Both the H.323 and SIP protocols specify the bearer capabilities that are required to support the service. We have considered basic call and session control as well as supplementary services, largely as end-to-end services. A general approach to allowing applications to utilise network connections, messaging and information is developed in Chapter 8. Specific networks using packet multimedia are not described. An important case, namely the 3G mobile communications system, is described in Chapter 7, where the basic and advanced service provision aspects are examined.

The H.323 architecture, with its defined functional entities maps readily to the NGN Framework in Figure 4.17. The basic SIP standard [188] identifies user agents in client and server roles that take part in transactions. Several types of servers are identified. The proxy, centralised (B2BUA) and registrar server roles are important building blocks in carrier-grade networks. SIP messages are concerned with both the association between users and gaining agreement on media stream details. The former is a service control issue and terminal user agents and servers are thus located in the Service Control Functionality layer. Media control is a Resource Control layer issue. Other building blocks for multiparty, multimedia services would be allocated to layers with appropriate separation of functionality. Figure 4.36 shows the mapping of a number of entities occurring in a SIP architectural context to the NGN Framework.

The media gateway has its companion control function. Other media processing devices may also be conveniently paired: mixer with media processing control function (MPC) and media server with media resource controller (MRC). These devices are best classified as edge devices.

This chapter started by setting out generalised notions of packet multimedia before examining specific protocols. While H.323 and SIP have different origins, philosophies and protocols, their generalised architectures become more similar as SIP takes on more and more features found in carrier networks, for example media gateways and IVRs. With these generalised notions, we map physical and functional entities in packet and switched circuit networks into the NGN Framework.

The architectures, messages and procedures for IP telephony described in this chapter are presented in a generalised or *canonical* form. When protocols such as SIP are used in a specific architecture and have to interwork with other standards, the VoIP protocol must be applied in a well-defined and restricted way. Such a specialised application is called a *profile*. A specific application of SIP in 3G mobile communication systems is described in Chapter 7.

Chapter 5

Integrated Enterprise ICT Systems

In Chapter 1, we indicated that legacy enterprise networks took the form of separate voice (PBX) and private data networks. Enterprise networks are evolving to a single infrastructure to support real-time services and information processing. Convergence occurs in the transport network and the method of supporting enterprise applications. In this chapter, we examine the effect of convergence on both communications support and information processing in enterprises. Convergence in the enterprise domain has several aspects. First, network convergence leads to an *integrated enterprise network*: a managed communications infrastructure serving needs of corporations, government or similar entities.

Second, synergy exists between the *common services* required in the enterprise's information processing systems and the control of the integrated communications infrastructure. The Service Oriented Architecture (SOA) described in Section 5.2.4 is a current enabler of this synergy. Third, end user applications may perform both conventional IT functions and make use of network connections and messaging, leading to *application convergence*.

This chapter is concerned with more than enterprise network convergence. Rather, we examine the supporting functionality required by both telecommunications and enterprise information services and describe an integrated architecture for integrated enterprise ICT systems. The enterprise domain is strongly influenced by dominant vendors, both in the networking and enterprise resource planning areas. We therefore present a vendor-neutral architecture for an *integrated enterprise ICT system* in the unifying NGN Framework developed in Chapter 2. This integrated view provides a basis for convergence at the network, common services and applications levels.

Section 5.1 examines the factors that drive convergence in enterprise ICT systems and the nature of convergence. Section 5.2 reviews four developments that contribute to convergence toward an integrated ICT system. The integrated enterprise network described in Section 5.3 is a specific case of a NGN. Common service control functionality for the integrated enterprise is identified in Section 5.4. We conclude in Section 5.5 by examining the implications of the integrated enterprise ICT system on development of public NGNs.

5.1 DRIVERS AND REQUIREMENTS

Figure 5.1 summarises the historical situation in which enterprises found it necessary to have two forms of private network: a voice private network using TDM technology and a packet-switched data communications network. Two forms of interconnection between sites were required and gateways were required to both the PSTN and the

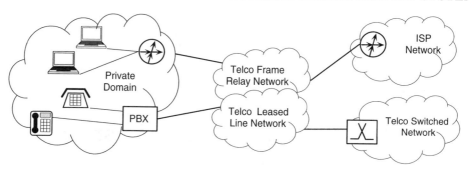

Figure 5.1. Duplication of intrastructure in legacy enterprise networks.

Internet. The legacy voice and data private networks are largely independent. A point of contact occurs between the two types of network in computer–telephony integration (CTI). There, information such as the calling party number is passed from the switch to an application where it is used to retrieve caller-related data for use by an agent.

5.1.1 CONVERGENCE IN THE ENTERPRISE DOMAIN

Development toward the integrated enterprise ICT system is more than moving to a multiservice transport network. Rather, the object is to enable convergence between telecommunications and IT applications in the enterprise context. In this context, we define a *telecommunications application* as computer-based logic that has as its primary purpose the provision of enhanced call, conferencing, messaging or data communications. By contrast, an *information technology application* has as its primary purpose the support of information processing to meet end-user requirements. Application convergence means that telecommunications-oriented applications could be created using the methods found in IT and IT applications would be able to use telecommunications services.

Figure 5.2 identifies the main streams of convergence leading to an integrated approach to enterprise ICT systems that support converged applications.

Enterprise network convergence results from the replacement of the dual legacy infrastructures shown in Figure 5.1 by a single packet network capable of supporting real-time services using packet multimedia standards such as SIP and H.323 and data communications. The packet network may have islands of users served by switched circuit PBXs interworking through media gateways. Similarly, analogue telephones working into media gateways may be used.

Service convergence results from the increasing commonality of generic support for communications, data and information services. In the telecommunications area, generic services such as call control are required. Given the integrated view of telecommunications and IT, generic support services for IT applications are equally important. An important influence in IT service convergence is the Service Oriented Architecture. The SOA seeks to support application developers by a standardised system of encapsulating software components in with an interface that describes the service and provides an XML document-based invocation method. The SOA is a

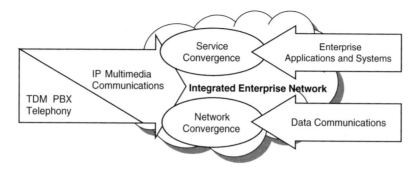

Figure 5.2. Convergence toward an integrated enterprise architecture.

way of capturing service control functionality. The SOA is increasingly adopted in enterprise resource planning systems. In enterprise telephony, a number of standards described in Section 5.2.3 provide generic telephony control for contexts such as PBXs and call centres. This pattern of generic service functionality increasingly pervades the IT and private telecommunications architectures.

While network convergence is an essential enabler, service convergence provides the basis for application convergence in the enterprise context. We define *application convergence* as the ability to design software applications for a range of purposes with the support of service capability functions (SCF) that provide real-time communications, messaging, management, access control, accounting, database, content management as well as business domain functions. Here, service capability function is used in the sense defined in Chapter 8: Service Control Layer functionality made accessible via an open API.

This form of convergence occurs in a private domain and allows the network and information systems platforms to be managed to provide adequate QoS to each of the variety of applications.

5.1.2   CORPORATE REQUIREMENTS

Historically, large firms and multinationals were the principal users of enterprise networking. Small firms were treated as a special category, for example the small office, home office (SOHO) type. Small firms look increasingly to enjoy the same benefits from enterprise networks as do large firms. With the ability to provision virtual private networks at different scales, small firms are able to enjoy the benefits of private networks and to partner with large firms.

Many companies operate globally. Some have global operations with sites and staff in different countries. Others market globally using both traditional methods and e-commerce to sell goods and services. In all cases communications and information systems are essential to business effectiveness. The general requirements of enterprises are expressed using the following themes.

- A *range of organisational models* must be supported. In the past, private communication systems mapped readily onto the multiple sites occupied by

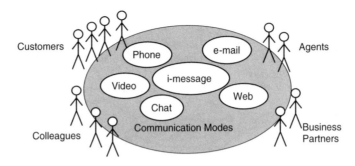

Figure 5.3. Roleplayers and communication modes in enterprise environment.

an enterprise. The future enterprise is seen as borderless: not demarcated by physical sites or national boundaries. For particular purposes enterprise ICT support must cross company boundaries seamlessly.

- A *range of ways of working* must be supported. The workforce is increasingly mobile with many workers spending only a short period in an office or, in some cases, having no office base. Teleworking is increasing. Collaborative working requires the support of services such as conferencing, whiteboarding and document sharing.

- Multiple modes of communication should be supported between roleplayers as shown in Figure 5.3: colleagues, customers and business partners should be able to communicate via various media: voice, messaging, Web-based interfaces and chat. The last named illustrates opportunities to exploit changed paradigms. Chat is a mass service available in the public Internet, allowing people to interact through text messages with a short time delay. Provision of a chat facility in a secure managed environment is potentially useful in business.

- *Diverse stakeholders* require support. Support is not restricted to the staff of the enterprise but may extend to partners, suppliers and customers. For example, supply chain management is a common IT application and its value is enhanced by exchanging information between computing nodes, both within an enterprise and between companies, using standardised electronic documents.

- A *range of applications* must be supported on a single infrastructure, for example the selection shown in Table 5.1.

- Support must be *continuously available and secure*. Members of an enterprise and its customers expect carrier-grade performance and availability from ICT infrastructure. Global operations require 24 hour per day and 7 days per week support across many time zones. Security must be maintained in the presence of various collaboration models. Conflicting requirements must be met. An enterprise ICT system must simultaneously be readily accessible to those with authorisation but provide a high degree of protection of important data and prevent abuse of communications.

Table 5.1. Broad applications in the enterprise environment

Telecommunications Applications	Enterprise Applications
Private branch exchange	Enterprise resource planning
Call centre	Supply chain management
Collaboration support	Finance
Teleworking support	Personnel
Unified messaging	Customer relationship management
	e-commerce
	Electronic data interchange

- Enterprises seek to become *ICT service providers*. With convergence, it is feasible to offer business services over communications networks as well as enhance communication services with IT applications. While such service may have been offered in-house, enterprises can now become service providers to outside customers over private and public network infrastructure.

- Various models must be supported for the ownership, provision, governance and use of facilities for the enterprise. Enterprises seek flexibility in choosing a model for provision of the ICT infrastructure. For example, the infrastructure may be owned by a telco, a service provider or the enterprise, or a combination of these players. The owner may not manage the infrastructure.

The goal of application convergence, supported by network convergence, is to achieve unification. The unified approach is embodied in the concept of an *integrated enterprise ICT system* having the following properties:

- A single, converged networking and computing infrastructure supports both communications and enterprise applications.

- Details of the infrastructure are hidden from users and appropriate abstractions are created, hiding unnecessary detail from application developers. The business logic must be independent of the underlying computing platforms, database systems and networks.

- The infrastructure supports a range of large and small business, partnership, collaboration and workgroup models, crossing physical, network, company and geographical boundaries.

- While this goal may be premature, a *unified approach* to all critical applications in the enterprise is beneficial.

These requirements form a general specification for an integrated enterprise ICT system.

5.1.3 MODELLING INTEGRATED ENTERPRISE ICT SYSTEMS

With convergence in the enterprise context, a common network infrastructure emerges. Common services that traditionally were either telecommunications-oriented or IT-oriented form a continuum. Applications that were similarly classified as telephony-oriented or IT-oriented will be distinguished only by the relative mix they provide of information processing, database transactions, content delivery and communication on behalf of end users. With convergence at the enterprise network, service capability and application layers, a consistent modelling approach is required. The NGN Framework layered model developed in Chapter 2 is used in the enterprise system context.

Services and application have nuanced meanings. Applications describe computing systems that meet a specific needs of the end user or enterprise. Services are more generic and support applications. A model of an enterprise ICT system may have sublayers to the application and service layers. Service capability functions are generic and may be applied to both telecommunications-oriented and IT applications. The range of service control functions is large and commonly required functions are listed in Table 5.1.

Applications and services in the enterprise environment are influenced strongly by the Internet paradigm. An end-to-end service model is favoured. This statement does not preclude particular service architectures. The end-to-end principle means that applications are decoupled from the underlying network and that, apart from restriction by enterprise policies, any entity may provide applications. The NGN Framework layered model accommodates an end-to-end service paradigm, with applications in the user domain.

The end-to-end model does not imply a simple client–server model in which each server must perform all common functions such as security and billing. The SOA described elsewhere provides these facilities as reusable components.

5.2 CONTRIBUTIONS TO CONVERGENCE IN ENTERPRISE ICT SYSTEMS

Network convergence, namely the adoption of a single multiservice packet network, is a significant contributor to integrated enterprise ICT systems. Four other significant factors influencing the enterprise environment are identified in the following sections: the use of commercial off-the-shelf building blocks; the adoption of the softswitch principle; computer–telephony integration; and the adoption of the Service Oriented Architecture for enterprise applications.

The development of softswitches for the enterprise environment, that is, call servers capable of supporting multimedia services tailored to enterprise users, is reviewed in Section 5.2.2. A form of application server long accepted in private networks in computer–telephony integration is described in Section 5.2.3.

5.2.1 *ADOPTION OF COMMERCIAL-OFF-THE-SHELF COMPONENTS*

Legacy private network infrastructure was often built using dedicated, single-purpose equipment such as PBXs. Such equipment required large investment, resulting in long amortisation periods. Apart from using standard signalling protocols, these systems were generally proprietary and closed. Coupled with their inflexibility with respect to new services, innovation was slow.

Emerging enterprise systems are based on commercial off-the-shelf (COTS) systems and components for communication as well as computing functions, reducing the need

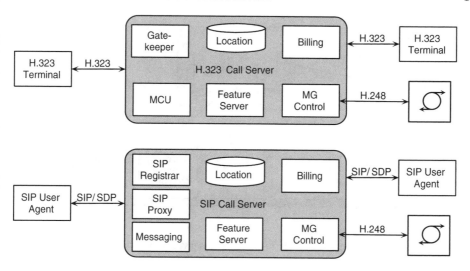

Figure 5.4. Telephony call server functionality.

for dedicated hardware, reducing cost and providing flexibility through programming using open standard interfaces. COTS computing equipment is capable of call and session control, information processing and media processing not only for adapting information to transmission media but also adding or extracting value.

The CTI standards allowed applications implemented on a COTS computer to invoke functions such as connecting a call in networks without the programmer having detailed network knowledge. The Service Oriented Architecture described in Section 5.2.4 provides a similar separation between generic or reusable functionality that supports a number of applications and the dedicated application. Such software functionality can be made available as COTS components.

Coupled with the move to COTS equipment, open standards for protocols and APIs have developed. While vendor systems have embraced open standards, system architectures are portrayed differently by individual vendors. Physical and functional architectures are therefore expressed in the NGN Framework layered model.

5.2.2 *THE MULTIMEDIA SOFTSWITCH*

An important building block of an enterprise network is a telephony call server or call manager. Such servers would be built on the principles of packet multimedia established in Chapter 4. A full-featured, fully managed call server would require billing and IN-like service features.

The H.323 and SIP standards define a number of system building blocks and protocols but leave implementation details undefined. In consequence, an equipment vendor might design a call server for the enterprise environment by integrating the necessary functionality onto a server as shown in Figure 5.4. Functionality includes protocol specific building blocks: the gatekeeper, gateway controller and multipoint control unit in H.323 and the proxy server, registrar and location server for SIP. Each

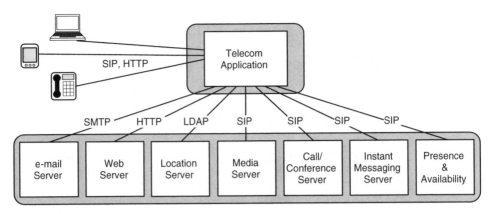

Figure 5.5. Telecommunications application server and supporting service control functionality.

server is classified as H.323 or SIP according to the protocols used by the terminals. Not shown in Figure 5.4 is the integration (glue) logic necessary for the server to function.

This form of server architecture lacks a clear structure required for an open integrated enterprise telecommunications and IT applications architecture. For example, enhancing telephony services using Web pages or messaging requires *ad hoc* addition and integration of servers. Similarly, applications may need to send, receive and interpret large volumes of short messages. In the enterprise environment the location and availability of workers to take part in communications requires management. We therefore use the layering principles of the NGN Framework to separate the application logic from the generic supporting functions.

Figure 5.5 shows a layered architecture of an enterprise telecommunications application server and its supporting functionality [177]. The descriptor telecommunications applied to the applications server indicates that applications use more than telephony: e-mail, Web pages, instant messaging, multicast media streaming, as well as user mobility, presence and availability. These supporting services are regarded as reusable functions that are accessible via standard protocols. The supporting services form of service control functionality layer in the NGN layered model. The set of supporting functions shown in Figure 5.5 is not exhaustive: others may be added for example billing using the Radius protocol.

End users interact with the application server using telephony (SIP) or Web (HTTP) protocols. The application must operate in different modes. For example, in a simple two-party call, it may act as a proxy, forwarding the call request to the call/conference server. In a more complex call, subsequent requests to add or drop parties would be forwarded similarly. In attempting to fulfill a call request, the application server may make use of the presence and availability server to determine whether the user is logged on to the network and is able to accept the call. If a call

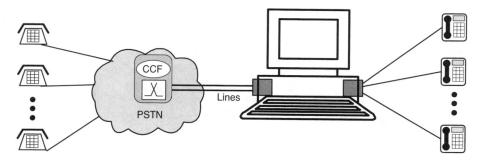

Figure 5.6. Physical architecture of a desktop computer–telephony integrated (CTI) system based on exchange lines.

request is not successful, the application may invoke a call completion service, for example by sending an e-mail to the unavailable called party.

The architecture shown in Figure 5.5 supports both first and third party services. The ability of end users to invoke the telecom application via HTTP enables control of telephony applications through a Web browser.

Figure 5.5 represents a distributed architecture based on protocols. From the application developer's point of view the distribution is achieved through the application layer protocols, for example SIP, LDAP, SMTP and HTTP, together with their transport protocols and DHCP and DNS. The various protocol-based servers act as service capability functions for the application. Application programmers must have knowledge of all these protocols or protocol-specific APIs.

This architecture exploits the flexibility of SIP to be deployed in different configurations and its ability to support not only real-time call control but also media streaming and instant messaging.

Within the application layer, various methods for creating applications are possible. The simplest form is a scripting approach. More complex forms may exploit operating system level APIs for the various protocols.

5.2.3 COMPUTER TELEPHONY INTEGRATION

An important influence in enterprise telecommunications and the integration with enterprise application is applications programming interfaces that allow applications to make use of telecommunications services. These APIs were first described as *computer–telephony integration* and were developed for to the private domain. These APIs support control of calls within the private domain, incoming calls to and outgoing calls from the private domain. With the emergence of powerful APIs, additional features such as third party call control, interaction with intelligent terminals and messaging have been added. The previous restriction to the private domain becomes blurred as advanced features are added to the APIs.

Systems based on telephony APIs are deployed at different scales. Some systems are implemented using a desktop computer or other single platform such as a PBX with connections to the public network using telephone lines or a single trunk as shown in

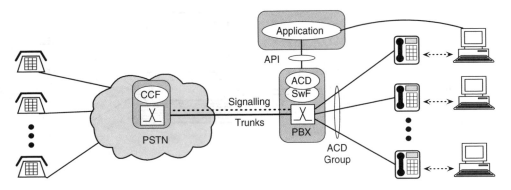

Figure 5.7. Physical and functional architecture of a computer–telephony integrated system based on exchange trunks.

Figure 5.6. The desktop computer performs the connection function between exchange line interfaces and local telephones.

The essential elements of larger scale or distributed systems, such as those used in a call centre, are shown in Figure 5.7 in the case of a TDM-based network. Connections to the public network are made using TDM trunks with a separate signalling channel. A PBX provides local switching and, in addition to switching control has an automatic call distribution (ACD) function that directs incoming calls to an appropriate agent who is logged into the system. An agent in the telephony API context is a device that is associated with a telephony application and is able to terminate calls associated with the device. The association would be established either by configuration parameters or a human agent logging on via the device. At its simplest, ACD operates as a hunting group with logged-on agents. The ACD function may seek the assistance of a telephony application to determine the group of agents or particular agent as the destination for the incoming call, for example by setting up an interaction with the caller or applying a policy.

While a device is logically a telephone, it may be implemented in different ways, for example a soft phone or a physical phone logically associated with a user's computer. The application generally provides a *screen pop*, that is information on the associated computer relating to the caller or service that has been called.

Two questions arise in evaluating a CTI system. First, what services are supported by the switching layer and what events can be reported to the application? Second, how are the underlying network and resources abstracted, for example in object models, connection models and state machines?

Four telephony APIs developed primarily for the private domain are described below: CSTA, TAPI, JTAPI and Asterisk. While their individual emphases vary, similar concepts and principles are used in their definitions. The API consists of services that may be invoked by an application on a provider entity and events that are supplied to the application. A *service* in this context is an operation, method call or message that invokes defined behaviour, including exception handling. An *event* is the notification of an occurrence in the network or call process that the application knows

Table 5.2. Summary of CTI standards

Standard	Environment	Definition	Network Adaptation
TAPI	Windows	C	Defined
JTAPI	Java	Java	Use existing API
CSTA	Neutral	ASN.1 and WSDL	Undefined
Asterisk	Linux	C/Open Source	Software implemented

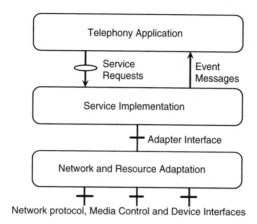

Figure 5.8. Generalising computer-telephony integration using selected NGN Framework layers.

how to receive, for example by having requested or registered for such notifications. The service provider entity is capable of accessing actual network protocols to control connections. The network interface is defined in some standards but not in others.

Service implementation is based on a defined state model that ensures predictable behaviour under various scenarios. Connections involve devices such as telephones in the private domain or trunks or lines in the public domain. Abstract connection models describe the connections and devices involved in the connection. Applications programmers must understand the call concepts, the state model, the connection model and models for devices such a telephones. No detailed knowledge of the networking technology should be required, for example whether the private domain is circuit-switched or packet-based or whether the public domain on which a call originated or terminates is the PSTN, an ISDN, a mobile network or an IP network.

CTI systems are described in terms of their ability to process first party calls and third party calls. In the enterprise context, a *first party call* is initiated by one of the parties to the call. Call initiation is either by loop signalling from a phone, by incoming signalling from the public network or by an action in the terminal, for example an ISDN or IP phone. A *third party call* by contrast is initiated by an agency other than one that will exchange media streams in the call. A third party call in an enterprise environment is generally initiated by an application via a telephony API.

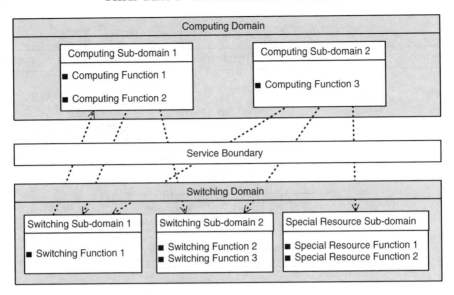

Figure 5.9. Domain model for Computer Supported Telephony Applications.

CTI interfaces assume a comprehensive model for a phone. While the capabilities of a phone may vary, a number of attributes can be set by the application, read by the application or can be programmed for state changes to initiate event reports. Phones can also upload or download user data. The most comprehensive phone model includes the hookswitch, speaker, microphone, display, ringer, buttons and lamps. Each of these can be interrogated and altered through API functions. The capability of a phone may be queried by the application.

We now review the four CTI standards listed in Table 5.2. This review is carried out using the NGN Framework adapted in Figure 5.8 to show the specific terminology used in some of the CTI architectures.

CSTA Standard

The Computer Supported Telephony Applications (CSTA) is an open standard developed by the European Computer Manufacturers Association (ECMA). Selected CSTA standards and technical reports have been adopted by both the ISO/IEC and ETSI. The core standard presents the modelling concepts, the definition of the services – in the sense of operations for invoking network services – and events [49]. The set of services defined in CSTA has grown through three phases from telephony call control to now include messaging and media processing. CSTA supports both first party (phone initiated) call control as well as third party (application initiated) call control.

The CSTA standard uses a domain model, summarised in Figure 5.9. Three types of domain are defined: computing, switching and special resource domains. The *computing domain* is the locus of application logic and is the *telecommunications application* in Figure 5.8. The *switching domain* contains the generic functionality

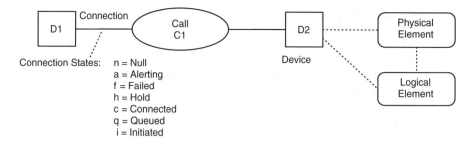

Figure 5.10. Modelling concepts used in CSTA.

that implements the network services defined in the API. This functionality ranges from low level to high level functions. The *special resource domain* can stand alone or be part of other domains. All three domains may have sub-domains in which particular services are implemented. The *service boundary* reflects the location of the API interface.

While the normal expectation would see the computing domain in a client role and the switching and special resource domains as servers, the computing domain may in certain cases act in a server role.

The application programmer has a view of the switching system and the user devices consisting of several abstractions. Devices are modelled logically and physically. The logical model represents the association of a device with a call. A finite state machine reflects the state of the device with respect to the call. A device also has a physical model representing the elements of a phone terminal that are observed or manipulated. The actual implementation of the switching system is of no interest to the application programmer.

Capabilities of switching domain systems may vary. The CSTA standard allows systems of different levels of capability to be CSTA-compliant. CSTA defines *profiles* containing sets of services that must be supported by a system that complies with the profile. For example, a system satisfying the basic telephony profile must support services Answer Call, Clear Connection and Make Call and must be able to start and stop monitoring events. The Connection Cleared, Delivered, Established, Failed, Network Reached, Originated and Service Initiated events must be supported.

The switching subsystem can be implemented in many ways. It may be a legacy processor-controlled PBX or an IP telephony call manager. The application in the computing subsystem must have a consistent view of the switching system irrespective of the implementation. Figure 5.10 summarises the modelling concepts used in CSTA that provide the consistent view of devices and a call.

A *device* represents a telephone. If the device is in another domain, for example in the public network, a network termination represents the device. A device within the domain has a *physical model* in which its various constituent parts, hookswitch, speaker, microphone, display, ringer, buttons and lamps, can be observed and controlled. For example, in handsfree operation, the hookswitch of the terminating phone is operated by the switching system and not by the act of lifting the handset.

A device also has a logical form that, for example, associates it with particular numbers or groups.

One or more *calls* are present in within a service session. In CSTA, a call is a relationship among two or more devices, or with one device during setup and cleardown. A connection represents the detailed relationship between a device and the call. In particular, seven connection states shown in Figure 5.10 describe the connection. Each party to the call thus has a connection state.

Before describing an illustrative use case, we note that a computing domain, that is an application, can control and monitor calls and devices. Monitoring tracks the progress of a call and changes of state at devices through event notifications. The switching function must implement monitoring of events such as dialled digits and local call control operations such as call forwarding. The computing domain enables and disables monitoring as required. Filters may be applied to allow and prevent specified event notification. CSTA allows call objects and device objects to be monitored. Monitoring can be enabled for the lifetime of the call object or while a device is associated with a call.

The relationships between devices and the call or calls in progress are depicted in a scenario, a type of connection view notation. Figure 5.11 shows a number of scenario diagrams at different stages during the progress of a call.

Figure 5.11 shows an illustrative sequence of services and events. The notation used in CSTA for depicting the call state is also shown. A call is initiated by a device D1 outside the private domain and this call is converted to a conference call. The call is connected by a device N1, a channel on an incoming trunk. Device N1 is monitored by the application in the computing domain. Device D2 is the called party. Phone D1 is first connected to D2. The user of D2 then makes a consultation call to a party with device D3 in the private domain. The three phones are then connected in a conference.

1. The switching mechanism detects an incoming call on trunk device N1 and uses the **RouteRequest** service to determine the destination number. For example, the conference access number must be translated to the number of the conference co-ordinator.

2. The **RouteSelect** service provides the routing number to the switching function. The conference co-ordinator may be provided with a screen pop over a data network.

3. The switch routes the call and, if successful, uses the **RouteEnd** service to complete the routing dialogue. Dialogue 1–3 illustrates the ability of the switch to play a client role in CSTA.

4, 5. Several notifications may be supplied to the application culminating in issuing an **EstablishedEvent** that notifies the application that the called party has answered.

Devices D1 and D2 now have an active connection. The conference co-ordinator now wishes to add the party using device D3 to the conference but first wishes to consult privately with the new party.

6, 7. The application is informed of this need by a mouse click on the conference control screen. A **ConsultationCallRequest** is made by the application to the switching function.

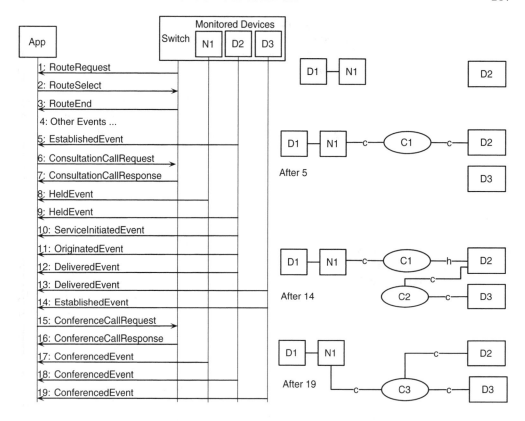

Figure 5.11. An illustrative sequence of CSTA service calls and events: routing and answering an external call, making a consultation call and combining calls into a conference.

8, 9. The switching function performs the detailed operations to provide the consultation call service. Device D2 is put on hold in call C1. Notifications may emanate from both devices.

10–14. Some notifications sent to the application while the D2–D3 call is made and answered at 13 are shown. A new call C2 now exists.

15, 16. Again using a mouse click on the screen pop, the conference co-ordinator requests the conference to be completed. The Application invokes the ConferenceCallRequest.

17–19. The switching mechanism ends call C1 and connects the three parties in a conference in a new call C3. The application is notified by each of the three devices that the conference connection has been made successfully.

The CSTA protocol has from its first form in Phase I been defined in ASN.1 and is therefore language-independent [48]. In Phase III a Web Services Description

Language (WSDL) form of the protocol is defined. This XML-based definition makes Web service forms of CSTA accessible to telephony application programmers. The XML Schema Definition (XSD) and the composition and interpretation of XML documents for invoking services and reporting events is considered easier to use than ASN.1 abstract definitions.

An example of a service request makeCall illustrates the ASN.1 type definition and the equivalent WSDL definition. This service request issued by a computing domain to a switching domain initiates a third party call. Part of the ASN.1 definition follows.

```
makeCall OPERATION ::=
    { ARGUMENT MakeCallArgument
      RESULT MakeCallResult
      ERRORS {universalFailure}
      CODE local: 10
}
MakeCallArgument ::= SEQUENCE
    { callingDevice                DeviceID,
      calledDirectoryNumber        DeviceID,
      accountCode          [0] IMPLICIT AccountInfo      OPTIONAL,
      authCode             [1] IMPLICIT AuthCode         OPTIONAL,
      .. }

MakeCallResult ::= SEQUENCE
    { callingDevice ConnectionID,
      mediaCallCharacteristics  [0] IMPLICIT MediaCallCharacteristics
                                                         OPTIONAL,
      initiatedCallInfo         [1] IMPLICIT ConnectionInformation
                                                         OPTIONAL,
      .. }
```

An extract from the corresponding WSDL definition follows. Like the ASN.1 definition, the schema relies on type definitions in other documents, for example DeviceID.

```
<xsd:element name="MakeCall">
  <xsd:complexType>
    <xsd:sequence>
      <xsd:element name="callingDevice" type="csta:DeviceID"/>
      <xsd:element name="calledDirectoryNumber" type="csta:DeviceID"/>
      <xsd:element ref="csta:accountCode" minOccurs="0"/>
      <xsd:element ref="csta:authCode" minOccurs="0"/>
      ..
    </xsd:sequence>
  </xsd:complexType>
</xsd:element>

<xsd:element name="MakeCallResponse">
  <xsd:complexType>
```

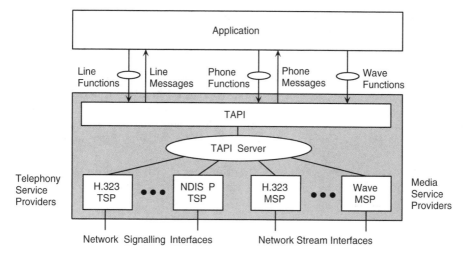

Figure 5.12. Outline of the Telephony Applications Programming Interface architecture.

```
<xsd:sequence>
  <xsd:element name="callingDevice" type="csta:ConnectionID"/>
  <xsd:element ref="csta:mediaCallCharacteristics" minOccurs="0"/>
  <xsd:element name="initiatedCallInfo"
               type="csta:ConnectionInformation" minOccurs="0"/>
  ..
</xsd:sequence>
</xsd:complexType>
</xsd:element>
```

In summary, the CSTA API is aimed at providing access to services implemented in the switching domain that are best described as supplementary services or service features. While CSTA supports low-level functions on physical devices, for example interrogating and setting the hookswitch status, call services are generally at a higher level. Several examples of CSTA service requests illustrate this observation: Conference Call and Consultation Call, illustrated in Figure 5.11, as well as Hold Call, Retrieve Call, Intrude Call and Park Call.

TAPI Standard

The Telephony Application Programming Interface (TAPI) is a Microsoft standard for CTI in the Windows environment [156]. A simplified view of the TAPI architecture is shown in Figure 5.12. The TAPI system allows a Windows-based telephony application to interact with software functions that in turn perform low-level telephone network functions in both TDM and packet networks. The telephony application occupies the NGN Framework Application Layer. Typical applications are PBX call control and call centre control.

The TAPI *service provider* layer represents the implementation of interfaces implementing functions defined in the TAPI API. The TAPI layer is also the source of event notifications that are sent to the application. The functions of the TAPI layer are generic call-related operations. The TAPI layer is therefore an SCF-type layer.

The TAPI server shown below the TPI layer is the central repository for user data, tracks the use of resources, registered applications and pending functions.

The *telephony service provider* (TSP) interfaces provide a defined way for network signalling to interact with software in the Windows environment. Similarly the *media service provider* (MSP) interfaces allow media operations to interact with network interfaces.

The TAPI interface has groups of functions as shown in Figure 5.12. The API provides functions that may be called by an application and messages for reporting events to the application.

- The *line functions* define a number of services for handling incoming calls. Examples of basic line functions are lineTranslateAddress for converting a canonical address to a routable number, function lineDial dials an outgoing call, while lineAnswer answers an incoming call.

- Similarly, *line messages* provide call-related event reports. For example the LINE_GATHERDIGITS message indicates that the requested number of dialled digits have been collected.

- *Phone functions* allow the application to control a phone with an assumed set of features in the private domain. For example the level of the microphone signal may be set using the function phoneSetGain.

- *Phone messages* report events related to an individual phone. For example, the PHONE_STATE message reports any change of state at the phone to the application.

- The *wave functions* support media-related operations. The characteristics of a stream can be interrogated and streams may be started, paused, or stopped.

- TAPI has a comprehensive set *supplementary call functions* such as lineSetup-Conference, lineForward, lineHold and linePark.

JTAPI

The Java-based Telephony API (JTAPI) was developed to support telephony applications in a Java language environment [139]. Like other Java-based APIs, a benefit of JTAPI is the portability of applications due to the inherent portability of Java programmes.

Two types of environments in which JTAPI is applied are illustrated in Figure 5.13. The JTAPI *implementation layer*, a type of SCF layer, abstracts the underlying network details. The implementation executes in a Java run-time environment, providing independence of the underlying system details.

Unlike TAPI, JTAPI does not define a network interface but expects the implementation to use one of the existing interfaces. In a desktop CTI implementation,

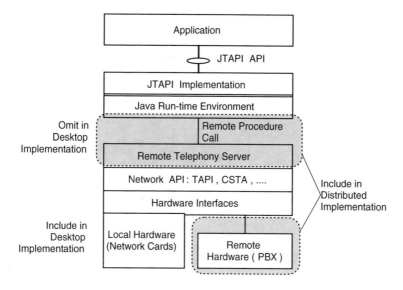

Figure 5.13. JTAPI architecture for desktop and large scale distributed system scenarios.

the network interface is local. In a distributed system, the network interface layer is accessed via a remote procedure call methodology.

Similarly, the physical interfaces to the network are either local or are accessed over a network using a suitable protocol or distribution mechanism.

The JTAPI specification is expressed in Java and is therefore object-oriented [199]. Application developers must have knowledge of the object model shown in Figure 5.14 and the associated state models used to represent a call and the parties to the call. A large number of methods are defined. Some are at a low level, for example creating calls, setting up event listeners and managing the containers in which logic executes. High level call control functions are provided, for example conference(), transfer(), drop() and consult(). Phone extension methods include setGain(), setHookswitchState() and setRingerPattern().

The Provider class is the specification for an object that is the point of access for the application for controlling calls. The Provider object keeps track of the terminals in its domain and their addresses. Terminals and addresses are represented by static objects. Terminals and addresses do not have fixed associations but become associated via connections being made in the course of a call. For example, in a call centre application only one address may be used, the number known to customers. To handle a particular incoming call, a physical connection is made to a particular terminal according to a call distribution rule.

Each call is managed by an instance of the Call class. The Call object maintains the state of the call as a whole. Each party to a call is termed a *call leg*. Each call leg is modelled by a Connection object that maintains the state of the leg and creates the logical association between an address and a terminal. A Terminal Connection object

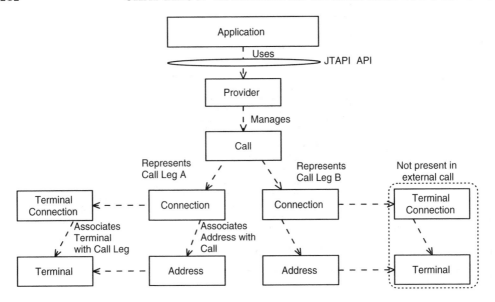

Figure 5.14. JTAPI object model.

represents the logical relationship between the **Connection** and **Terminal** objects. When a call party is outside the domain of the CTI system, the terminal connection details are unknown to the application. The call leg is represented by the **Connection** and **Address** objects only in this case, as shown in Figure 5.14.

The object model defined in JTAPI provides an important basis for representing calls in Open Service Access systems for public networks discussed in Chapter 8.

Asterisk: an Open Source PBX

Asterisk differs from the three CTI standards described above in that it is an *open source* implementation of a PBX rather than an *open standard* API as in the case of CSTA, TAPI and JTAPI. Asterisk is described as a single C-language application that has a modular structure [196]. The principal architectural constituents of Asterisk are shown in Figure 5.15, organised by the layers in the NGN Framework.

While Asterisk is described as a PBX, its basic hardware is the host computer and network interface cards, for example to terminate analogue or ISDN lines and E1 or T1 trunks. These cards terminate both signalling and media streams. Similarly, a standard network interface connects the host computer to an IP network router. This interface carries both signalling (SIP or H.323) and media streams. Media processing functions may be implemented entirely in software in the host computer. Hardware components that have on-board DSPs for media processing reduce the load on the host processor. The media gateway function between TDM and IP network is performed in the computing node. For example, media transcoding is performed by terminating

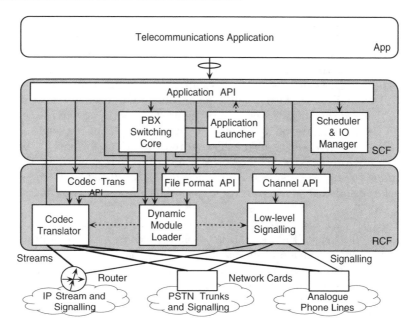

Figure 5.15. Architecture of Asterisk open source PBX.

the physical channel or trunk and performing the signal processing operations in the computing node.

The *application API* does not represent a clear domain boundary as reflected by the CSTA, TAPI or JTAPI interfaces. The Application Layer is the locus of the more complex PBX features or *telecommunications applications* in Asterix such as conferencing, voicemail, paging and directory service. Custom applications developed by third parties also reside in the application layer. Such applications can be invoked from the SCF layer below and can invoke features implemented in the SCF layer. The application API is defined via the open source programme listing.

Three lower level APIs, the *channel* API, the *codec translator API* and the *file format API*, are provided to access the functionality of loadable modules.

The *PBX switching core* is the principal constituent of the Service Control Functionality Layer. The switching core controls connections between callers in the various types of networks.

Asterix deals with user signalling and streams by loading the required modules. Loadable modules fall into two types. First, Asterisk handles different types of signalling by loading low level signalling modules as required. Several types of call signalling are accommodated including IP telephony protocols SIP and H.323 and those from TDM networks. Second, user media streams are processed by loading *codec translator* modules. The codec translator performs any required media transcoding, for example between a TDM channel and a packet voice encoding.

The channel API hides the individual network protocols from the PBX switching core. The channel API provides a metaprotocol that abstracts the detail of specific protocols associated with individual users or channels on a trunk.

The codec translator API hides the detail of the media coding from the switching core. The codec translator represents a resource control functionality. Audio signals in various TDM and packet formats may be encoded and decoded and transcoding is supported. The codec translator represents the media processing and connections that are made in the media gateway between a switched circuit and packet-based system.

The *file format API* allows media files stored in various formats to be read and written. The file format API depends on the codec translator module. For example, if an incoming stream to be stored is encoded in G.723 format, it is decoded by the codec translator and then written in the appropriate format on the disk, say the wav format.

Unlike other CTI standards, Asterisk is not a distributed application: the entire system reflected in Figure 5.15 is implemented on a single computing node. Custom telephony functions are implemented within the Asterisk application. While Figure 5.15 shows evidence of layering, namely application, SCF and RCF, these layers do not represent possibilities for physical disaggregation of the Asterisk system.

An important part of the Asterisk design is the *dialplan*. The dialplan identifies the users, allows users to be placed in groups and stores information on treatments to be applied to calls to specific users.

Asterix supports networked PBXs through its proprietary internode protocol, IAX. This protocol transfers both signalling and media streams between computing nodes. Because IAX is designed for this specific context, it has low protocol overheads that the more general open standard Real-time Protocol described in Chapter 4.

5.2.4   *ENTERPRISE SYSTEM SOFTWARE ARCHITECTURE*

Enterprise software system architecture is influenced by major vendor products. Individual vendors of enterprise resource planning (ERP) systems have developed their own architectures. Such architectures rely on application servers, relational databases and provide user interaction interfaces. Increasingly such systems make use of communications services.

Enterprise IT applications are, in general, transaction-oriented and database-intensive. The three-tier model shown in Figure 5.16 is commonly found in enterprise applications. The first tier includes the user interface and end-user access mechanism (EU). The middle tier contains the business logic and is hosted on an application server (AS). The third tier is the database. Typical enterprise systems operate in a transaction oriented manner, making synchronous invocations on databases.

The increasing need to create applications to meet enterprise needs also benefits from a an approach that divides an application into reusable components and application-specific logic. The reusable components perform generic functions including user interface, database, communication and print spooling. The Service Oriented Architecture is used increasingly as an architectural principle and to provide the reusable service components [183].

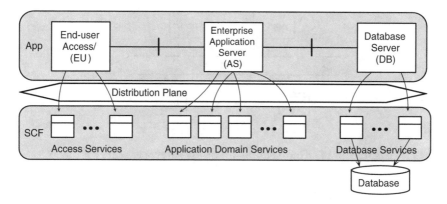

Figure 5.16. Enterprise resource planning three-tier model disaggregated using reusable SOA components viewed as SCF elements.

The Service Oriented Architecture

The SOA defined by the World Wide Web consortium (W3C) is introduced in Section 3.8.3. Figure 3.33 shows the triangular relationship between the service requestor, the service provider and the repository in which the service description is published. During the application development phase, the developer finds services that perform functions required by the application using the discovery mechanism. The completed application invokes services as required by exchanging documents conforming to the WSDL interface definition.

The objective of this chapter is to develop an integrated approach to ICT systems. This objective is related to the more general need to exploit software reuse in creating applications. While various software reuse strategies could be used when designing the application, the resulting application was monolithic. The adoption of the SOA allows the identification of decoupled, encapsulated components with well-defined interfaces. Applications in the three-tier architecture are being disaggregated into services according to the SOA model. To place this process in the NGN framework we allocate the components to a SCF layer as shown in Figure 5.16.

Some distribution plane mechanism is required that allows applications to invoke services. The concept of an *enterprise service bus* (ESB) is gaining currency [40] for applying the Service Oriented Architecture to IT applications. Figure 5.17 represents the essential elements of an ESB. The heart of the ESB is a message transfer mechanism, commonly referred to as *message-oriented middleware* (MOM). MOM provides reliable communication between applications and services using asynchronous messages. A built-in queueing system deals with requests that cannot be serviced immediately and protection against failure of components.

The enterprise service bus provides services in addition to the message-oriented middleware. Transformation of message formats and content dependent routing are supported. Applications are shown above the messaging system while services are shown below. Connections between applications and services are supported at the OSI

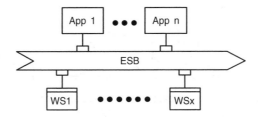

Figure 5.17. Representation of an enterprise service bus.

application and transport layers, for example the combination of SOAP and HTTP. The ESB is based on a loosely coupled model. An application, say App1, may invoke a variety of services without prior configuration. An ESB allows messages to be passed between components in response to events, connections to be made and data formats to be translated as required.

The layered form of the Web services model that complies with the NGN Framework Layers, as shown in Figure 5.16, is therefore useful. Common services made available as components provide generic, stable functionality. We therefore broaden the interpretation of the service control functionality layer to include functions other than communications functions. Other classifications of services include user access functions, database service and content services. Within particular application domains, sets of services perform domain-specific business logic.

Some IT applications use event-driven notifications to initiate necessary action, for example, a change of state of a variable. For example, the balance in a trading account falling below a predetermined level, causes a notification to be sent to a programme module that requests a transfer of funds into the account.

5.3   NETWORK LEVEL CONVERGENCE

An integrated ICT system supports converged applications. Converged applications are able to call on reusable functionality from telecommunications, IT and specialised business domains. The ability to invoke a wide range of reusable functions is enabled by extending the notion of the service control functionality layer from telecommunications into information technology and the support of distribution technologies such as CORBA, RPC and Web services. A multiservice packet network must in turn support the service control functionality. The following sections summarise the converged enterprise ICT system, starting with the converged enterprise network.

The enterprise network is a specific case of a multiservice packet network that is characterised by a defined geographical layout as well as clear ownership, management and usage roles. Unlike the Internet, a multisite enterprise network is well-defined and can be engineered and managed to meet the corporation's requirements, including high-availability of connectivity and services.

An enterprise network is usually defined with three main parts: access network, distribution network and core network [38]. The *access network* is concerned with connecting and concentrating the end user traffic, for example from workstations, IP phones or legacy phones via an access gateway. The access network is also responsible

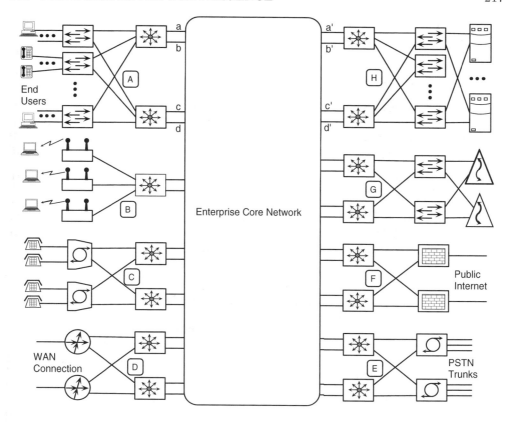

Figure 5.18. Enterprise network configurations for end user access, server connection and network interconnection. See Figure 5.19 for typical core network configurations.

for the connections to servers and specialist nodes such a call managers, and DNS and DHCP servers. Other forms of access provide interconnection to other networks: the PSTN via a media gateway, a remote island of the enterprise network via a WAN connection or to an external IP network, for example the public Internet via a firewall.

The network architecture in an enterprise responds to the requirement for high availability. Single points of failure are avoided except in the final connection to the end user. Access devices such as Ethernet switches, media gateways, firewalls or WAN routers generally gave more than one connection to the distribution layer.

The *distribution network* carries concentrated traffic and is generally constructed using high-performance switched routers [38]. The *layer 3 switch* or *switched router*, also called a routing switch, is an essential element in enterprise networks. The switched router, in addition to exploiting hardware-implemented forwarding algorithms, is able, because of the limited extent of the enterprise network, to process routing information efficiently and to deal with changes in network configuration and state rapidly.

Six access and distribution configurations are shown in Figure 5.18.

A. *Wired-connection end user*: this is generally provided by Ethernet connections. terminals are typically desktop computers or IP phones.

B. *Wireless-connected end user*: wireless access points support connections to a number of devices.

C. *Media gateway access to analogue phones within the enterprise*: this configuration arises either to integrate groups of legacy analogue phones into the packet-based enterprise network or to exploit the low cost of POTS phones for users that do not require data connections.

D. *Wide area network connection*: this interface provides the connection to remote sites that are part of the enterprise network.

E. *Media gateway to PSTN*: the media gateway has the characteristics described in Chapter 4. Interconnection with a legacy public switched telecommunications network is provided via these gateways. Signalling gateways, not shown in the figure, are required to transfer PSTN signalling to the call manager in the private domain.

F. *Access to Internet*: physical access from the public Internet requires firewalls.

G. *Call manager*: control of telephone calls and multimedia sessions in the private domain requires a call manager. The duplication of call managers is intended to create high availability. A single call manager may control calls involving parties on all physical sites of the private network.

H. *Server or server farm*: corporate servers are generally configured in a duplicated, high availability configuration. Depending on the number of servers, a server farm configuration may be used.

End systems not shown in Figure 5.18 include a network management centre (essentially a set of servers and terminals) and DHCP and DNS servers.

Figure 5.18 indicates the points of connection of two typical access and distribution groups to the *core network* by a, b, c and d and a′, b′, c′ and d′. The corresponding points of connection to the core network are shown in the various cases in Figure 5.19 [45]. Several common core network configurations are shown in Figure 5.19.

a. *Small network based on a collapsed backbone*: in the case of a small network, different types of access elements, for example end users and servers, are connected to a single pair of switched routers that double as the core routers.

b. *Small network with full mesh interconnection*: where the number of access networks is small, for example two or three, core switching may not be requires and it is sufficient to interconnect the distribution switched routers in a full mesh. The connections may be physical or virtual.

c. *Medium scale network with switched routers*: as the network becomes larger, switching is required in the core. Circuit (c) shows the use of layer 3 switched routers in the core.

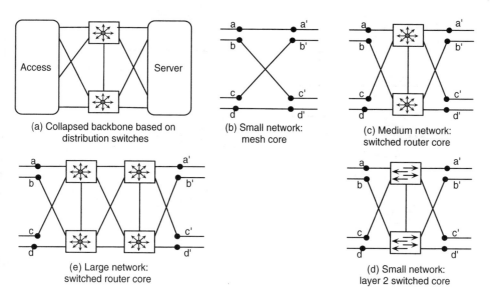

Figure 5.19. Core network configurations for enterprise networks at different scales.

d. *Medium scale network with layer 2 switching*: this configuration is an alternative to (c) that uses layer 2 switches in the core.

e. *Large scale network with switched routers*: this configuration differs from (c) only in the scale of the network.

Figure 5.20 is an example of the combination of access and core networks to create an enterprise transport network. The elements involved in a traditional three-tier model are shown. The physical connections to end users and servers are shown.

The objective of an enterprise ICT system is to support mobile workers and teleworkers outside the physical coverage of the enterprise network. Figure 5.21 shows a number of access methods to workers who are off-net. Five cases are shown.

- The PSTN is used for both voice and data access. Voice connections are made via a media gateway. Dial-up data access makes use of a point of presence (PoP) and routes packets via the Internet. Access using an ADSL line reaches the enterprise network via the Internet. Secure connections on the Internet section are achieved using various security mechanisms. The enterprise may have special billing arrangements for calls leaving and entering the private domain in general and for mobile and teleworkers operating from remote sites.

- Second generation mobile networks function similarly to the PSTN when providing access to the enterprise network. Voice connection and billing considerations are similar. Data connections via GPRS must pass through a GGSN.

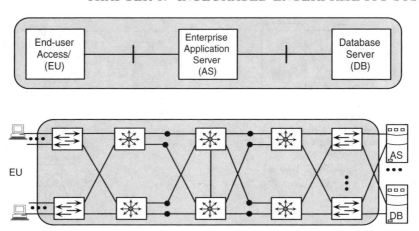

Figure 5.20. Enterprise resource planning three-tier components and underlying network.

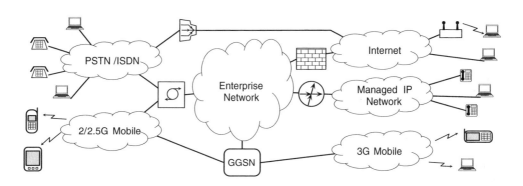

Figure 5.21. External access mechanism to enterprise network.

- The third generation mobile network operates in packet mode and therefore does not require a media gateway. Speech transcoding may be required.

- Access via the Internet to the enterprise network is supported by creating a virtual private network (VPN) connection. The VPN restricts access to authorised users, ensures that data is authentic, protects the data from tampering and prevents unauthorised parties from reading the data. Mobile workers and telecommuters are accommodated using an *access VPN*. By extending the VPN to the customer premises, the end user can enjoy rights and privileges as if logging on from inside the enterprise network. Users at fixed stations may be accommodated using an intranet or extranet VPN.

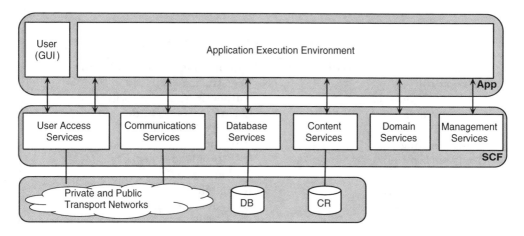

Figure 5.22. Generalised architecture of an enterprise system in a layered model.

5.4 APPLICATION AND SERVICE LEVEL CONVERGENCE

The thesis of this chapter is that application convergence reflects, on one hand, the increased use of network services, connection, messaging and network and user data. On the other hand, the software architectural principles underpinning telecommunications applications and enterprise IT applications develop an increased degree of commonality. While the SOA is a significant unifying influence, it is not the sole architecture. The key to convergence between telecommunications and IT applications is the adoption of a layered approach to software functionality in both areas. In telecommunications, the common call, messaging and network data functions are located in an SCF layer. In the CTI architectures, this common functionality is accessible through an open, secure API. Implementations are possible on several platforms. One CTI implementation, CSTA, follows the Web services or SOA model with the open interface defined in WSDL. In enterprise IT, the widespread adoption of the SOA allows a layered decomposition in NGN Framework terms: application specific logic in an Applications Layer with supporting service components in a Service Control Functionality Layer [54]. The latter abstracts network and database resources from the application.

While separation of unique application logic and reuseable services into layers provides a basis for convergence, differences exist in the inter-layer interactions and the distribution mechanisms used. Telecommunications applications and the supporting service control functionality are implemented using objects that are created and deleted as required. An application consists of several objects that communicate frequently using application layer protocols or method calls. For example in the JTAPI call model, a call object is created when a call starts. As legs are joined to the call and terminal connections are made, the corresponding objects are created. As legs are removed from the call, related objects are deleted and the call object does not exist after the call. Execution is computing intensive and generally must be completed within short deadlines. Invocations of methods on objects are in general asynchronous.

Table 5.3. Selected service control functions for the enterprise environment

Call and Session Control	**User management and control**
Call control	Registration
Conference control	Authentication, authorisation
Call handling	Location (contact address)
Videoconferencing	Mobility (geographical)
Whiteboarding	Directory
Event notification	User profile and history
Data session	Presence
Least-cost routing	Availability
Transaction co-ordination	Diary (calendar)
Accounting, call data record	Contact lists
	Subscription management
Call Centre Services	Service discovery
Automatic call distribution	
Caller-related data pop	**System and network management**
Outbound dialling	VPN management
	QoS
Messaging Services	Policy management
E-mail	Application management
Chat	Load balancing
SMS	Account management
Unified messaging	Storage management
Voice messaging	DHCP/DNS/ENUM
MMS	Lifecycle management
Instant messaging	Diagnostics
Fax	Statistics
Device capability	
	Personal and Workgroup Support
Media Services	Portals
User prompting and response	Video conferencing
Speech recognition	Transaction support
Speech synthesis	Web form support
Text to speech	
	Domain Services
Content Operations	e-commerce
Media streaming	Supply chain management
Multicast conferencing	Finance
Media adaptation	
Content indexing	
Licence and rights management	
Content presentation	
Content transformation	
Print queueing	
Access control	
Version control	

Similarly, transactions are asynchronous and large numbers of transactions may be concurrent, requiring a co-ordination mechanism. The invoking entity need not wait for the response to a request. Telecommunications processes usually have defined state machines where state transitions occur due to event or requests. Data is both static and dynamic. The first category includes user and service profiles. Dynamic data include user presence and availability information.

The application logic controlling lower level interactions in the telecommunications application case, for example JTAPI call control, may be wrapped as a Web service, presenting a single interface to the invoking application. For example, a high level function similar to the CSTA makeCall service hides the detailed Java invocations on the JTAPI interface. An enterprise application invokes the wrapped application using a WSDL-defined message.

The ease with which converged enterprise applications can be created depends on the range of service components available in telecommunications and IT-oriented SCFs. Figure 5.22 reflects the broad types of SCFs that would be required. A selection of possible SCFs in various categories is given in Table 5.3.

The view of convergence presented above suggests that telecommunications and IT-oriented SCFs should be regarded as a continuum in the enterprise context.

5.5 CONCLUSIONS

The integrated enterprise ICT system is more than a converged network. Rather, we view the generic reusable service control functionality that supports both telecommunications and IT applications as a common pool of resources. The resulting configuration is consistent with the Service Oriented Architecture based on Web services. The telecommunications-oriented service control functions hide the detail of the underlying network and the applications programmer view abstractions such as calls, messages and locations rather than the actual networks serving the user. Distribution support technologies including the enterprise service bus and CORBA allow seamless access to services and co-operating applications.

The integrated enterprise ICT system is a form of NGN and conforms to the NGN Framework layering. Several features are relevant to public NGNs. The enterprise network infrastructure is a multiservice network that serves users in a defined environment. The network management spans the users and facilities in the private domain. The softswitch principle is applied in the integrated enterprise network: call control is implemented in the SCF layer. The softswitch principle is extended in various CTI architectures to accommodate an application layer. This principle is examined in the context of public networks in Chapter 8.

Chapter 6

Legacies and Lessons: Broadband ISDN, TINA and TIPHON

6.1 LEARNING FROM HISTORY

At any stage in the development of ICT, several technologies vie for adoption to fulfill infrastructure and service needs. Some gain acceptance while others do not, even after adoption as standards. Since the development and deployment of the narrowband ISDN in the 1980s, the telecommunications world has always had a promised next generation network on the horizon. Initiatives ranged in scope from incremental developments to all-encompassing architectures. Candidate networks had a low rate of adoption, including the narrowband ISDN itself. Three successes of the 1990s were the Intelligent Network, adding a range of services to the existing circuit-mode voice network, the second generation mobile network and the Internet.

In this chapter, we examine the lessons from three networks that were not adopted into general use in public networks. Each has at least some of the objectives and characteristics of NGNs and provides an informative example of an architecture. All three contribute to thinking about the development of NGNs and, with the benefit of hindsight, hold important lessons and are worthy of study.

First, the *Broadband Integrated Services Digital Network* (B-ISDN) was essentially an extension of the narrowband ISDN to support services requiring higher bandwidths and multiple media. The principal contribution of B-ISDN was Asynchronous Transfer Mode (ATM), a packet switching protocol capable of supporting a range of types of real-time and asynchronous data streams with appropriate QoS. The Broadband ISDN held out the promise of a multiservice, multimedia network with broadband connections to the user terminal. B-ISDN standardisation took place in the later 1980s and early 1990s. Apart from the use of ATM as a switching technology, the B-ISDN was not deployed.

Second, the *Telecommunications Information Networking Architecture* (TINA) initiative from 1993 to 2000 sought to define an architecture that would support existing and emerging telecommunications and information services and allow management functionality to be built naturally into the architecture. The TINA Architecture is a set of concepts and principles for the constituent architectures of a complete infrastructure for the control and management of services and network resources. Management is integral to the architecture. A computing architecture provides a uniform approach to specification and design of software and exploits distributed object computing.

TINA sought to produce an implementation-free definition of all aspects of a future network: no implementation specifications were given. While many successful trials

Network Convergence: Services, Applications, Transport, and Operations Support
Hu Hanrahan © 2007 John Wiley & Sons, Ltd

of the TINA architecture were carried out, TINA has not been adopted as the basis for a next generation network. We examine the principal reasons for TINA's lack of adoption as a whole. Aspects of TINA have, however, been adopted by subsequent NGN standards, for example, Parlay, discussed in Chapter 8, draws on TINA concepts.

Third, the ETSI *Telecommunications and Internet Protocol Harmonisation over Networks* (TIPHON) initiative, which sought to harmonise packet voice standards, was seen as a next generation network in the late 1990s. Several packet voice standards had emerged, principally SIP, H.323 and BICC. TIPHON aimed to support interworking with switched circuit networks and the consistent use of these protocols. Partial harmonisation was achieved through the decomposition of the H.323 gateway discussed in Chapter 4. TIPHON extended harmonisation of packet voice networks further by introducing a greater degree of structuring.

After reviewing key aspects of the B-ISDN, TINA and TIPHON architectures using the NGN Framework developed in Chapter 2, we identify important principles they contribute for NGNs. We then identify the contributions and limitations of each and the possible reasons for not finding acceptance as a NGN.

6.2   THE BROADBAND ISDN

The Broadband Integrated Services Digital Network (B-ISDN) is an evolution of its narrowband counterpart. The narrowband ISDN was conceived as a multiservice network offering both real-time and data services through a single digital interface, the S-interface, at the customer premises [100]. The S-interface supports 64 kbit/s bearer or B-channels and a packet-oriented channel, the D-channel, for user-to-network signalling by the terminal equipment (TE). The transport network of the N-ISDN is based on circuit-switched B-channels. While it is possible to combine two or more channels to give higher bit rates, the N-ISDN is not suited to broadband services, that is services requiring bit rates in excess of 1.5 or 2 Mbit/s. The B-ISDN sought to overcome this limitations by introducing a packet-switched transport network that could carry several types of user traffic with appropriate performance. Apart from the change to end-to-end packet transport, other aspects of the ISDN architecture remained largely unchanged in attempting to meet the objective of supporting broadband, multimedia services. In the following sections we summarise the general architecture of the B-ISDN and the packet transport network before evaluating the architecture in the light of current understanding of the requirements for a next generation network.

6.2.1   B-ISDN ARCHITECTURE

The architectural concept of B-ISDN is adapted from N-ISDN taking the broadband multiservice objectives into account. Figure 6.1 is a composite form of several diagrams used to describe the ISDN architecture. The reference configuration for user access to the B-ISDN as well as the core network capabilities are summarised.

Access to the narrowband ISDN is based on two reference points. The S-reference point is implemented as a S-interface that provides the connection to the ISDN terminal equipment (TE). An S-interface is concerned with a single source and sink of information.

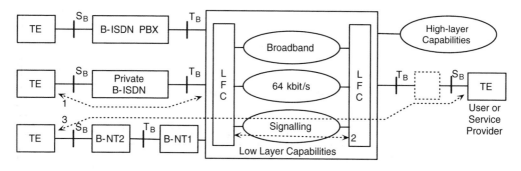

Figure 6.1. B-ISDN architectural model with access reference points S_B and T_B.

The T-reference point supports the multiplexed information and signalling flows of several terminals. For example, an ISDN private branch exchange (PBX) presents S-interfaces to the individual ISDN phones and a T-interface toward the public network. The B-ISDN maintains this access architecture. The S- and T-interfaces carry only packet traffic and are designated S_B and T_B, respectively. Bearer signals are carried in packet mode and user-to-network signalling has the same basic form as in the ISDN.

The core network capabilities are enhanced by being able to make broadband, multimedia connections. Signalling patterns remain the same while the detailed method of signalling transport is changed. Several virtual circuits are carried on the user-to-network interface (UNI) (marked 1 in Figure 6.1); one of these virtual circuits carries access signalling messages. User-to-network signalling is an extension to Q.931. Inter-exchange signalling is carried in ATM virtual circuit connections (2 in Figure 6.1). User-to-user signalling is carried from end-to-end, also in a virtual circuit connection (3 in Figure 6.1).

As in the N-ISDN, teleservices and supplementary services are envisaged in the B-ISDN. The concept of a call remains largely unchanged, as described in Section 6.2.3. End-user services envisaged for B-ISDN are categorised as *interactive* or *distribution* services. Interactive services may be *conversational* involving real-time signals, perform *messaging* (as evidenced by a store-and-forward mode of operation) or provide information *retrieval*. Distribution services provide large-scale point-to-multipoint transfer of information. This transfer may have a degree of user control.

6.2.2 ASYNCHRONOUS TRANSFER MODE

The principle chosen to create a multiservice network is asynchronous time division multiplexing. Short, fixed-length packets, called *cells*, are used to transport both signalling and user information. The short cell length, 53 octets, and high link speeds give low latencies for competing packets at a user-to-network interface. The standard is entitled Asynchronous Transfer Mode (ATM) [131]. Choice of ATM gives considerable flexibility in multiplexing user streams with a wide range of bandwidths, constant and variable bit rates, timing requirements and quality of service requirements. With the ATM transport network, a range of services could be supported.

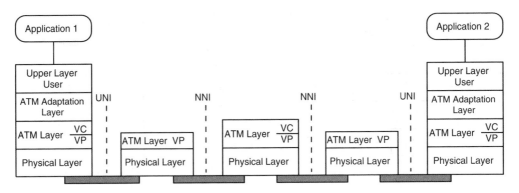

Figure 6.2. B-ISDN protocol model for UNI and NNI. Note the VP and VP/VC switches.

The ATM transport network developed for B-ISDN is based on OSI principles with a specific layered model. The B-ISDN protocol stack for transport of user streams and data is shown in Figure 6.2 at end and intermediate stations. The physical layer is not prescribed. The ATM switching mechanism, based on cells and virtual circuits is independent of the physical layer. Different upper layer protocols are adapted to the ATM cells using a set of ATM Adaptation Layers, each suited to a particular type of user [114]. The ATM layer is therefore independent of the serving and served layers. Connection-oriented and connectionless services are supported.

Six AALs have been defined of which four have found use:

- *ATM Adaptation Layer 1* (AAL1): designed to adapt constant bit-rate streams with real-time timing requirements to a stream of ATM cells.

- *ATM Adaptation Layer 2* (AAL2): designed to adapt variable bit-rate streams with real-time timing requirements to a stream of ATM cells. AAL2 is suited to transport of low bit-rate encoded speech or encoded video.

- *ATM Adaptation Layer 5* (AAL5): designed to adapt asynchronous streams without strict timing requirements to a stream of ATM cells. AAL5 is agnostic of the payload protocol and readily adapts frames of different types and layers to the stream of ATM cells.

- *Signalling ATM Adaptation Layer* (SAAL): designed to adapt SS7 signalling application layer protocol messages to a stream of ATM cells. The SAAL provides signalling co-ordination and application specific support and uses AAL5 for actual mapping of messages.

The B-ISDN standards also incorporate operations and maintenance principles. Like narrowband ISDN, these focus on management of protocol stacks through a system of layer and plane management.

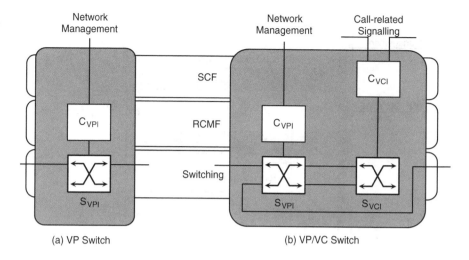

Figure 6.3. B-ISDN switches showing the logically separate but associated switching blocks and control blocks.

6.2.3 CALL AND CONNECTION CONTROL

The ATM network provides end-to-end transfer of cells between user access points by concatenation of links at the ATM layer. On a given physical layer, the streams belonging to different end-to end connections are distinguished by identifiers called the Virtual Path Identifier (VPI) and Virtual Channel Identifier (VCI). An end-to-end *ATM connection* is formed by two types of virtual connections. A *virtual channel* (VC) carries a single stream of cells related to an end-to end transfer requirement. This stream is identified by a VCI. On a particular link, a number of VCs can be grouped into a *virtual path* (VP), identified by a VPI. A physical link may carry a number of virtual paths. Each switch is responsible for the concatenation of VCs and VPs such that the end-to-end connections are made.

B-ISDN retains the PSTN concept of an *exchange*, that is co-located switching and connection control, for the nodes in the network. Each exchange is seen as having *switching blocks* and *control blocks*, associated in the same way as the CCF and the switch fabric of a circuit switch. Separate switching and control blocks are envisaged for VC and VP switching functions. Virtual path control configures the virtual paths and is assumed to be exercised by a management system. Virtual circuit control uses SCN-like connection setup via user-to-network signalling or internodal signalling.

An ATM switch may switch only virtual paths, and in this case, is the known as a *VP-switch* or an *ATM crossconnect*. An *ATM exchange* performs switching at both the VP and VC levels.

Figure 6.3 shows the building blocks of an ATM switch classified using three of the NGN Framework layers. The VP and VC switching blocks belong to the switching layers. Their respective control blocks are of two types. The VP control block is a RCMF function that participates in a management process for setting up paths across

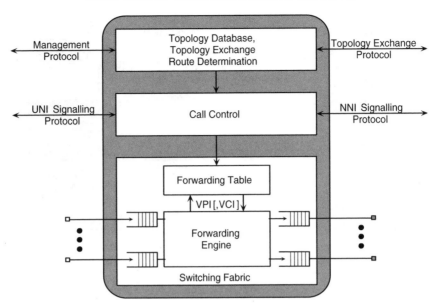

Figure 6.4. Reference model for an ATM switch. Adapted from [96].

an ATM network. The Virtual Path Identifiers may be rewritten as the cell leaves a switch. The VC control block is capable of on-demand switching and is therefore classified as a Service Control Function.

The switch fabric shown in Figure 6.4 switches cells according to the VPI and VCI fields in the cell header and information contained in the forwarding table. The forwarding table contains the outgoing link to be used for each combination of incoming VPI and VCI. The VPI and VCI may be translated before emission of the cell onto the outgoing link. The detailed architecture of the forwarding engine is an implementation decision.

6.2.4 ROUTING AND CALL CONTROL PROTOCOLS

The protocols used in an ATM network are of two types. First, as in other packet networks, a routing protocol is required to determine topological and state information on the network links. Second, the retention of the exchange concept in ATM networks requires a call or connection protocol for use between switches (NNI) and at the user-to-network interface (UNI). A protocol that includes both types of functions is the Private Network–network Interface (PNNI) protocol [96]. The ATM Forum reference model used in the PNNI specification for an ATM switch illustrates these relationships in Figure 6.4.

The B-ISDN assumes a teleservice model with intelligence in the customer premises equipment. Transport for the streams involved in the service is provided by the ATM network. While some services are connectionless, ATM supports connection-oriented operation. Virtual circuit connections are set up on a per-call basis. Each switch therefore has a call/connection control element. At the user-to-network interface, an access signalling protocol is used. PNNI includes Q.931 operations extended to support multimedia and point-to-multipoint connections. The call setup protocol is also used at the NNI between switches.

The PNNI routing protocol distributes topological information between switches. To support routing in very large networks, PNNI allows a hierarchical partitioning of the network. Link state information is exchanged by all members of a particular layer in the hierarchy. Information is exchanged between adjacent layers on links between border elements on layers. Since a large number of hierarchical layers is possible, very large networks can be supported efficiently.

The path followed by a particular connection is specified by means of a Designated Transfer List (DTL) of logical nodes to be traversed.

6.2.5   B-ISDN AND IN

The ITU-T Q.1200 series Intelligent Network Recommendations extend beyond the switched circuit network. Capability Set 3 sought to support advanced service control, mobility and services in the B-ISDN [132]. Recommendation Q.1237 describes the call model, application protocol and procedures when a B-ISDN transport network is used [134]. The basic call state models are extended to deal not only with the state transitions but also to accommodate bearer modification. The INAP protocol is enhanced to accommodate ATM-related parameters. Like B-ISDN itself, this extension of IN has not been deployed. This approach has the limitation of the closed, vertically integrated IN architecture.

6.2.6   APPRAISAL OF B-ISDN

The principal contribution of B-ISDN to telecommunications is the ATM network now used for other purposes. ATM finds several applications in the transport and access networks. For example, ATM is used to create virtual links in other networks, for example for interconnecting mobile switching centres, as the backbone for Frame Relay networks and as Internet service provider backbone networks. Such networks are semi-permanently configured and on-demand connections are not supported. ATM cells are used in Digital Subscriber Line Access Modules (DSLAM) to multiplex user data streams.

The complete B-ISDN has not however contributed to the provision of services to end users. Contributory factors are:

- The B-ISDN architecture does not conform to the softswitch model. Call and connection control are not separate.

- While the ATM network provides important support for quality of service, admission control and flow control at the UNI, it is seen as complex and expensive in relation to the rapidly expanding Internet Protocol (IP) networks.

6.3 TELECOMMUNICATIONS INFORMATION NETWORKING ARCHITECTURE

The need for a new telecommunications architecture to succeed the switched circuit network was recognised in the late 1980s. The TINA Consortium was founded in 1993 with a membership representative of the major telcos and equipment vendors to define the architecture and standards. The main standards development effort lasted until 1997. Several trials continued until 2000. While some TINA standards were accepted as ITU-T Recommendations, no general deployment of TINA-compliant products took place. An appreciation of the TINA architecture is, however, important to the study of NGNs. We review TINA but flavour the treatment with the benefit of hindsight.

The TINA Consortium aimed 'at defining and validating an open architecture for telecommunications systems for the broadband, multi-media, and information era' [35]. Behind this statement was a recognition that telcos would need to not only serve telecommunications needs but would also provide or at least support information services: this objective is reflected in the *Information Networking* element in TINA's full name. The architecture addresses both voice-based services as well as possible new services: interactive multimedia, information services and management services [24]. The standard was to be technology-independent and allow implementations using different networking standards and computing technologies.

A second set of objectives of TINA addresses a number of computing considerations. OSI-based communications protocols have long provided the means for computing elements distributed across computer nodes to communicate. The OSI-reference model is a simple peer-to-peer paradigm. In arbitrary distributed systems, the OSI-RM protocol stacks require significant management effort to ensure knowledge of transport addresses, especially if software is relocated, for example in the case of node failure or load balancing. Distribution transparency was therefore a requirement in the TINA environment. The TINA initiative also sought to exploit software reuse. To pursue these objectives, the TINA standards exploit the concepts of object-orientation, Open Distributed Processing (ODP) and a distributed processing environment (DPE).

Because of its comprehensive coverage, the TINA architecture and detailed specifications are very complex. Several methodologies described in following sections were used to manage the complexity: architectural separations, layers and planes, and the RM-ODP viewpoints.

6.3.1 *THE TINA ARCHITECTURE*

Because it sought to be technology independent, the TINA architecture concentrates on software for service control, network resource control and the management of services, network resources and the software itself. Such software is complex and is inevitably distributed over several computing platforms.

At a high level, the TINA architecture was defined as an *overall architecture* that is divided into four constituent architectures with defined interrelationships. In the TINA context, an architecture is a set of concepts and principles that guide the development of the standard and in the design of services. Because of its anticipated complexity, the overall TINA architecture defines the four constituent architectures, each with a specific purposes and defined relationships among the architectures.

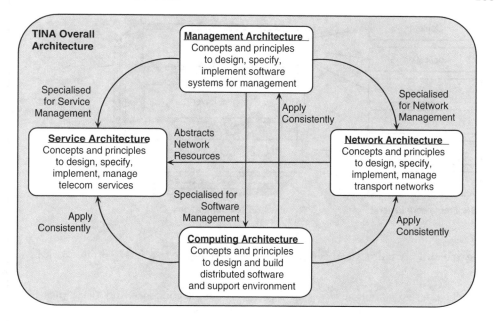

Figure 6.5. The TINA architecture with its four constituent architectures and their relationships.

Figure 6.5 shows the four constituent architectures and their principal relationships. Distinct service and network architectures implement the softswitch principle: the separation of service control from control of network resources.

- The *service architecture* defines a set of concepts and principles for the design, specification, implementation and management of telecommunications and information services. Important concepts in the service architecture are sessions, service components and reference points.

- The *network architecture* defines a set of concepts and principles for the design, specification, implementation and management of transport networks, including connection management.

The service architecture accommodates the logic needed to implement services for use by end users. Such services generally require the establishment of network connections for transport of streams and data. The network architecture is required to present an abstract view of network resources to the service architecture. Network technology, routing topology and other details must be hidden from the service architecture. One or more reference points are therefore needed between service and network architectures.

The TINA Consortium sought *ab initio* to incorporate management of network and computing resources into the standards and therefore made the management architecture an integral part of the overall architecture:

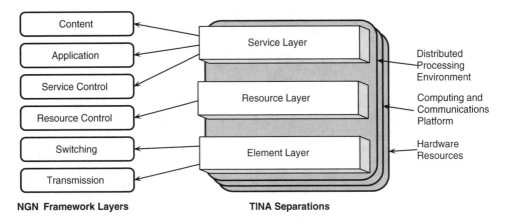

Figure 6.6. Architectural separations in TINA related to NGN Framework layers.

- The *management architecture* defines a set of concepts and principles for the design, specification and implementation of software systems to manage services, resources, software and the underlying technology.

The management architecture guides the management of service software, computing platforms and network resources. The common principles of the management architecture are specialised for each of the other architectures: service, network and computing.

The service and network architectures are the concepts and principles for the software elements for control of services and network connections. The management architecture similarly supports the definition of software elements for managing various resources. Because of the great complexity and distributed nature of such software systems, a computing architecture was defined.

- The *computing architecture* defines a set of concepts and principles for designing and building distributed software and the software support environment, based on object-oriented principles.

The computing architecture must be applied consistently when designing software for service or network resource control and for management. A single approach to software specification and implementation is used across all architectures.

6.3.2   TINA LAYERS

Figure 6.6 illustrates further separations identified in the TINA Overall Architecture. Three layers are distinguished: service, resource and element layers. A network infrastructure to support a range of services generally straddles all three layers.

The *service layer* contains objects that are required for the provision of services to end users. The service layer corresponds largely but not exclusively to the service architecture. Service Layer objects provide generic functions such as access control, user profile data and service management as well as service-specific logic and data.

The *resource layer* is concerned with objects that control and manage elements or collections of elements. The resource layer also provides abstract views of the element layer to the service layer.

The *element layer* contains objects that represent physical or logical resources. Such resources generally consist of a functional part and a physical processing part, as observed in Chapter 2. For example, an ATM switch shown in Figure 6.4 exhibits such a division. The functional part can be represented by an object with interfaces for control and management. Such an object is often referred to as a *proxy* for the physical element.

The three TINA layers are called *management layers* because they corresponds to three of the Telecommunications Management Network (TMN) layers described in Chapter 9. The relationship of the TINA overall architectural layers to the layers defined in the NGN Framework in Chapter 2 is shown in Figure 6.6. The *element layer* corresponds to the Switching and Transmission Layers since the TINA network resource model contemplates the control of both switching elements such as ATM switches and transmission elements such as SDH multiplexers. TINA's well-developed resource layer corresponds to the NGN Framework's Resource Control Function Layer. The services layer in TINA contains both generic and specific service logic and therefore spans the NGN Framework's Service Control Function and Application Layers. TINA does not develop the treatment of content, other than services involving content streaming. Content servers would be accommodated in an unspecified way in the service architecture.

6.3.3 PLANES: HIDING PHYSICAL RESOURCES

The three planes shown in Figure 6.6 reflect computing and networking concerns and hide different levels of detail. The first plane is concerned with the distributed processing environment that supports communications among objects in the service, resource and element layer proxies, without regard to their location or computing platform. The next plane, termed the *native computing and communications environment* (NCCE) collects concerns about the computing and communications platforms, for example operating systems and protocols. The third plane reflects the actual computing and communications hardware. The planes provide a form of abstraction: each plane hides the detail of the plane below.

The DPE is conveniently viewed as a plane that provides support for inter-object communication that does not require detailed knowledge of the computing platform or underlying network. The DPE plane insulates the service programmer from technological and topological detail of the computing and networking resources.

6.3.4 TINA AND RM-ODP

The TINA architecture and resulting specifications represent a large distributed software system. The TINA standardisation process was therefore guided by the principles of the RM-ODP. Table 6.1 summarises the alignment of aspects of the TINA architecture with RM-ODP viewpoints.

The enterprise viewpoint is relevant at two levels. First, for the architecture itself, the generic roleplayers in the provision and use of services are identified. Their identities and relationships are captured in the TINA Business Model described in

Table 6.1. Alignment of TINA architectural features with the RM-ODP

RM-ODP Viewpoint	Aspect of TINA architecture
Enterprise	TINA business model
	Inter-domain reference points
Information	Comprehensive information modelling
	Session models
Computational	Service components and interfaces in the service architecture and network resource architecture,
Engineering	Aspects of the distributed processing environment
Technology	Not defined: an implementation consideration only

Section 6.4. Second, when designing and implementing an individual service, for example a video-on-demand service, the business model is used as a framework and the individual roles are developed in detail.

Developing the information viewpoint is an important stage in the definition of each of the service, network and management architectures. Detailed information modelling is used to define the object classes and relationships.

Similarly each of the service, network and management architectures requires the development of the computational viewpoint. Agreed service components with defined interfaces give the computational architecture. Where the interfaces are accessed across a domain boundary, they constitute a reference point.

The engineering viewpoint defines a common method for deploying software components belonging to all architectures.

The RM-ODP viewpoints, with the exception of the technology viewpoint, pervade the definition of the constituent parts of the TINA architecture. Similarly, the viewpoints would provide an organising principle for the subsequent design of individual services within a TINA-based system.

6.4 BUSINESS MODEL AND REFERENCE POINTS

An important objective of the TINA initiative was to create an open architecture. At the simplest level, openness is obtained by having defined standard interfaces at which systems in different domains can interwork. At a more comprehensive level, the interdomain relationship is specified from various viewpoints. One such scheme is to use the RM-ODP viewpoints. The business roles of interacting domains emerge from the enterprise viewpoint definition. In a standardised architecture such as TINA, the reference points must suit typical ICT business interactions. This consideration lead to the formulation of the TINA business model and reference points [52].

6.4.1 GENERIC BUSINESS DOMAINS

The *TINA business model* identifies a number of generic business roles applicable in the ICT context. These business roles could be specialised or supplemented by other roles. The generic roles are therefore described as *initial business* roles. The business roles are defined around the concept of a *TINA system* that we define as a telecommunications and information service infrastructure that exposes TINA reference points to stakeholders. Six initial business roles are defined:

- *Consumer*: a party that benefits by being able to use the services provided by other stakeholders in the TINA system. The consumer has no business interest in the TINA system. The consumer in general pays for the use of services and is therefore the source of revenue for other stakeholders in the TINA system. Different types of consumer may be identified. The *subscriber* is a party that has a contract with the retailer, described below. A *user* is a party that the subscriber authorises to use the subscribed services.

- *Retailer*: the essential service provider to the consumer. The retailer registers the consumer for various services that it either provides itself or makes available through third party or connectivity providers. The retailer functions as the home network for the consumer by maintaining subscriber profiles. Subscription management services are provided to the consumer. Subscriptions may be long lasting or short-lived. All charges are levied by the retailer irrespective of whether other providers actually supply services or network connections. The retailer must therefore collect and log accounting information and distribute payments to other service providers. The retailer controls access to the TINA system and to individual services. Once a service has been started, the retailer manages the service session. The TINA architecture therefore assumes a single point of contact between the consumer and the retailer. Retailers may co-operate with other retailers to provide services.

- *Connectivity provider*: an infrastructure operator that owns or operates networks that can provide connections between users. Service-related requests for connections are issued either by a retailer or a third party operator. The connectivity provider must make, and if requested, modify physical connections between endpoints in response to a request by the retailer (or third party provider). Connections must be managed and accounting information collected.

- *Third party service provider*: provides services on the request of a retailer acting on behalf of a consumer. These services may involve service logic, for example conference control, or may provide content, for example a video on demand service. Depending on the service, the third party provider may request connections from the connectivity provider.

- *Broker*: this party's function is to provide stakeholders with information on services that are available from other stakeholders. Underlying the concept of the broker is the assumption that the DPE on which the TINA system is based allows a client to contact a serving object provided that the object reference to the server is known. The broker function could be offered on a white pages basis: the requestor names the requested service and the broker returns an object reference through which the server can be contacted. Alternatively, the yellow pages-like mechanism allows the requestor to describe the required service and the broker returns an object reference.

TINA systems may be constructed using federation of providers and composition of services. *Federation* is a process in which service providers agree to operate jointly

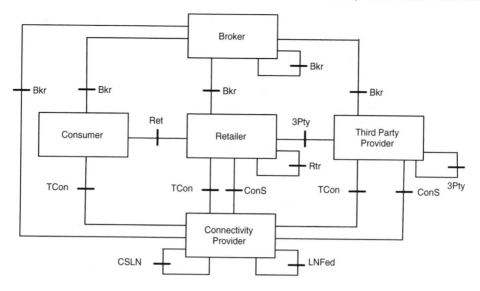

Figure 6.7. The TINA business model showing the five generic roles and reference points.

for defined purposes, making identified resources available and following specified procedures. *Service composition* is the process of creating a service by using existing service components or services where the services may be provided in different domains.

The retailer and third party (3Pty) domains are able to federate, that is work together to perform specified functions according to an agreement. Similarly, a service that appears as a single service to the consumer may be implemented by composition of service logic in the retailer and third party domains. Network providers in separate administrative domains are able to federate to provide end-to-end connections with specified quality of service.

The TINA architecture follows the general softswitch principle of separating service logic from the network connectivity function. The connectivity provider role in the TINA business model encapsulates all connection-related functions.

The TINA architecture does not draw a clear distinction between the retailer and the third party service provider in terms of the functions that each performs, for example whether the service logic in the retailer domain should be light-weight and generic while complex service logic should be restricted to a third party domain.

6.4.2   INTERDOMAIN REFERENCE POINTS

The inter-domain reference points shown in Figure 6.7 are identified to support a number of business relationships: Ret, 3Pty, TCon, ConS and Bkr. Reference points for federating domains are also identified: Rtr, 3Pty, CSLN, LNFed. In addition, intra-domain reference points may be defined, for example between management and service logic subsystems is the service architecture.

Table 6.2. Interdomain reference points in the TINA business model

RP	Name	Status
Ret	Retailer	Defined in [92]
3Pty	Third party	Undefined: parts of Ret usable
TCon	Terminal connection	Defined in [50]
ConS	Connectivity services	Defined in [162]
Bkr	Broker	Undefined: use CORBA services
Rtr	Retailer-to-retailer	Undefined: use Ret interfaces
CSLN	Client-server layer network	Undefined
LNFed	Layer network federation	Undefined

A full definition of a reference point involves the RM-ODP viewpoints. The enterprise viewpoint describes the relationship between the domains that interact across the reference point. This description is in natural language. The information viewpoint describes the information shared across the reference point. The computational viewpoint identifies the client and server service components that interact across the reference point and the interfaces used. The interfaces are usually grouped in two sets. The *access part* is concerned with gaining access, authentication, service discovery and service initiation. The *usage part* is a subset of the interfaces required during service execution. The engineering viewpoint provides the technology independent definition of how the components are deployed and communicate across a reference point.

Not all the reference points are defined. The retailer reference point (Ret) has a comprehensive, validated definition with access and usage parts. Interfaces at the TCon and ConS reference points are defined in the TINA standards. The third party reference point has not been defined. However some retailer RP interfaces are often used at the 3Pty interface. Table 6.2 summarises the status of the TINA reference points.

The broker role has not been developed and the broker RP interface has not been defined. Where required, the white and yellow pages functions have been implemented using CORBA services. For example, the CORBA trading service provides for location of services by description. CORBA services do not provide a complete substitute for the full broker role. The broker was not just a technical function but was envisaged as a commercial role, offering subscriptions to and gaining revenue from other roleplayers.

6.4.3 BUSINESS SCENARIOS

The four business model domains, together with the reference points, allow a number of scenarios. Figure 6.8 shows four illustrative relationships.

In the two-party call scenario in (a), one consumer is assumed to have a subscription with the retailer. The retailer domain contains the necessary service logic. The retailer accesses the connectivity provider to obtain the required connections. The retailer is able to invite other end users to the service.

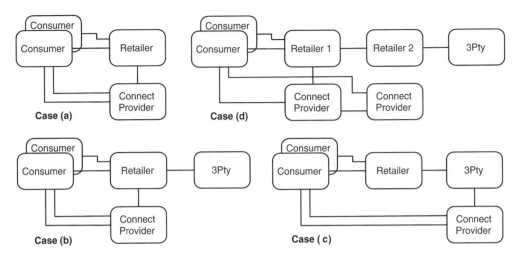

Figure 6.8. Four possible business scenarios enabled by the TINA business model and its reference points.

In (b), the two-party call is enhanced by service logic residing in a third party domain. Service logic is invoked by the retailer on behalf of the client. The retailer requests network connection from the connectivity provider and invites other users to the service.

Case (c) would apply when a retailer has no network facilities but obtains connections for the consumer through a third party. The TINA business model permits this scenario since the third party provider may use the connectivity service (ConS) reference point to request connections from a connectivity provider. An example of this scenario is a video-on-demand (VoD) service. The user initiates the service by contacting the retailer. The retailer invokes the VoD service of a 3Pty provider. The logic to control the VoD service, for example rights management, streaming control, pause and rewind, are in the 3Pty domain. The connection to the end user(s) is initiated from the VoD provider.

Case (d) elaborates case (b) to illustrate further possibilities within the business model. First, the retailer serving the consumer accesses the third party provider through a retailer-to-retailer federation arrangement. Second, the network connection is made by federating connectivity providers.

6.5   TINA SERVICE ARCHITECTURE

The *TINA Service Architecture* (SA) defines a set of concepts and principles leading to a set of computational objects and interfaces that allow individual services to be designed and implemented. The SA is intended to support a wide range of services, both telecommunications and information services. Services must be manageable and capable of being tailored to user needs. Access to services must be ubiquitous, that is accessible by the user from various locations. Service instances must be independent

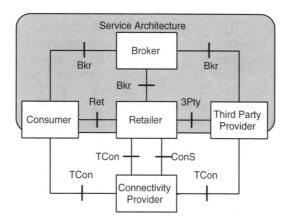

Figure 6.9. Enterprise viewpoint location of service architecture in relationship to the TINA Business Model.

of the underlying network resources and interoperable across platforms provided by different vendors. As in the case of all service architectures, service creation and deployment should have short lead times.

The TINA service architecture is defined according to RM-ODP principles [22]. The enterprise view states the objectives of the service architecture, summarised above, and locates the service architecture in relation to the business model. The service architecture spans most of the consumer, retailer and third party domains as well as the broker, as shown in Figure 6.9. The broker role may not be implemented explicitly; CORBA services may be used.

The service architecture is the location of the service logic, encapsulated in particular computational objects. The service architecture also contains generic functions such as authentication and service initiation.

6.5.1   SERVICE ARCHITECTURE: INFORMATION VIEWPOINT

Information modelling concepts used in the TINA specifications follow the RM-ODP information viewpoint principles: the purpose is to express what the system must do, not how it does it or the packaging of objects or the nodes on which they execute. Information modelling defines the following: What are the information objects? What types of objects assist with the definition of information objects? What relationships exist between the objects/types? What constraints and rules apply to the entities? In addition, the way in which objects are created and destroyed must be defined. The TINA specifications predate the widespread adoption of the Unified Modelling Language. Information models are expressed in a graphical notation called Object Modelling Technique (OMT). The use of OMT is not mandatory. Specifications using OMT diagrams are readily intelligible to those familiar with UML.

Session Model

The TINA standards uses the concept of *sessions* as an organising principle. Within the service architecture, three types of sessions are defined: access, service and communication. A fourth type of session, the connectivity session is defined in the Network resource Architecture.

The information objects that comprise the information viewpoint are defined in [22]. A core concept is the *session model*. In general, a session is a relationship between entities that exists for a period and provides the context for interaction between the entities. The session in TINA is more comprehensive than both the PSTN call concept and the media streaming session in SIP or H.323.

Several types of sessions and their interrelationship are defined. A session generally involves interacting entities in different domains. Within each entity, however, there are state variables and policies applicable to that entity alone. The service architecture specification therefore defines a *domain session* as the basic form of session. An interdomain session is related to the domain sessions in each domain and their binding to perform functions across the two domains.

An *access session* results from the secure binding of two domain access sessions. An access session is initiated when one domain contacts the other. Authentication may be required. Once an access session is in existence, one domain may the request operations to be performed by another domain. Within an access session, a party may:

- make initial contact to establish interfaces to use for subsequent interactions;

- authenticate with the provider;

- discover available services;

- initiate and terminate services;

- terminate the access session.

An access session must be set up when a consumer wishes to use services offered by a retailer. Similarly, an access session must be set up when a retailer accesses a service offered by the third party provider.

A *service session* represents the information and interactions required to execute a service. A service session depends on the existence of at least one access session. A service session is launched from an access session. For example, the core logic of a multiparty conference service is represented by a service session. One party must have an access session with the provider to launch the conference service logic. Within a service session, domain sessions have specialised roles. A domain user session may be a *user* or a *provider* of services. Domains may have a peer relationship or may work together to *compose* an overall service from their respective service capabilities.

The service session is concerned with the logic to perform a service. When parties need to transfer real-time or data streams, a new type of session, the *communications session*, comes into existence. The communications session represents a view of the connections invoked by the service. This view of the connection is abstracted from the actual method of making the connections by the network resources. The communications session is independent of both the network topology

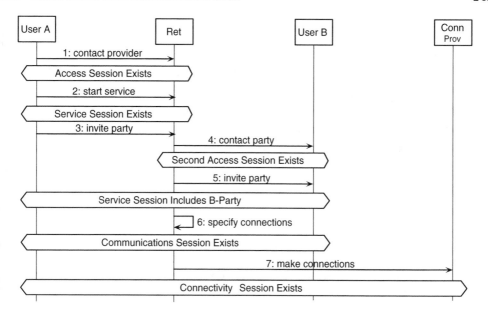

Figure 6.10. Illustrating four types of session defined in TINA. The operations 1–7 are hypothetical.

and technology. Rather, the communications session contains a representation of the endpoints involved in the connections, their format, for example point-to-multipoint, connection properties and required quality of service. A communications session is controlled by a single service session and may be initiated, modified and terminated within a service session.

The fourth type of session, found in the network architecture, is the *connectivity session*. This type of session represents the processes required to convert the abstract view of the communications session into concrete network connections.

Figure 6.10 illustrates the relationships among the different types of sessions. An access session is initiated between user A and the retailer at message 1. Within this access session, the user requests a service, for example a two-party call. A service session involving the retailer and user A is set up after message 2. User A then issues message 3 to invite party B to the call. At message 4, the retailer contacts user B and sets up a second access session. The retailer then invites user B to the service session (message 5). The requirement to connect users A and B by a stream connection requires a communications session to be set up (6). The communications session spawns a connectivity session resulting in the actual physical connections being made between users A and B (message 7).

6.5.2 *COMPUTATIONAL VIEWPOINT: THE SERVICE COMPONENTS*

In RM-ODP terms, the computational viewpoint defines how the objects defined in an implementation-independent manner in the information viewpoint are encapsulated.

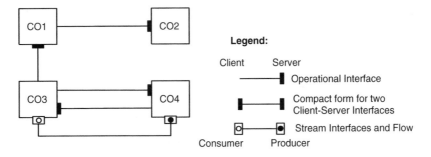

Figure 6.11. TINA computational model.

The units of encapsulation are termed *computational objects*, also referred to as *service components*. In TINA, a computational object has one or more interfaces. An interface is a named collection of operations (method calls) located on a defined computational object. Generally, the computational object from which the methods on an interface may be called are defined.

Figure 6.11 shows the computational model used in the TINA standards. Computational objects have defined interfaces that can be of two types. The most common interface is an *operational interface*, that is one comprising a defined set of operations (method calls). Each interface therefore has a defined client (requestor or user) role and an identified server (provider) role. Apart from parameters, no other data crosses an operational interface. In the service architecture, components have the descriptors Req(uest), Ind(ication), Exe(cution) or Info(rmation) to distinguish the role of the interface.

Where two computational objects exchange data with their own characteristic structure, for example files, encoded voice and images, the interfaces are called *stream interfaces* [123]. A stream is unidirectional from a producer to a consumer. A bidirectional flow is simply two contradirectional stream flows. No operations are transferred on a stream interface. Stream interfaces are generally abstract representations of actual user data flows.

The computational view spans the services, resource and element layers. Starting with the services layer, we build up the set of computational objects that makes up the TINA architecture. In the following sections, we describe the service architecture by identifying patterns of component for different groups of actions.

Service Component Overview

A full implementation of the TINA architecture uses a significant number of service components spanning the service architecture and the network resource architecture. Figure 6.12 shows the majority of the service components defined in the TINA standards. Management components are not shown. The set of components shown in Figure 6.12 supports all operations in a service instance: enabling user access control, service initiation and execution, invocation of services from a third party provider

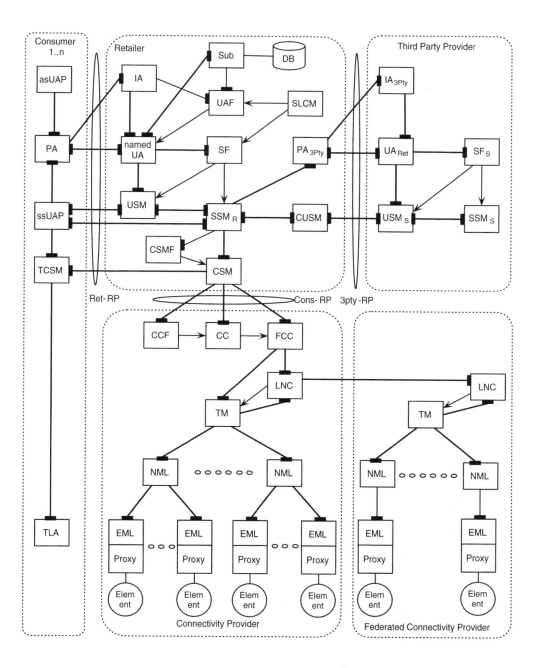

Figure 6.12. TINA service and network resource architecture components.

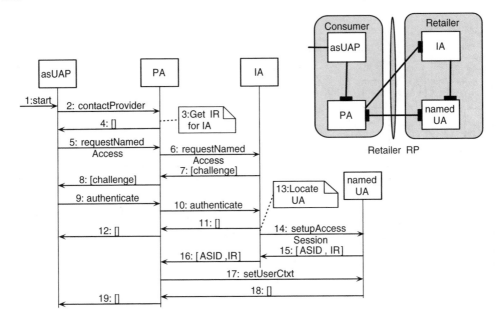

Figure 6.13. Use Case 1: pattern of components and messages involved in starting a TINA access session.

and the completion of connections in a connection-oriented network. Individual components are named and described in the use cases given below.

Figure 6.12 represents a complex system. The TINA service component [106] and network resource architecture [51] specifications describe the interfaces and typical interactions between components. Our purpose is to give an appreciation of the role of the service components and how the TINA session model is used to access and execute services and to make network connections. We therefore define use cases and identify the participating groups of components in Figure 6.12. We also identify recurrent patterns of operations that can be re-used for different functions.

Six illustrative use cases are identified to illustrate the initiation of different types of sessions: access, service and communications. The six use cases together employ all components identified in Figure 6.12. Each component is defined and its characteristic interactions are illustrated.

Use Case 1: User Gains Access to Provider

The first use case describes a user that is known to a service provider (retailer) through an existing subscription contacting the provider, being authenticated and starting an access session. The use case also illustrates how service component interaction is supported by DPE services.

The components involved in the user gaining access to the provider are extracted from Figure 6.12 into Figure 6.13. The accompanying message sequence chart illustrates the component interactions. A short description of the participating service components follows.

- *Access Session User Application Part* (asUAP): a component described as a *User Application Part* (UAP) represents logic in the consumer domain that is part of the overall application performed by interaction across the domains. A UAP-type component must interact with other components through defined interfaces and through interfaces, not part of the standard, toward the [human] user. The UAP is divided into two parts. The asUAP is concerned with starting a service. The ssUAP, concerned with service execution, is described in the second use case.

- *Provider Agent* (PA): exists in the user domain to represent the provider domain. Interdomain access in TINA is based on the use of agents, including the Provider Agent. Similarly, a User Agent (UA, described below) exists in the provider domain to represent the user as shown in Use Case 3. The PA may, on request of the asUAP, start an access session. Similarly, the PA may receive a request from the provider (retailer) to invite the user to join an access session. Thus, the PA shown in Figure 6.13 is an essential participant in an access session. The PA is service independent: the same PA is used irrespective of the service invoked by the user.

- *Initial Agent* (IA): an initial point of contact is required in the retailer domain. The Initial Agent therefore presents interfaces that can be accessed before an access session is set up by a user that has not yet been authenticated. The reference to these interfaces would be published or readily discovered using DPE services. The IA performs authentication of the requesting user and initiates an access session. The IA is service independent.

- *Named User Agent* (namedUA): this component represents the user in the provider domain. (An *Anonymous User Agent* is also defined for users that are not required to disclose their identities and be authenticated to access specific services.) The named UA is the point of contact for the PA once an access session is set up. User contexts can be negotiated and specific service invoked through interfaces defined on the namedUA. The namedUA has knowledge of the user's profile. The namedUA is service independent. The namedUA can be thought of as a component that implements generic user-related and service invocation operations and holds the profile data for a particular user.

- *Subscription Management* (Sub): this component is the provider's single point of control of subscribers and users as they access services and the management of subscriptions. The Sub component requires a database that is not part of the TINA standard. The Sub component maintains the user-specific data of each namedUA, for example updating the data when a subscription management operation is completed. The Sub component is service-independent.

- *User Agent Factory* (UAF): the factory design pattern is used in the TINA architecture for creating objects required for instances of a service or user. The UAF creates the namedUA (or anonymousUA) when an access session is commenced.

Use Case 1 involves a sequence of operations shown in Figure 6.13.

1. The user initiates the logon, for example by switching the terminal on. The asUAP obtains the reference to the Provider Agent.

2–4. The PA is requested to contact the provider. The interface reference (IR) to the provider's Initial Agent is obtained, for example by using the DPE naming service. The asUAP is notified of success. The notation [ ] signifies returned results or simply an acknowledgement.

5, 6. The asUAP issues a requestNamedAccess operation on the PA to gain access to the provider as a known (named) user. The PA, acting as a proxy for the provider, forwards the requestNamedAccess request to the provider's IA.

7–12. The response to requestNamedAccess holds an authentication challenge. The user completes the authentication with the IA using one of the available methods.

13. The IA locates the UA and creates the UA. The exact method is not specified but the Sub and UAF components are involved.

14–16. The IA then requests the namedUA to start an access session. The namedUA returns to the PA via the IA a session identifier (ASID) and a reference to its interface (IR) that can be used for invoking services.

17–19. The PA uses the setUserCtxt method to provide information on the user domain to the namedUA.

The existence of the access session allows the user to execute further user cases, for example discover available services, list the services to which the user has a subscription, manage the subscription, query billing data or start a service. The last is the subject of Use Case 2.

Use Case 2: User Initiates a Service

Successful completion of the actions in use case 1 is prerequisite to starting a service. The additional service components involved in this user case are:

Service Factory (SF): this factory object is specific to a particular service. On starting a service via the UA, the SF creates the required SSM and USM(s) for the service and returns interface references to the user domain.

Service Session Manager (SSM): this component contains both generic service logic, for example multiparty session control, and logic specific to the service, for example a mailbox service. Services provided by the SSM are grouped into feature sets, allowing SSMs of different complexity to be designed.

User Session Manager (USM): this component represents a single user at the service usage level. It therefore contains both generic session control and service-specific logic.

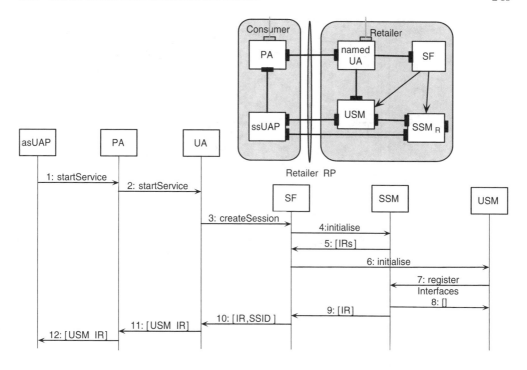

Figure 6.14. Use Case 2: pattern of components and operations involved initiating the TINA service session.

Service Session User Application Part (ssUAP): this component, residing in the consumer domain, is service specific and therefore has many potential manifestations. It may contain part of the service logic or may simply invoke service logic (feature sets) contained in the SSM. The ssUAP may participate in the service session. The ssUAP may be depicted with stream interfaces. This gives an abstract representation of the need to source and sink information streams by the user. (Other components, TCSM, TLA discussed in further use cases elaborate the actual streaming control.)

The message sequence shown in Figure 6.14 illustrates how a service session is started. Completion of Use Case 1 is a prerequisite to this sequence, in particular the existence of the user's UA.

1, 2. The user interacts with the asUAP in an unspecified way to request a service to be started. The service is identified by a service ID. The startService operation is invoked on the PA in its role as proxy for the provider and forwarded to the namedUA.

3. The UA has an access session active and starts a service session by invoking a createSession operation on the Service Factory for the desired service.

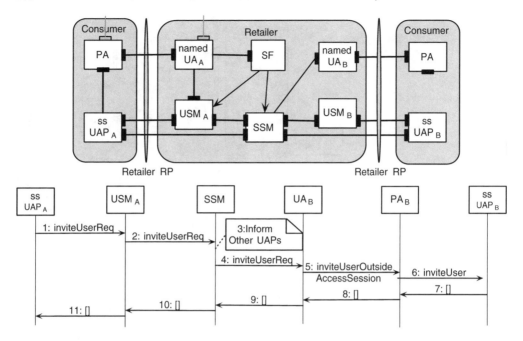

Figure 6.15. Use Case 3: inviting a party to an existing session.

The Service Factory could have been located by using the service ID and other information to query the naming service.

4, 5. The Service Factory creates the SSM for the requested service. The SSM returns its interface reference to the SF.

6, 7. The Service Factory creates a USM to represent the user. The SSM's interface reference is given to the USM. The USM registers its interfaces with the SSM.

9–12. The USM interfaces are returned to the PA and the asUAP. This enables the ssUAP object to be created in the user domain by a method not shown here. The ssUAP is now able to contact the USM and the SSM during execution of the service.

The service logic contained in the SSM and the ssUAP is now able to perform a range of functions. Two such functions are described in Use Cases 3 and 4.

Use Case 3: Invite a Party to a Service Session

This use case is typical of a provider inviting a user to join an existing session. No new types of service component are required. The ability of the PA to receive a request from a retailer is used to invite the party to the session. The invited user, B, is assumed to be known to the retailer, that is B's UA exists in the retailer domain.

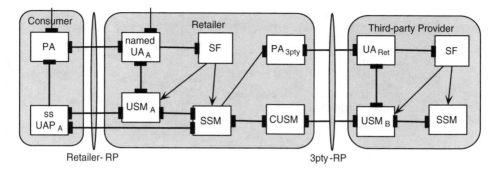

Figure 6.16. Use Case 4: accessing and invoking services of a third party provider.

1, 2. The invitation is initiated by user A's ssUAP and forwarded to the SSM by the USM.

3. If there are other parties to the service, they may be notified of the impending invitation to B to join the service.

4. The UA for B is created and the invitation is sent to the UA.

5. The UA sends the invitation to the PA of user B. This form of invitation from the provider to a user may be issued if no access session exists.

6. The ssUAP is created in user B's domain.

7–11. Acknowledgments and interface references are returned as required.

A variation on this use case would occur when the B-party is registered to another retailer. The retailer to party A would identify the retailer serving party B using DPE services. A peer-to-peer relationship is set up between the two retailers using the Rtr reference point. A special USM, the PeerUSM, and a special form of the UA called the PeerAgent, support federation of domains with a peer-to-peer relationship.

Use Case 4: Retailer Invokes a Third Party Service

Having established a service session as in use case 2, for example a multiparty call service, the service logic may determine that a control function available though a third party provider is needed, for example a complex interactive voice response unit interaction. The retailer then accesses the third party provider, using similar mechanisms to those user by the user to access the retailer using PA_{3Pty} to represent the service provider and the IA_{3Pty}. The retailer is known to the third party provider and has a namedUA (UA_{Ret}) available in the third party domain.

The service logic is instantiated in the third party domain by using the mechanism already seen in the retailer domain. The service factory for service X is used to create SSM_X and USM_X. One further component is required in the retailer domain.

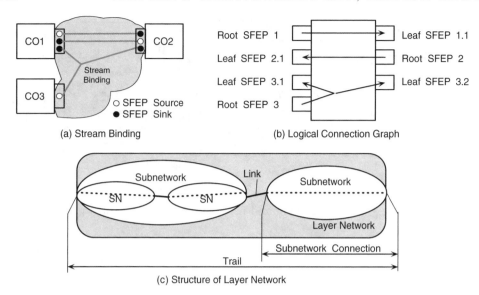

(a) Stream Binding

(b) Logical Connection Graph

(c) Structure of Layer Network

Figure 6.17. Stream binding and logical connection graph.

- *Composer Usage Session Manager* (compUSM or CUSM): this is a specialised form of the USM that supports composition of services. It resides in the domain of the user of the composed service, the retailer in Use Case 4.

The message sequences are similar to those in use cases 1 (gaining access) and 2 (starting the third party SSM).

Use Case 5: Stream Binding is Set Up

Separation of functions between the service and network architectures is supported by using an abstract representation for the desired stream flow connection in the service architecture. At a point in the execution of service logic in the SSM, say the retailer SSM, the requirement for a network connection arises, for example between two endpoints. The first step is to set up a *stream binding*. A *stream binding* is a modeling concept that represents a set of related stream flows that serve a purpose for an application. An information transfer requirement is represented by a *stream flow* between *steam flow endpoints*. The number of endpoints to a particular flow depends on the required topology, for example point-to-multipoint. A bidirectional flow is modelled as two contradirectional undirectional flows.

Figure 6.17(a) shows an example of a stream binding. The stream flow end points are stream interfaces between stream interfaces on a computational object. In this example, two parties, CO1 and CO2, have a bidirectional channel and the third CO3 streams media to the other two. The stream binding encompasses the complete set of stream flows.

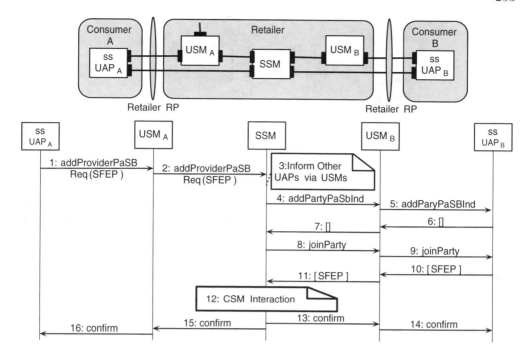

Figure 6.18. Use Case 5: setting up a stream binding.

The desired stream flows are represented by an information object in the service architecture called a *logical connection graph* (LCG). Figure 6.17(b) shows a representation of the LCG. Each unidirectional flow has a *root stream flow end point* or source and one or more *leaf stream flow end points* (sinks). Bidirectional and multicast stream flows are defined in terms of unidirectional flows as shown in Figure 6.17(b). The logical view of a connection is translated into a physical connection as shown in Figure 6.17(c). the connection may be arbitrarily complex involving subnetworks, internetwork links and tandem networks. Networks may be in different administrative domains. The task of converting the logical connection to a physical connection is discussed in Use Case 6.

The components involved in setting up a stream binding are all service session components shown in Figure 6.15. Part of the message sequence required to set up a stream binding is shown in Figure 6.18. The initiating ssUAP is assumed to have already ascertained the details of the SFEP from the TCSM in its domain.

1, 2. The ssUAP requests the SSM to set up a stream binding via the USM for user A.

3. If there are other parties they may optionally be notified at this stage.

4, 5. The B-party is already involved in the service session from use case 4. The SSM requests the B-party ssUAP to join the stream binding.

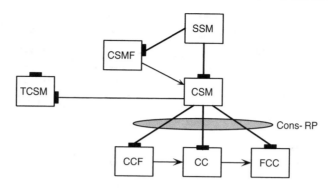

Figure 6.19.   Use Case 6:   pattern of TINA components for controlling a communication session and initiating actual connections.

6, 7. The participant(s) who agree to join the stream binding confirm their intention.

8–11. The joinParty operation is issued and the parties return their SFEPs. The SSM has all the information about stream flow endpoints and constructs the logical connection graph.

12. The interaction with the CSM described in Use Case 6 takes place to effect the connection.

13–16. A confirmation of establishment of connection is sent to the parties.

The physical connection that fulfills the abstract specification of the stream binding is performed by the connectivity providers network and the customer premises equipment.

6.6   NETWORK RESOURCE ARCHITECTURE

The *TINA network resource architecture* (NRA) is identified in Figure 6.12. We use the sixth Use Case to give a brief introduction to the NRA. The treatment is brief because the NRA is strongly oriented toward ATM transport networks and requires detailed consideration for application to other connection-oriented technologies.

6.6.1   USE CASE 6: CONVERT STEAM BINDING TO NETWORK CONNECTION

Use Case 5 illustrates the process of establishing the connection requirements by setting up a stream binding and recording the requirements in an information structure called a logical connection graph. Use Case 6 explains how the actual connections are realised using the TINA network resource architecture.

The underlying network in which the physical connections are to be made is assumed to be connection-oriented.   The connection network is assumed to be structured using the methods used in ITU-T recommendation G.805 [133].   The elements in a connection are shown in Figure 6.19.   A *layer network* provides a connection of a particular type between two termination points. A layer network may in turn be composed of subnetworks. Subnetworks are interconnected via links. A link may be a simple point-to-point link or may be created by means of another layer network, for example an SDH network or an ATM virtual path network. Subnetworks may be decomposed into further subnetworks. The lowest order subnetwork is a single switch.

The physical connection is initiated using further components, defined in the network resource architecture.

- *Communication Session Manager* (CSM): this component manages the bindings between stream interfaces on computational objects that represent end-to-end stream flows at the application level. The CSM presents interfaces to the SSM for controlling stream flow connections.

- *Communication Session Manager Factory* (CSMF): this object creates a CSM on the request of a SSM.

- *Terminal Communication Session Manager* (TCSM): manages flow connections within the customer premises equipment.

The SSM requests the CSMF to create a CSM. A communications session is started and is linked to the current service session. A terminal CSM (TCSM) must be in existence. The CSM contacts the TCSMs in each user domain to correlate the core connection requirements with the local connections in the customer premises equipment.

Further components then come into play:

- *Connection Co-ordinator* (CC): this object controls a connectivity session, that is all the end to end connections required by the service. The CC controls principally at the session level.

- *Connection Co-ordinator Factory* (CCF): this object creates the Connection Co-ordinator.

- *Flow Connection Co-ordinator* (FCC): this object is concerned with controlling an individual flow required by the service. If the connection involves multimedia flows, one FCC is required per flow.

The CCF, CC and FCC present interfaces that can be invoked by the CSM. These interfaces constitute the connectivity service reference point (Cons-RP).

The components described in this Use Case are the high-level controllers at the session and flow level. Each flow must be created through a layer network that consists in general of subnetworks and individual switching elements. Actual connections are controlled by connection performers, arranged in a hierarchical structure, matching the structure of the subnetworks that make up the layer network shown in Figure 6.20.

The TINA NRA was influenced by the ATM network developed as the transport network for the B-ISDN. Each switch is controlled by an element-level connection performer. The proxy associated with each element level CP represents adaptation from the standards-based CORBA communications to the control protocol used by the network element. The switches within a subnetwork are controlled by a network level

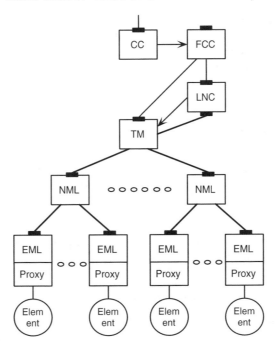

Figure 6.20. Pattern of TINA network resource architecture components for control of connections.

connection performer. Subnetworks make up tandem networks under the control of a tandem manager. Not shown in Figure 6.20 is the mechanism for federating networks in different administrative domains to make end-to-end connections that meet the requirements of the individual flows.

6.7   LESSONS FROM TINA FOR NGNS

Was the TINA initiative a failure? Clearly there are no TINA-compliant networks in general use. The TINA initiative, however, pioneered many architectural concepts and principles and several of these have influenced other NGN initiatives. Reviews of TINA stress both achievements and shortcomings [149, 41]. We follow this approach to learning from TINA by first reviewing its successes and then analysing factors explaining why it did not become widely adopted as an NGN standard.

6.7.1   ACHIEVEMENTS OF TINA AS AN NGN

The architectural innovations and achievements of the TINA architecture are listed in Table 6.3. In general, the TINA architecture is a softswitch architecture with separation of service control from network resource control. A second general contribution is the demonstration of the RM-ODP method applied to a complex system. Several specific contributions emerged in the TINA architecture.

Table 6.3. Important Contributions of the TINA architecture

Overall Architecture	Service Architecture
Business Model and Reference Points	Access Session
TMN Layering	Service Session
Computing Architecture and DPE	Third party Service provider
	Subscription Management
Network Resource Architecture	Accounting Management
Communications Session	Communications Session
TMN-like layering	Virtual Home Environment
Layer Network Federation	

- The *TINA Business Model* introduced the notion of business modelling to NGN architectures.

- The TINA *Session Model*, with its linked access, service, communication and connectivity sessions provides a comprehensive model that is applicable in all types of services.

- The TINA architecture also introduced the concept of a *third party service provider*.

- Despite important gaps in the architecture such as the lack of a defined third party reference point, other aspects of the standard were well advanced. *Accounting management* and *subscription management* are two such areas.

The TINA standards contain a comprehensive definition of information structures, operations and interfaces for many aspects of telecommunications service control and management. These range from multiparty session control (adding, suspending resuming and removing parties), through subscription management to usage accounting. These TINA information structures have contributed significantly to the OSA/Parlay standard described in Chapter 8. The TINA architecture is based on concepts and principles that are generally applicable in NGNs. The detailed definitions of infrastructures, methods and interfaces is a treasure house for developers of NGN standards.

6.7.2 WEAKNESSES OF TINA

Architectural Limitations and Flaws

While TINA was in many ways a development not hampered by legacy networks, it was nevertheless influenced by B-ISDN thinking. The consumer domain is analogous to the ISDN terminal equipment (TE). The TINA NRA is strongly influenced by the ATM transport network developed for B-ISDN. TINA was influenced by B-ISDN, which suffered from some of the technology-push shortcomings of N-ISDN. The focus fell on the capability to deliver services while the marketable services remained unproven. For example, the provision of ATM to the desktop was mooted before the need for broadband access was widely perceived.

The significant rise in usage of Internet services and the emergence of commercial Internet service providers occurred midway through the TINA development process and the process did not respond to the acceptance of the Internet model and the growing importance of the intelligence-free IP network with end-to-end services.

The TINA standards did not engage pragmatically with interworking between legacy switched circuit network and packet networks. The problem of interconnecting circuit-switched and IP networks was not addressed. Rather, it fell to the IETF to develop the concept of the media gateway.

While the TINA architecture introduced the third party service provider, the standards do not contemplate third party initiated services. While TINA has a powerful inter-domain access and service invocation mechanism, for example as illustrated by Use Cases 3 and 4, the consumer must always initiate a service by contacting a retailer. Considerable interest exists in applications that are initiated by a third party application. A simple application is a wake-up call service while a more complex service would be a pre-arranged, managed video conference.

The TINA service architecture defines SSM components in both the retailer and third party provider domains. The SSM is the locus of the service logic. No guidance is given on how the overall logic for an application should be split between the retailer and 3Pty domains, other than that dictated by the retailer's role as the point of contact for the user. The retailer and 3Pty SSMs are therefore likely to contain logic that is reusable from application to application and logic that is specific to an application. A clear division corresponding to the SCF and Application Layers in the NGN framework does not exist in the TINA service architecture. The 3Pty reference point is in any event undefined and service implementers have reused interfaces and operations in the retailer reference point for interaction between the retailer and 3Pty SSMs.

The TINA network architecture maps to the RCF and switching layers of the NGN Framework. The Cons reference point is a possible basis for an RCF to SCMF interface. To obtain a mapping in the service architecture, we impose a constraint on the SSMs not contained in the TINA standard. The retailer SSM is restricted to generic service logic, for example call session control functionality. Service-specific logic, for example the core logic of a video-on-demand service is placed in a third party SSM. With this restriction, the TINA service architecture can be expressed in the NGN Framework layered form. The operations defined in the retailer reference point include multiparty session control. If these operations are used at the 3Pty reference point, a multiparty, multimedia API is presented by the retailer (SCF) domains to the 3Pty domain.

TINA as a Disruptive Technology

TINA was a bold idea, attempting to address both future multiparty, multimedia services while dealing with mobility in a subtle way as well as seeking to integrate management into the architecture. TINA also sought to exploit distributed processing, an idea without robust, proven CORBA implementations at the time. The boldness of TINA may have contributed to its lack of acceptance.

Conventional wisdom on sustainable convergence holds that technologies should meet two requirements. First, technologies should evolve from their predecessors and, second, new technologies must interwork with legacy technologies. The thinking underlying TINA was novel in most respects. The approach was therefore revolutionary and has the characteristics of a disruptive technology. While interworking with legacy technologies was an objective of TINA, this aspect was not developed and proven. Unlike the Internet telephony standards, no concept of a media gateway or signalling gateway developed in TINA.

The distributed processing concepts on which TINA relied require a conceptual leap for those accustomed to working with communications protocols. The TINA DPE was initially an abstract definition and it was only later that the CORBA standards developed sufficiently to give a practical realisation of the DPE. With a workable DPE, subtle support for architectural and service development became possible. For example, the broker (Bkr) reference point was not developed. The CORBA trading service provided the type of services envisaged for the broker. Similarly, support for mobility could be enhanced through CORBA services.

TINA breaks the ISDN line of thinking that centred on the call and access (UNI) signalling using a protocol such as Q.931 or Q.2931. Similarly, the notion of a media-related session that had developed in packet-switched networks, that is perhaps closest to the connectivity session, was supplanted by the more powerful and hence complex notion of interlinked sessions in TINA. Protocol-based interactions are replaced by with interactions via interfaces supported by a DPE was another disruptive effect. The UNI is replaced by a rich reference point and API-type interface.

In addition to having characteristics of a disruptive technology, TINA was too early for the marketplace: while vendors could have delivered TINA-based products – and a few did so – telcos were not ready for the significant step from circuit-switched, IN-based networks to the revolutionary TINA architecture. Unlike the later introduction of managed packet voice networks, there was no simple, self-evident evolution path.

A more general reason for TINA being seen as disruptive may be added and a lesson learned. There is a natural reluctance among telecommunications engineers to embrace seemingly complex architectures and standards. Abstraction and modelling are used to manage the complexity of such standards. A significant effort is often required to get to grips with the abstraction and modelling concepts before the different parts of the architecture can be understood. While the TINA architecture contains a number of separations that are effective in dividing and conquering its complexity, these were fully understood only by those involved in the development or prepared to make a significant commitment to understanding the standards. Newcomers often found the four sub-architectures, each with its own separations, daunting. By contrast, Internet standards, while large in number, represent incremental developments, and seem easier to assimilate.

TINA's Legacy

Was TINA a failure? Clearly TINA failed to become the NGN to succeed the PTSN/IN. However, most of the concepts, principles, information structures and interfaces of TINA were sound. Also, any perception that implementation of the TINA architecture was an all-or-nothing matter was erroneous. For example, parts of the service architecture, for example the access session, could be implemented in another context. Many aspects of TINA have been adopted into the OSA/Parlay standard described in Chapter 8. The TINA standards are a treasure house of useful elements for solving problems in the development of new architectures. We conclude, therefore, that the study of next generation networks should involve learning from history, and that an important part of that history is the TINA architecture.

6.8 TIPHON

6.8.1 OBJECTIVES OF TIPHON

Implementation of networks based on the packet multimedia protocols described in Chapter 4 raise a number of issues.

- How can services appear comparable from an end user perspective if implemented using different protocols?

- How to ensure interworking between packet-based and switched circuit networks?

- How to integrate the suite of protocols required to implement the multimedia communications system? Not only must call signalling be performed, but media negotiation is required. The Real-time Transport protocol as well as the transport protocol for signalling messages must be incorporated. A media gateway protocol must be used. Signalling transfer with switched circuit networks must be handled.

In response to such problems, the ETSI Telecommunications and Internet Protocol Harmonisation over Networks (TIPHON) initiative set out to provide a framework for harmonising the operation of various types of packet networks and switched circuit networks and ensuring their interoperability.

TIPHON's early achievement was the development of a reference model for the H.323 network in which the monolithic H.323 gateway was decomposed into three elements: media gateway, media gateway controller and signalling gateway [58]. TIPHON then developed a technology-neutral architectural framework that would ease the use of the single-technology, multiple services protocols and the interworking between different types of networks. Interactions within the architectural framework were defined using a metaprotocol. Specific protocols would be mapped into the framework and the behavioural description of the metaprotocol.

TIPHON was initially concerned with simple calls [62]. Objectives included developing voice services over IP networks and ensuring interworking between IP networks and switched circuit networks: PSTN, ISDN and mobile networks. TIPHON also sought to enrich the service features available in an H.323 IP telephony network. The IP networks envisaged in TIPHON would compete in the general voice market and therefore need accounting to enable charging, QoS, privacy of voice channels and security of user data. Services based on number and address translation must be possible. Roaming is to be supported, requiring registration, authentication and authorisation of users, both in home and visited networks. These networks would also operate in regulated markets and must therefore be capable of supporting mandated services: emergency (911/112/999) services, legal intercept and number portability. Subsequently, the requirements on TIPHON were expanded to include multimedia services [65]. The TIPHON initiative has been subsumed into the ETSI TISPAN programme, ETSI's next generation fixed network initiative described in Chapter 10.

TIPHON's approach to meeting its objectives is captured in several constructs. The layering used in the network and environment model is presented in Section 6.8.2.

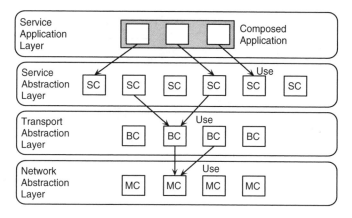

Figure 6.21. Application and abstraction layers in TIPHON model.

The business model is presented in Section 6.8.3. The TIPHON domain model is described in Section 6.8.4.

6.8.2 TIPHON LAYERED MODEL

The layering used in the TIPHON network and environment model is shown in Figure 6.21. The end objective, namely the provision of services, is expressed in the *service application layer*. *Application services* residing in this layer are composed by using capabilities in other layers through reference points.

The *service abstraction layer* is the location of service capabilities. A *service capability* is a discrete and unique set of functionality accessible through a defined interface. Functionality includes control of calls, access to transport services and terminal functionality. Examples of service capabilities are the registration of a terminal with a network and the establishment of an association between the entities in a call. An application service is executed by invoking service capabilities. For example, the application service could be a script or programme that invokes a number of service capabilities in sequence.

Two further abstraction layers are defined that contain capabilities that support the service capabilities. The *transport abstraction layer* is concerned with end-to-end connectivity requirements. These capabilities are called bearer functions. The network abstraction layer is concerned with communications capabilities, that is specific types of media that end users require to be transported such as voice, image, video and data as well as media processing such as the choice of video or audio codec. The capabilities of these two layers are available to applications but only through the intervening layers in a client–server manner.

6.8.3 TIPHON BUSINESS MODELS

TIPHON seeks to support connections across multiple domains, including switched circuit networks. TIPHON also supports user mobility in that a user may register in

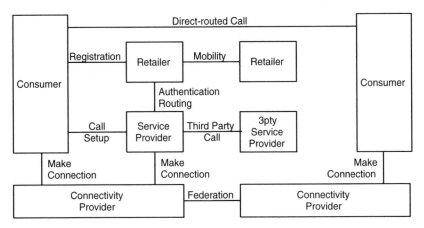

Figure 6.22. Possible business relationships enabled by TIPHON.

a visited or serving network and can receive services based on the user profile stored in the home network and a service-specific profile held in the terminal.

Possible modes of usage of a TIPHON-based network are encapsulated in the TIPHON business model, defined in [62] and expanded in Figure 6.22. The TIPHON business model is based on the TINA business model described in Section 6.4. The retailer role focusses on managing the registration and authorisation of users and their access to services. The retailer and consumer interact in terms of a pre-existing contract. The *service provider* makes services available to subscribers. The service provider is the point of contact during service execution, for example in setting up a call. The service provider does not have a contractual relationship with the end user, but access to services may rely on authorisation from the retailer. The softswitch principle is entrenched by the separation of service provision from connectivity provision.

6.8.4 TIPHON DOMAIN AND INTERWORKING MODEL

The TIPHON architecture uses a number of domains to demarcate a set of functional entities that are under the administrative control of a single operator. Several types of domain are identified:

- the terminal, containing a number of functions;

- origination networks, which may be the home *network* of the subscriber but may also be a *serving network* providing access to a roaming user;

- *transit* or *intermediate networks*, providing both transport of user streams but also passing signalling for registration of users;

- *terminating networks* to which a terminal domain is connected, containing the non-initiating party.

- A *terminating gateway network* is a packet-based network that provides a gateway to a switched circuit network.

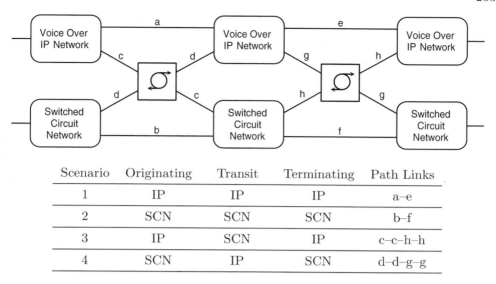

Scenario	Originating	Transit	Terminating	Path Links
1	IP	IP	IP	a–e
2	SCN	SCN	SCN	b–f
3	IP	SCN	IP	c–c–h–h
4	SCN	IP	SCN	d–d–g–g

Figure 6.23. Scenarios for application of TIPHON.

On a larger scale, TIPHON is concerned with interworking of packet and switched circuit networks via media gateways. Several scenarios are possible. Important combinations are shown in Figure 6.23. For example, a packet network is used for trunking between switched circuit networks in scenario 4. The four scenarios represent test cases for TIPHON.

While the TIPHON telephony application supports a number of services, its present release does not show how external SCPs or application servers could interact with the telephony application.

6.8.5 TIPHON FUNCTIONAL ARCHITECTURE

TIPHON is a functional architecture that is defined in terms of functional entities and information flows across reference points [60]. Implementation of an actual network is achieved by various information flows embodied in specific protocols. While using a functional entity method, TIPHON departs from the approach to previous ITU-T and ETSI standards. The functional entity approach recognises the existence of many existing protocols which perform control and management actions required in TIPHON. No existing protocol performs all the functions required in TIPHON. Protocol operations are defined at an abstract level in ASN.1: the *TIPHON meta-protocol*. Mapping from the metaprotocol into existing protocols such as H.323, SIP and H.248 is standardised. Thus, the extent of conformance of a SIP implementation of call control signalling in TIPHON would be known. The choice of functional entities provides decomposition of all the functions required for multimedia call control at a fine grain. This fine grain is necessary to allow existing protocols to be applied.

TIPHON is assumed to have a management plane, details of which are not standardised. TIPHON is concerned mainly with the telephony application but

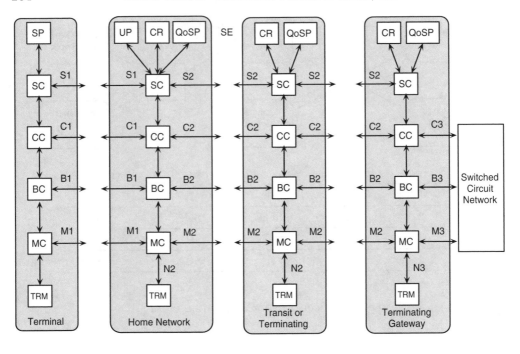

Figure 6.24. Layered functional entities in TIPHON in different domains showing interdomain reference points.

also defines a suitably abstracted network control interface. The *transport resource manager* (TRM) encapsulates transport layer functions needed to support QoS. The functional entities in the telephony application are layered as shown in Figure 6.24.

The *Services Functional Layer* (SE): is a collection of data required to execute service functions. The various data types are defined for different functions:

- *User-service profile function* (SP): contains information used and acquired during registration of a user. This function is resident in the terminal domain.

- *User profile function* (UP): holds information about the user and is located in the home network.

- *Call routing* (CR): function supports address translation and telephony routing functions.

- *Accounting function* (not shown), handles and stores call and service data for use by appropriate operators.

- *QoS policy function* (QoSP): manages policies for QoS in the IP network and authorises requests for QoS at a specified level.

The *Service Control Functional Layer* is concerned with the management of user registration and supporting the Call Control layer (described below) through the following functions:

- *Service control function* (SC): allows access to the services layer for actions such as authorisation, registration and call routing.

- *Terminal registration* (TREG): registers a user at a terminal with a TIPHON telephony provider.

- Three functions for user/terminal registration are provided:

  - *Service network registration Function* (SREG): accepts the registration of a user at a terminal by the serving network, allowing the terminal to connect to an IP network.

  - *Intermediate network registration function* (IREG): accepts requests from a user on a serving network and passes (proxies) registration requests toward the home network, applicable only when Serving and Home networks cannot communicate directly.

  - *Home Network Functional Group* (HREG): accepts the registration of a user at a terminal by the home network, is aware of the services subscribed by the user.

The *Call Control Functional Layer* has only one functional entity:

- Call control function (CC): is responsible for maintaining the call state and managing changes to the call state. The CC signals to other CCs involved in the call.

The *Bearer Control Functional Layer* is concerned with the logical association of entities in an IP telephony call that are involved in an end-to-end-media flow.

- Bearer control function (BC): controls admission to media streaming, negotiates with other media control functions and accesses media resources from the Media Control function.

- Aggregate bearer admission control (ABAC): keeps track of the state of aggregate flows in the transport layer through management information and controls admission of additional flows to existing aggregate bearer. (An aggregate bearer is an end-to-end media flow between endpoints.)

The *Media Control Functional Layer* is concerned with individual media flows, the encoding capability and reservation of paths that provide QoS.

- *Media control functions* (MC): include transport addresses, control of resources such as signal processors, reserves paths and controls firewalls.

- *Aggregate bearer measurement function* (ABM): determines that capacity used and available in an aggregate bearer, taking into account flows requested but not yet admitted.

The arrangement of the functional entities by domain and reference points is shown in Figure 6.24. Four horizontal reference points are shown: S, C, B and M. Variants for different interdomain boundaries are indicated by a digit: 1 for the terminal interface, 2 for packet network to packet network and 3 for packet to switched circuit

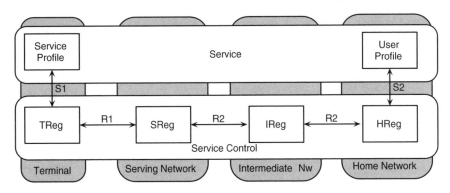

Figure 6.25. Generalised registration mechanism in TIPHON.

network. The vertical N-reference points communicate with the control functions in the underlying network. Only the S and C reference points carry signalling messages. Bearer information is embedded in the call signalling messages at the C-reference points and is passed down to the bearer control within the domain. A typical signalling sequence is given in Figure 6.26.

6.8.6 REGISTRATION AND MOBILITY IN TIPHON

TIPHON allows for registration of users in their home network as well as when roaming in another network. In the home network a user can attach to a service node, giving the required authentication information, for example a username and password or key. Once attached, the user may invoke services. The sequence of actions shown in Figure 6.25 is as follows:

1. The user (registrant) requests registration, indicating the service to be used and providing authorisation information.

2. The registrar in the serving network, having determined the service node to be used for the requested service, notifies the service node of the identity of the user, the service to be used and provides an authorisation ticket for the user. The service node is usually in the home network.

3. The service node accepts the client.

4. The registrar confirms the registration and supplies credentials to the user.

5. The user then sends a request to the service provider node, supplying the authorisation ticket and asking to be attached to the service.

6. The service provider responds with a service offer.

   Figure 6.25 shows the functional entities involved in registration in a roaming situation. The user attempts to register with a serving network in the roaming area. As in the mobile network case, the user profile is in the home network. The roaming

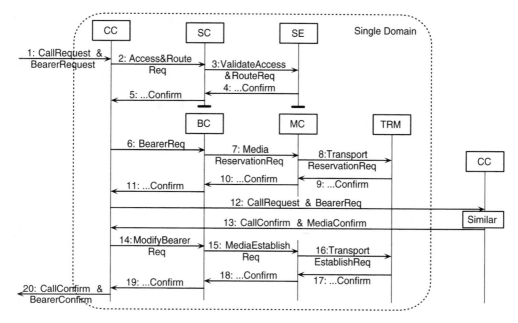

Figure 6.26. TIPHON metaprotocol call setup operations within a single domain.

user discovers the access interface, the registration point of attachment (RpoA), to be used in the local network. The terminal sends a registration request to the serving RPoA. The serving RPoA checks whether a roaming agreement exists with the caller's home network. If such an agreement exists, the serving RPoA acts as a proxy and sends the registration request via intermediate networks if necessary to the home RPoA in the home network. If the user is registered for roaming services, the home RPoA conducts verification and authentication procedures as required. If successful, the user is attached to a service point of attachment (SpoA) in the home network and notified. The user may then request a service. The SPoA used may be either in the serving network or in the home network.

6.8.7 TIPHON METAPROTOCOL

The TIPHON metaprotocol defines the messages and parameters to support registration and call control. Supporting data structures are also defined. The messages are grouped by reference point and are supported by behavioural descriptions, SDL diagrams and message sequence charts [63]. In addition, a number of use cases are described that confirm that the metaprotocol supports the required service capabilities [61].

During call setup and clear-down, domains interact via the CC functions. The CC carries out information exchanges with functional entities in the same network. Figure 6.26 shows the metaprotocol interactions within a single domain.

1. A Call Request is received from the CC of another domain via the C1 or C2 reference point. A related Bearer Request is received.

2. The CC makes a Service Access request to the Service Control Function.

3. The SC makes a request to the Service Layer (SE) to validate the service request and provide the required routing information. If necessary, the called party address is translated.

4, 5. The notation ...Confirm indicates the confirmation message relating to the requests 2 and 3 respectively.

6. The CC then issues the Bearer Request to the Bearer Control (BC).

7. The BC requests the Media Controller to reserve capacity for the required flow.

8. The MC requests the Transport Resource Manager to reserve capacity on an aggregate bearer.

9–11. Confirmation is returned of successful transport and bearer reservation in 9 and confirmation of bearer availability.

12. The Call Request is now sent to the CC of the next network toward the called party. The Bearer Request accompanies the Call Request. Steps 2–11 detailed above are now performed in the next network.

13. Confirmation of the call request and bearer request in 13 is received from the next network.

14. The CC issues a Modify Bearer Request to the BC to take up the bearer reserved in 7.

15. The BC issues a Media Establish Request to the MC, taking up the reservation made in 8.

16. The MC issues a Transport Establish Request to the TRM, to make use of the reserved capacity in the network layer.

17–19. Represent the confirmation of successful completion of 14, 15 and 16.

20. The CC confirms to the upstream requester that the call and bearer setup are proceeding successfully.

6.8.8   *TIPHON IMPLEMENTATION USING ESTABLISHED PROTOCOLS*

The functional entities, reference points and metaprotocol define the structure and operation of a TIPHON-based system. This definition is technology-independent. Each instance of a TIPHON-based network is implemented using specific protocols that have mappings with the TIPHON metaprotocol. Mappings have been defined for H.323, SIP and Megaco (H.248). At each reference point, an existing protocol is used to implement the metaprotocol operations and to elicit the behaviour that an

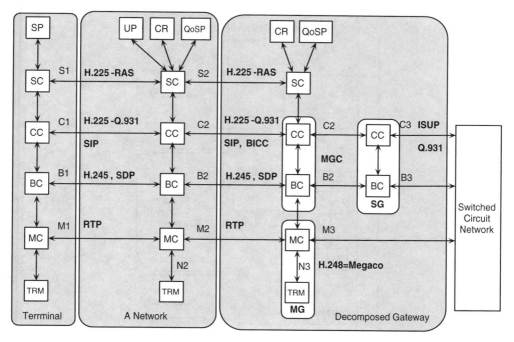

Figure 6.27. Implementation of the TIPHON architecture using existing protocols at various reference points.

operation expects. Figure 6.27 shows a number of candidate protocols for various reference points.

Registration is supported by the H.225 RAS protocol and by the REGISTER method in SIP to implement the R1 and R2 interfaces. The C1 and C2 reference point meta-protocol operations carry call setup, modification and cleardown messages. H.225/Q.931 and SIP provide realisations of the metaprotocol methods. For example the CallRequest message in Figure 6.26 is realised using a H.225 Setup message or a SIP INVITE method. Rules for mapping of parameters, for example from SIP headers are defined [64]. The Bearer Request corresponds to the SDP in the SIP INVITE message. BearerConfirm corresponds to the SDP in the SIP 200 OK message

Not all capabilities defined in the TIPHON specifications are supported by actual protocols. For example SIP has no data structure for transferring information that indicates the allowed services, the caller location or controls caller number presentation [64]. SDP lacks QoS-related fields such as delay budget and packet delay variation defined as parameters in the metaprotocol Bearer Request operation.

The TIPHON specifications concentrate on functional entities, information flows and expected behaviour. A number of physical configurations are possible to implement a TIPHON-based network. For example, a particular network, say the home network, may be controlled by means of a softswitch implementing all functional layers except MC, that is SE, SC, CC and BC. MC is a media gateway function.

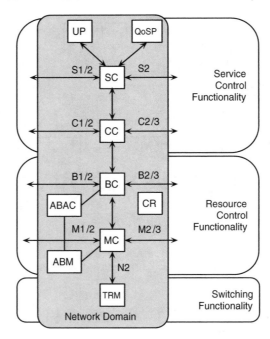

Figure 6.28. Mapping of functional entities in TIPHON to NGN Framework layers.

Alternatively, two physical elements may be used: a limited softswitch or media gateway controller, implementing CC and BC, while the SC and SE are implemented on a back-end server.

The gateway to a switched circuit network reflects the decomposition of the monolithic H.323 gateway. A media gateway implements the Media Control functions. The media gateway controller (MGC) implements the SC, CC and BC functions. The signalling gateway (SG) provides signal transfer/translation of two kinds: service related (INAP) and connection related (ISUP, Q.931).

The alignment of the TIPHON layers with those of the NGN Framework is substantial across the Service Control, Resource Control and Switching functionality layers with the grouping of functional entities shown in Figure 6.28.

6.8.9   TIPHON AND MULTIMEDIA SERVICES

TIPHON's initial objectives were aimed at telephony in IP networks and interworking with switched circuit networks. Subsequently, the requirements for multimedia communications were defined by expanding the service capabilities relative to those needed for a simple call [65].

6.8.10   APPRAISAL OF TIPHON

A starting point for appraising TIPHON's contribution to the long-term development of NGNs is the mapping of the various IP telephony protocols onto the TIPHON model.

The SIP and H.323 protocols described in Chapter 4 have simple architectures that are predominantly end-to-end service models with servers having limited roles. TIPHON provides a mapping and disaggregation of IP multimedia protocols. For example, in SIP a user agent in a terminal interacts with a user agent in a proxy server. The SIP message carries address and media information. The TIPHON layering emphasizes the separate operations in setting up an association between call parties, defining the bearer between the endpoint addresses, and the media specification. The TIPHON architecture, with its service layers, supports quality of service, policies, profiles and routing information. Roaming of users across networks is supported.

TIPHON, however, lacks two architectural features that support value-added services. There is no triggering method that would allow a condition detected in processing a signalling protocol to invoke an application server. Also, there is no open interface that allows an application to invoke network connections, for example to initiate a third party call as in the CTI architectures reviewed in Chapter 5.

6.9 CONCLUSION

The three architectures that failed to achieve widespread deployment in public networks reviewed in this chapter hold lessons for those concerned with the evolution of next generation networks. These lessons have technological aspects. For example, the B-ISDN perpetuated the service paradigm of the PSTN with switching control linked to the switching mechanism. The softswitch principle came later and was embraced readily.

These lessons also identify factors that work against the acceptance of a proposed technology or standard and deployment in networks:

- The scope or complexity of the architecture is excessive; abstractions may be available but getting to grips with the methods of abstraction could also be daunting.

- Developments are not incremental.

- Long standardisation processes cause flagging interest and sidelining of technologies by subsequent developments.

- There is a lack of agility and adaptability.

- The technology is disruptive rather than sustaining.

Chapter 7

Important NGNs: Third Generation Mobile Communication Systems

Two significant areas of growth in ICT over the past decade have been mobile telephone networks and the Internet. Growth of mobile networks has been driven by voice services and messaging, but, with the addition of packet-switched data services, these networks now show strong convergence trends. Mobile networks are seen as multimedia communication systems with multimedia messaging and Internet access as initial applications. Mobile communications networks therefore represent a major thread of evolution toward next generation networks.

Third generation (3G) *mobile communication systems* represent an evolution from second generation networks that provides enhanced packet-mode radio access networks, separation of service control from bearer transport, a unified packet core network, enhanced services including multimedia services and the opportunity for operators to provide distinctive services and applications. The point of departure for discussion of 3G communication systems is the second generation GSM mobile network described in Section 1.2.2, together with is packet enhancement, the General Packet Radio Service (GPRS) described briefly in Section 1.2.3.

Two main standardisation efforts define 3G mobile networks. The Third Generation Partnership Project (3GPP) has developed the Universal Mobile Telecommunications System (UMTS) by a process of evolution from the GSM standards and convergence with Internet standards. Standards evolving from second generation networks based on a CDMA-air interface are the product of the Third Generation Partnership Project 2 (3GPP2). These two sets of standards, while differing in air interfaces, are intended to have a common IP Multimedia Subsystem core network described in Section 7.5.

This chapter examines third generation mobile communication systems, concentrating on the 3GPP standards. Mobile communications systems are complex: they have all the service control functions of fixed networks as well as cellular radio access networks. In addition to call control, mobile networks have functionality for controlling radio resources and managing the locations of mobile subscribers. Common functionality such as mobility management serves both circuit and packet-mode services. Mobile networks, their standards and enabling technologies are constantly evolving. An in-depth study of mobile networks is therefore complex. In this chapter, we make use of the NGN-Framework as an organising principle. We overlay architectural subdivisions and functional entities used in the standards for each type of 3G network on the NGN Framework.

Network Convergence: Services, Applications, Transport, and Operations Support
Hu Hanrahan © 2007 John Wiley & Sons, Ltd

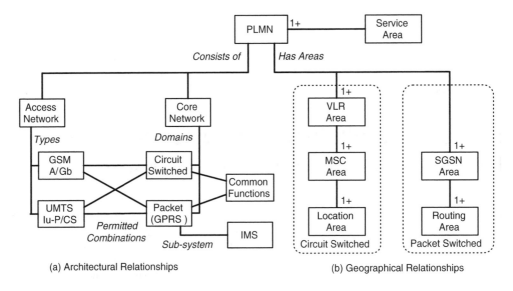

Figure 7.1. Concepts used in defining a Public Land Mobile Network.

Section 7.1 reviews the high-level architectural concepts used in the GSM and 3GPP standards for defining mobile communication systems. Section 7.2 sketches the evolution of mobile communication systems. Four areas of evolution are discussed: the access network, the core network, transport network and service provision. The various formally defined and informal generations of mobile communication system are identified. Section 7.3 describes the circuit-switched domain of the core network, its relationship to access networks, mobility management, call handling and the provision of IN-type services. Section 7.4 describes GPRS-based systems and their support for IP packet traffic between end users. Section 7.5 introduces the IP Multimedia Subsystem and describes the services offered, the use of the SIP protocol and relationships with the packet and circuit-switched domains.

7.1   ARCHITECTURAL CONCEPTS IN MOBILE COMMUNICATION SYSTEMS

7.1.1   CONCEPT OF A PUBLIC LAND MOBILE NETWORK

The principal architectural features of a second generation (2G) mobile network are outlined in Figure 1.3. The method of addition of packet-mode communications to 2G networks is shown in Figure 1.12, creating the 2.5G network. These diagrams give a simple view of 2G and 2.5G networks. Second generation networks provide the base for evolution to third generation networks.

GSM and 3GPP mobile communication systems conform to a set of architectural concepts and principles summarised in Figure 7.1. The term *Public Land Mobile Network* (PLMN) used in the 3GPP standards encapsulates all forms of network serving mobile users. An understanding of the PLMN architecture is based on both functional groupings and geographic relationships. The former are defined in the 3GPP standards using functional entities, reference points and interfaces. The latter provides a consistent method of managing communications with mobile users in different geographic areas.

7.1.2 STRUCTURAL FEATURES OF A PLNM

The highest level structural or functional groupings of a PLMN are the *access network* (AN) and the *core network* (CN) shown in Figure 7.1(a). Any PLNM must have a core network and at least one type of access network. The access and core networks are logical divisions of functionality and admit of different physical implementations.

Access Network

The *access network* is defined and understood under three broad headings. First, the access network architecture defines the functional blocks, their broad functions and the reference points between blocks. The reference points in turn are specified in terms of the protocols used at OSI-RM layers 2 and 3. The physical layer interface is a significant consideration at the air interface. *Air interface access* in second generation networks is narrowband. Encoded voice crosses the air interface in assigned time slots. Circuit-mode data also use time slots and data rates are limited to tens of kbit/s. In the so-called 2.5G network, packet-mode access is provided at the air interface to support the General Packet Radio Service (GPRS). In the Third generation mobile network, radio access is packet oriented and a range of air interfaces are standardised. Depending on the type of air interface, increased bit rates are available to the 3G user. Two types of access network are identified in Figure 7.1:

- The second generation GSM circuit-switched access network, was enhanced in two ways to provide packet-switched access. First, the GPRS was introduced. The packet-enhanced access network presents two reference points to the core network: the original A circuit-mode interface and an additional packet-mode interface denoted Gb. In the GSM and 3GPP standards the appellation A/Gb is used for this type of network. The second enhancement is the increase of data rates at the air interface. The enhanced interface remains classified as second generation: it remains a circuit-mode interface adapted to carry packets.

- The third generation access network operates in packet-mode, but also presents a circuit-mode reference point to the core network for interworking with legacy equipment. This network also presents two interfaces to the core network, one packet-switched denoted IuPS and the second circuit-switched, denoted IuCS. While GSM has a single form of air interface, the 3G standards admit of several implementations of low-level communications across the air interface.

Table 7.1 lists some of the commonly used identifiers of second and third generation communication systems introduced here or later in this chapter.

Table 7.1. Indentifiers for mobile network generations

Term	2G	3G
System	GSM	UMTS
Interfaces	A/Gb	IuCS/IuPS
Access network subsystem	BSS	RNS
Evolved access network	GERAN	UTRAN

Core Network

The *core network* contains the transport network, call, session and service control functions as well as functions particular to mobile networks such as mobility management.

The 3GPP standards define two types of core network domains shown in Figure 7.1. The *circuit-switched domain* (CS) represents an evolution from 2G GSM networks. The CS domain contains functionality required for the control of CS-type connections and the transfer of user traffic. Second generation GSM networks use TDM circuit switching. In 3G networks, CS-type connections refer to circuit oriented traffic where an A or IuCS access network interface connects a mobile station to the core. As core transport in 3G networks is packet-oriented, use of a media gateway is essential. The connection may terminate in a packet or circuit-mode network. In the latter case, a media gateway is required at the interface to the circuit-switched network.

The *packet-switched domain* (PS) provides packet-switched types of connections for control and transfer of user data. Functional entities in the PS domain are the elements making up the GPRS network. With the introduction of GPRS in 2G networks, packet streams are extracted at the Base Station Controller and are fed into the GPRS packet-based core network. This same core network forms the basis of the 3G network, supporting real-time services as well as data services. The packet core network provides transport for both traffic to and from a packet-mode access network interface and for trunking between media gateways. The PS domain is therefore capable of interworking with both circuit-switched core networks and other data networks including the Internet.

A subsystem of the PS domain called the *IP Multimedia Subsystem* (IMS) is defined to accommodate the elements necessary to support multimedia services based on IETF standards such as SIP. The relationship of the IMS to the GPRS-based core network is explained in Section 7.5. The IMS is a core network standard and requires an IP-capable access network. The 3GPP standards define permitted combinations of access and core network.

The changing nature of the core transport network from legacy circuit-mode to GPRS and IP multimedia networks brings about changes in the control of bearer and value added services. The MSC must therefore evolve both in call session functionality and protocols. Common functions such as mobility management also evolves.

Common Functions

The PS and CS domains rely on common functions, principally database functions. The common functions are:

- *Home Subscriber Server* (HSS): a database containing information related to users having subscriptions to the network. Data include user profiles, user identification, numbering and addressing, security information and user location information. The last relates to the user's registration and information received from other entities such as VLRs. The HSS consolidates the 2G Home Location Register (HLR), with data enhancements for 3G services, and the Authentication Centre (AuC) into a single logical unit.

- *Visitor Location Register* (VLR): a database containing information relating to a mobile station registered in the area served by the VLR.

We note that the VLR is essential in the CS domain but is not explicitly required in the PS domain. The corresponding function in the PS domain is the *Location* function in the GPRS SGSN described in Section 7.4.3.

Enhanced services in 2G networks are provided using an IN overlay, tailored to the needs of mobile networks. With the advent of 3G networks three approaches to advanced services are standardised. The IN approach is developed further, the network is opened to application providers and SIP application servers are integral parts of the IM Subsystem.

7.1.3   GEOGRAPHICAL RELATIONSHIPS IN A PLNM

A PLNM services mobile users in a particular geographic area. Operators of PLNMs enter into roaming agreements that allow each other's subscribers to connect to their networks. The most general description of a PLNM therefore identifies the main architectural features and defines geographical relationships. Figure 7.1 defines these geographical relationships.

The circuit-switched and packet-switched domains use the similar but not identical area definitions shown in Figure 7.1. The lowest level areas are the location area and routing area. A mobile station served by a CS domain may move within a designated set of cells, the *location area*, without updating location information in the VLR. A *routing area* in the PS context is a set of cells through which the mobile station may move with updating location information in the SGSN.

The next level area defined in the CS context is the *MSC area*: the set of location areas associated with base station systems connected to the MSC. A *VLR area* consists of one or more MSC areas. In the PS context, the *SGSN area* contains one or more routing areas.

The *service area* is the total area in which mobile stations may be reached. A service area may span several PLMNs with roaming agreements.

7.2   MOBILE COMMUNICATION SYSTEM EVOLUTION

Figure 7.1 identifies the main architectural features of a PLMN. Most elements evolve from second to third generation while the IMS is introduced in 3G. The evolution of

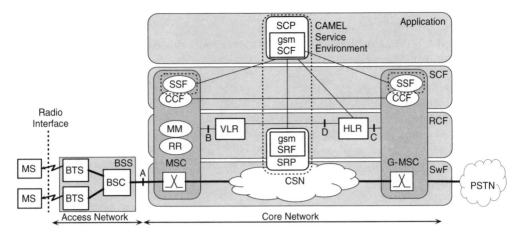

Figure 7.2. Second generation GSM network elements in NGN Framework.

mobile networks is made up of several threads, illustrated in Figure 7.3. We examine
the evolution of the access, MSC and mobile terminals using the GSM network as the
baseline.

7.2.1   BASELINE: THE SECOND GENERATION MOBILE NETWORK

The GSM network is a widely applied 2G network that serves as a starting point
for discussion of the evolution of mobile communication systems. We apply the
NGN Framework developed in Chapter 2 to the 2G and subsequent networks.
Figure 7.2 locates the principal architectural elements of a 2G GSM network against
the framework layers.

The access network consists of a Base Station Sub-system (BSS) using an air
interface based on time and frequency division multiplexing. The BSS consists of
Base Transceiver Stations (BTS) providing the air interface in each cell. The BTS are
connected back to a Base Station Controller (BSC). The BSC presents a signalling
and bearer traffic interface to the core network designated A.

The Mobile-system Switching Centre (MSC) is the main switching and control
node in the core network. Two MSC roles are shown in Figure 7.2. In its general
role an MSC has base station subsystems connected to it and serves mobile users.
Such an MSC has call control functionality (CCF) similar to that in the PSTN switch.
Inter-MSC signalling uses ISUP and a similar call model applies. In addition, mobility
management (MM) functionality is requires to track mobile users. Control over the
radio resources (RR) is also needed. The 2G MSC is therefore a large logical grouping
with functions that span the SCF, RCF and Switching layers of the NGN Framework,
as shown in Figure 7.2.

The second MSC role is as a *Gateway MSC* (G-MSC). The Gateway MSC
provides the point of interconnection with other mobile networks and fixed networks.

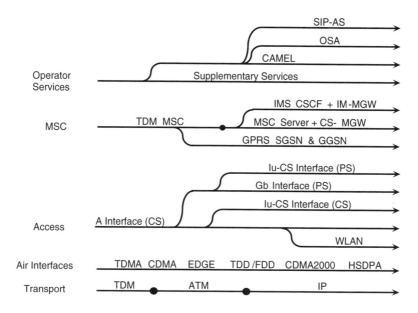

Figure 7.3. Threads in the evolution of mobile communication systems.

The G-MSC does not interface with access networks. The principal G-MSC functions are call control and switching.

The databases required in the 2G mobile network are placed in the resource control layer of the NGN Framework. Two of the databases are shown in Figure 7.2, the HLR and the VLR. The HLR contains two types of information: subscriber information and location information sufficient to route incoming calls to the MSC in whose area the subscriber is registered. The VLR stores subscriber location information for all mobile stations registered in the MSC area. Interfaces B, C and D are defined in the GSM standards for interaction among the MSC, HLR and VLR. These interfaces use the Mobile Application Part protocol defined for GSM [57].

Second generation mobile networks were developed in a narrowband ISDN service paradigm. Several switch-based supplementary services were defined in the standards and are implemented by switch vendors. As the Intelligent Network developed in the PSTN context, it was also adapted for 2G mobile networks. Functional entities corresponding to those in the PSTN/IN were adopted: Service Switching Function, Service Control Function and Specialised Resource Function as shown in Figure 7.2. Operator-specific services implemented as service logic hosted on the SCF in mobile networks must take the location of the mobile into account. Additional data is therefore required in the HLR to support invocation of SCF-based services. Additional parameters were added to the INAP protocol. This adaptation of the IN standards to support services in mobile networks became known as *Customised Applications for Mobile network Enhanced Logic* (CAMEL). The CAMEL-compliant functional entities are identified by the prefix gsm, namely gsmSSF, gsmSCF and gsmSRF. The term

CAMEL Service Environment (CSE) is used to denote the logical entity that supports operator's services.

No domains are shown in Figure 7.2. Several types of network domains are however used in mobile system standards. The *home network* is defined technically as the network where the Country Code and Network Code are the same as in the subscriber's International Mobile Subscriber Identity (IMSI). Less formally, the home network is the network to which the user subscribes and which holds the subscriber information in its HLR (or one of its HLRs). A *visited network* is a network where a mobile station is registered. The visited network may be the home network or another network that has a roaming agreement with the home network. The visited network is also referred to as the *serving network*. An *interrogating network* is a network containing a Gateway-MSC that interrogates the HLR of the home network to route a call toward a mobile subscriber of the home network. The interrogating network may be the home network. These network domains are used in describing mobility management in Section 7.3.1.

7.2.2 IDENTIFICATION OF MOBILE NETWORK GENERATIONS

The second generation network represents a phase in an evolutionary process toward several third generation system goals [1]. Voice, data and messaging services that co-exist in 2G should converge and be extended to provide multimedia services. The mobile terminal must provide seamless access to the Internet. Increasing user data rates must be supported.

Mobile network standards identify only two generations of network: GSM *second generation* networks, enhanced by GPRS and GERAN access networks and *third generation* networks with packet core networks and UTRAN access networks. Several informal identifiers have arisen for intermediate evolution stages. Addition of the GPRS packet transport to GSM was called 2.5G. The designation *Converged 3G* (C3G) indicates the goal of seamless interworking of wireless and wired IP networks, and wireless networks and the Internet. C3G is often equated to the IP Multimedia Subsystem described in Section 7.5. *Beyond 3G* (B3G) identifies a stage of development in mobile networks in which a number of heterogeneous access networks, for example UTRAN, WiFi and WiMax, are in use while a common IP network layer with IP-level mobility management is in use. Further increases in air interface speed are loosely designated 3.5G. The fourth generation, or 4G, represents a vision for the future, especially the quest for very high data rates on the air interface, a single IP core network and seamless handover between different access network technologies.

There is no single 3G communications system. The 3G standards represent a phase in evolution. Within 3G, the circuit-switched core network supports circuit-mode access network interfaces and interworking with the PSTN. packet-switched core networks support packet-mode access network interfaces. Multimedia services are enabled by adopting IETF standards.

Each aspect of 3G systems is complex and is at its own stage of maturity and implementation. In this chapter, we examine key aspects of the 3G mobile communication system standards and, where helpful, we identify relationships between parts of the system that are in fact logically or physically associated but are often treated in isolation.

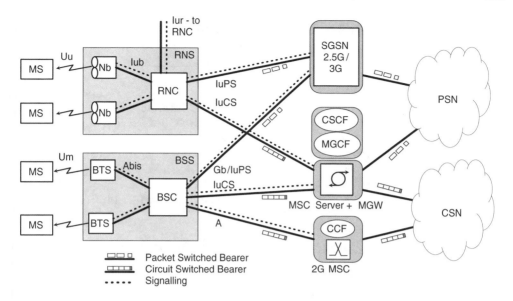

Figure 7.4. Evolution of access architecture from GSM 2G to 3G.

7.2.3 EVOLUTION OF THE RADIO ACCESS NETWORK

The GSM standards developed to a level denoted Phase 2+ before standardisation attention focussed on third generation systems. Circuit-switched voice and data and packet-switched data modes of operation are available in 2.5G GSM. Packet data service is implemented as the General Packet Radio Service. Access networks are based on the Base Station Subsystem concept, shown in Figure 7.4 with enhancements to support GPRS. The availability of GPRS is the principal milestone that marks GSM Phase 2+. Within Phase 2+, the packet data rates are increased by improved coding and modulation in the *Enhanced Data rates for GSM Evolution* (EDGE) standard. With the addition of EDGE, the 2G access network is called the *GSM/EDGE Radio Access Network* (GERAN). Two interfaces are presented to the core network: the original circuit-mode A-interface and the Gb packet-mode interface introduced with GPRS.

The GSM air interface uses frequency division duplexing (FDD) and time division duplexing (TDD). The frequency band available to an operator is divided into 200 kHz wide bands. Paired bands are allocated to the uplink and downlink, hence the FDD attribute. Each band carries a time division multiplex structure with eight time slots giving a maximum user data rate per time slot of 13 kbit/s. Circuit-mode bearer signals, both speech and circuit-switched data, are carried in time slots. GPRS packet-mode traffic is carried by allocating one or more time slot for this purpose. The TDD attribute arises because time slots are allocated to a user to avoid having to transmit and receive simultaneously.

With the move to 3G, the architecture, air interface and mode of data transfer change. The fixed part of the radio access network is called the *Radio Network*

Table 7.2. Interfaces in a PLMN for packet and circuit-mode services

Interface	Generation	Description
Um	GERAN	Air Interface
Uu	UTRAN	Air Interface
Abis	GERAN	Internal BTS–BSC Interface
Iub	UTRAN	Internal Node B–RNC Interface
IuR	UTRAN	RNC–RNC interface
A	GERAN	circuit-switched BSC–MSC Interface
IuCS	UTRAN	circuit-switched BSC/RNC–MSC Interface
Gb	GERAN	2G GPRS BSC-SGSN Interface
IuPS	UTRAN	Packet-mode BSC/RNC–MSC Interface

Subsystem (RNS) shown in Figure 7.4. The node responsible for transmitting and receiving radio signals to and from the mobile station (MS) is called *Node B*. The radio resources of several such nodes are under the control of a *Radio Network Controller* (RNC). A collection of RNCs and their associated Node Bs is termed a *Universal Terrestrial Radio Access Network* (UTRAN). The term UTRAN distinguishes this type of third generation radio access network from the enhanced GSM/EDGE Radio access network.

The 3G *Radio Access Network* (RAN) is designed specifically for packet-mode data transfer both across the air interface and the core network interface. Circuit-oriented data can be presented to legacy core network switches from the RNC. The lower layers of the radio access network may be implemented in several ways and develop over time to offer increasing data rates.

The RNC has two types of interface that support signalling and bearer channels to different types of core network. Figure 7.4 anticipates the discussion of core network evolution in Section 7.2.4. The IuPS reference point defines the interworking of an RNS access network with a SGSN, the edge element of a GPRS core network. The SGSN has a Gb interface from its 2.5G incarnation allowing packet-mode connections via a BSS. The RNS also has a circuit-mode reference point IuCS that allows an RNC to be connected to a 3G circuit-mode core network via a media gateway. The IuCS interface may also be implemented on the BSS, allowing 2G access networks to connect to the core with a 3G interface.

Table 7.2 summarises the RNS and BSS interfaces defined for the enhanced 2G and 3G contexts.

7.2.4   DEVELOPMENT OF THE MSC

The core of the GSM Phase 2+ system contains separate circuit-switched and packet-switched domains as identified in Figure 7.1. In the circuit-switched domain, the main building block is the Mobile-system Switching Centre (MSC). The packet-switched (PS) domain is based on GPRS Support Nodes: the Support GSN (SGSN) and the Gateway GSN (GGSN). Elements in both the CS and PS domains have access to location registers and service logic resident on a service control point via the Signalling System No. 7 network. In this section, we examine the evolution of the MSC from 2G to 3G circuit- and packet-mode versions. The role of the GPRS nodes in 3G networks is discussed in Section 7.4.

The MSC in second generation mobile networks is a monolithic node containing a number of functional entities: the call control function, mobility control and a GSM service switching function (gsmSSF) for invoking IN services hosted on a service control point. The MSC is able to interrogate the Home and Visitor Location Registers. The MSC switching function is circuit-oriented. Low bit rate coded speech transmitted over air interface A is transcoded in the BSC to and from 64 kbit/s A- or μ-Law pulse code modulation. The 64 kbit/s time division multiplexed signals are switched in the MSC. All circuit-oriented bearer signals, voice and data are switched in this way.

The MSC originated as a switched circuit element. The 3G standardisation process carried out in the Third Generation Partnership project aimed to move to a multiservice, multimedia type of network and therefore introduced a packet-mode core network. Interworking with legacy networks was, as always, a requirement. The MSC was therefore developed in the standards through a number of stages to meet these requirements.

Disaggregating the MSC

Three stages of evolution of the MSC are shown in Figure 7.5. The point of departure is the dual MSC and SGSN arrangement based on the GSM 2+ standards. This form MSC was defined in the 1999 or R99 version of the 3GPP standards. Services offered are at the same level as in GSM systems. The initial core network for transport of voice and data was an ATM network using ATM Adaptation Layer 2 transport. ATM adapters were added to the 2G-type MSC. The ATM adapter performs transcoding of speech and therefore functions as a form of media gateway.

The next revision of the standards, now referred to as Revision 4 (R4), took a first step in disaggregating the monolithic MSC toward a softswitch configuration as shown in Figure 7.5. The media gateway functions, together with ATM switching, are logically separated from the MSC. The media gateway element performs transcoding between circuit and packet formats and switches packets. The remaining MSC functionality, including call control and mobility management, became the *MSC Server*. Disaggregation allows the control functions to be located away from the physical gateways.

By Revision 5 of the standards, disaggregation and redefinition of the MSC Server and the new distinct entity, the Media Gateway, were influenced by two factors. First, the objective was to adopt an IP network in the core and to provide end-to-end connectivity based in IP. Second, there was a move to adopt SIP as the call session control signalling protocol to support future multimedia services. The media gateway took on the form discussed in Chapter 4. Megaco (H.248) was adopted as the control protocol for the media gateway. The MSC functions were repackaged into two main functional entities, in addition to mobility management as before. The *Call Session Control Function* (CSCF) contains all functions related to the call state. The *Media Gateway Control Function* (MGCF) encapsulates functionality required to control connections at the media gateway. The call model in the CSCF for circuit-switched services changed little as call signalling continued to be based on ISUP.

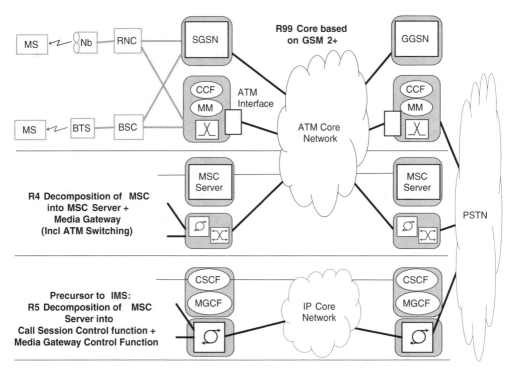

Figure 7.5. Evolution of 3G core network architecture from GSM 2+.

This decomposition foreshadows the corresponding elements in the IP Multimedia Subsystem described in Section 7.5 where a new form of CSCF is introduced.

Revision 6 also introduces the IP Multimedia Subsystem as a set of core network elements that support multimedia services, that is services involving audio, video, text and other forms of information in combination. The configuration of the IP Multimedia Subsystem is defined in Figure 7.19.

Circuit-switched Domain Architecture

The Revision 6 network architecture of a Public Land Mobile Network is defined in [9]. The principal elements of the CS-domain and associated access network and service platform are shown in Figure 7.6, arranged by layer according to the NGN Framework. Two possible forms of access network are GERAN and UTRAN. Figure 7.6 shows only the 3G UTRAN access network presenting the IuCS interface to the core network. In the case of a GERAN access network, the A interface would be presented to the core in two ways. Bearer channels connect with the media gateway while signalling channels terminate on the MSC.

In the core network, the MSC Server and Media Gateway for the CS domain (CS-MGW) are as in the R4 definition, shown in Figure 7.5. The MSC Server uses the

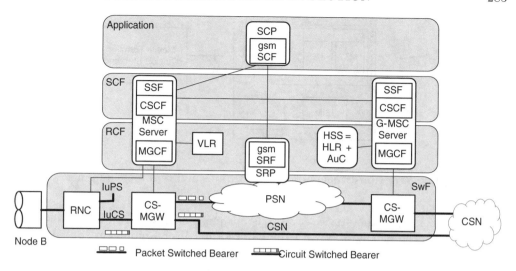

Figure 7.6. 3G Circuit-switched domain architecture evolved from GSM 2G concepts shown against NGN Framework layers.

Q.931 protocol on the terminal side and ISUP for call signalling toward another MSC server.

The switching layer contains a packet transport network. A reference point Nb is defined for the transport of user data, for example voice streams. Two forms of network are defined for transport of real time streams [21]. First, ATM Adaptation Layer 2 adapts user data to ATM cells for transport in permanent or switched virtual circuits. Second, user data is transported over an IPv4 or Ipv6 network using the Real-time Transport Protocol described in Chapter 4.

7.2.5 TERMINALS IN THE 3G CONTEXT

Complex terminology describes equipment that enables the mobile user to access the network and make use of services. The various entities identified in the 3GPP standards are shown in Figure 7.7. The *User Equipment* (UE) is 'a device allowing a user to access services' [2]. The UE is linked to the network via the radio interface (Uu). The UE consists of the Mobile Equipment (ME) and the Universal Subscriber information Module (USIM). The USIM is an application that resides on an insertable card that allows the user to register securely with the network and to access services provided by the mobile network. Several functional groupings occur in the ME. The Mobile Termination (MT) contains the radio interface functions while the Terminal Equipment (TE) is the group of low layer protocols that allow access to the network. The R-interface, similar to the ISDN terminal adapter interface, provides access to the MS for various mobile terminal equipments that are standardised elsewhere. Such terminals may implement teleservices and provide a human–machine interface. The terminal equipment may be physically integrated with the MS or may be separate.

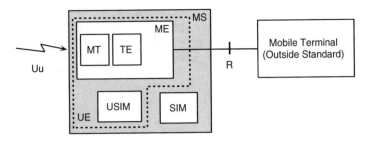

Figure 7.7. Constituents of mobile stations and terminal equipment.

When working at the service level it is adequate in most cases to use the term mobile station interchangeably with the more formal User Equipment to denote the equipment that provides and controls the basic bearer connections.

7.3   SERVICES IN THE CS DOMAIN

Figure 7.6 locates the principal elements of the CS-domain architecture on the layers of the NGN Framework. The Revision 5 decomposition of the MSC into an MSC Server is used. Inter-CSCF signalling continues to use ISUP.

The CS-domain contains no circuit-mode (TDM) switches. Rather, CS channels coming from the access network are converted to packet-switched format in the media gateway. Similarly, a media gateway converts the packet format to circuit-mode channels at the interface to a legacy CS network. The CS-domain core network is a packet trunking network under the control of legacy call controllers. Service functionality is limited by both the circuit mode bearer interfaces and the legacy call control.

7.3.1   MOBILITY MANAGEMENT IN THE CS DOMAIN

In the second generation network, mobility management relies on permanent user data stored in the Home Location Register and temporary data stored in the Visitor Location Register of the MSC area currently occupied by the mobile station. The mobile station is identified by the MSISDN, an E.164 number. The HLR and VLR are identified by Global Title Addresses, also E.164 numbers. A temporary E.164 number is allocated to a mobile for routing calls.

Mobility management in the 3G CS-domain is based on the same elements. The interaction diagram in Figure 7.8 shows two main cases in mobility management. In (a), the mobile user is registered in the VLR, for example as a result of a location update, and has a temporary number allocated. Steps to satisfy this precondition are shown by dotted arrows in Figure 7.8(a). The temporary number allows other switches to route calls toward the MSC to which the mobile is registered. The HLR is also updated with the identity of the VLR containing the subscriber's temporary data. The mobile user may be in its home network or in any network into which it is permitted to roam. When the user initiates a call request (1), the MSC retrieves information required to process the outgoing call from the VLR (2). The MSC then

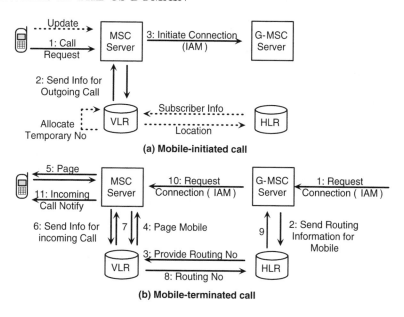

Figure 7.8. Interaction diagram showing essential steps in initiating or terminating a call in the CS domain (prepaging scenario).

starts the call setup toward the destination by sending an ISUP IAM message to the relevant MSC or GMSC (3).

The mobile termination case is shown in Figure 7.8(b). Preconditions shown dotted in case (a) must be fulfilled: the mobile station has a temporary number allocated and the identity of the VLR is stored in the HLR. The terminating mobile station may be in any network where it is permitted to register.

Case (b) shows a call request (IAM) coming from another network to the G-MSC Server (1). The G-MSC retrieves information for routing the call to the mobile from the HLR (2). This data includes the identity of the VLR to which the mobile is registered. In the process, the HLR requests the VLR to supply the temporary routing number allocated to the mobile station (3). The VLR requests the MSC to page the mobile station (4 and 5). The visited MSC retrieves information it needs to terminate the call (6 and 7). The temporary routing number is returned to the HLR and the G-MSC (8 and 9). The G-MSC now continues routing the call toward the visited MSC (10).

7.3.2 CALL HANDLING IN 3G SYSTEMS

In the circuit-switched context, the term call refers to a voice call, usually involving two parties, but also relates to supplementary services for call forwarding, call waiting and other services involving an extra party. The principles of call processing were established in the GSM specifications following a narrowband ISDN model: call control and connection control occur in the MSC. The 3GPP specifications introduce a softswitch approach. Call control involves the MSC servers and bearer

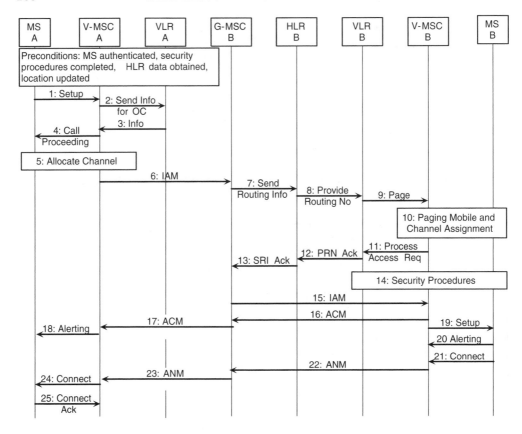

Figure 7.9. Principal information flows in a CS-domain mobile-to-mobile call (prepaging scenario).

connections are made in the underlying packet network using media gateways as required. Call handling procedures are applicable in both GSM and the UMTS CS-domain systems [19]. These procedures are based on ISUP signalling in the core network and would therefore be applicable to voice calls in GSM and calls handled in CS mode in UMTS. Call handling is not independent of mobility management: interaction between the MSCs and the HLR and VLR is essential to call handling.

Several cases arise, depending on the originating and terminating parties to the call. Parties include UMTS or GSM mobile subscribers as well as parties in a fixed network. The latter case is referred to as trunk-originated or trunk-terminated calls. Numerous call setup and cleardown scenarios exist. We examine a selected case.

The information flows for the case of a mobile-originating, mobile-terminating call is shown in Figure 7.9. The originating party, A, and the terminating party, B, may in general be in any allowed visiting networks. At the start of the call, mobile station A (MS-A) is registered in the area of MSC-A and has current information stored in the associated Visitor Location Register, VLR-A.

1. Mobile station A initiates the call by sending a Q.931 Setup message to its visited MSC. The called party is identified by its Mobile Subscriber ISDN number, an E.164 number.

2, 3. The MSC consults the associated VLR-A to obtain information that is required to initiate an outgoing call.

4. A Q.931 Call Proceeding message returns to MS-A.

5. A bearer channel is allocated on the air interface, linking the MS to MSC-A by signalling not shown here.

6. MSC-A starts routing the call by sending an ISUP Initial Address Message (IAM) containing the MSISDN of the called party. The IAM is routed to a Gateway MSC that has access to a Home Location Register containing subscriber profile information of party B, as well as the identity of the VLR to which MS-B is currently registered. The HLR and VLR are identified by Global Title Addresses.

7. On receipt of the IAM, the G-MSC interrogates the HLR. The MAP protocol Send Routing Information (SRI) flow is sent to the HLR of network B.

8. The HLR sends the MAP Provide Routing Number request to VLR-B. The object is to obtain the temporary number that VLR-B has assigned to MS-B while it is in the MSC area.

9–11. The VLR requests the MSC to initiate paging of the mobile station. The paging process elicits a response from the targeted mobile and a bearer channel is assigned between the mobile station and the MSC.

12, 13. On successful completion of paging, the routing number is returned to the Gateway MSC in the responses to the Provide Routing Number/Send Routing Information requests.

14. The air interface to MS-B is now secured by signalling messages not shown here.

15. The IAM message, using the temporary routing number as the B-party address, is sent to the visited MSC, MSC-B.

16–18. The ISUP Address Complete Message (ACM), returned via the G-MSC indicates successful routing in the core network. The Alerting message informs the calling MS of reaching the alerting state.

19, 20. The incoming call is signalled to mobile station B by the Setup message. The Alerting message confirms that the user is being alerted.

21–24. When the B-party answers, the Connect message signals this event to MSC-B, the ISUP Answer Message (ANM) relays the event back to the originating MSC and the originating side Connect informs the calling mobile station that the call is now ready.

25. The call is now active.

This sequence uses the ISUP signalling protocol between MSCs. Other protocols may be used.

7.3.3 *SETTING UP BEARER CONNECTIONS IN THE 3G CS DOMAIN*

The signalling sequence above is concerned with setting up the call. In addition, a bearer connection must be made in the transport network. On the core network side three cases contemplated by the standard [9] are shown in Figure 7.6.

First, the packet-switched network (PSN) acts as a trunking network between two circuit-mode interfaces, for example an A or IuCS access interface and a PSTN. The packet network functions as a trunking network similar to the VoIP case shown in Figure 4.5. A media gateway (CS-MGW) controlled by an MSC Server is located at every point of ingress and egress to the trunking network. All CS domain bearer connections pass through media gateways, as shown in Figure 7.6.

The second case shows the media gateway that terminates the CS channel from the access network making a connection directly to a circuit-switched network via a time-division multiplexed 64 kbit/s channels.

Third, not shown in Figure 7.6, the packet-switched network transports transcoded speech signals within the mobile network to a mobile user via a second media gateway and access network, under the control of a MSC server.

Two realisations of the core network use IP or ATM to create a transit network between two circuit-mode access network interfaces. An IP network uses the Real-time Transport Protocol described in Chapter 4. An ATM-based packet transport using ATM Adaptation Layer 2 (AAL2) to adapt encoded signals to the ATM bearer carries the voice signals to the distant media gateway.

The media gateway has resources such as echo-cancellers, codecs and conference bridges. Connections and media processing in the gateway are controlled via the Mc interface, implemented using the H.248 Gateway Control Protocol. The media gateway is also involved in changes in bearer connections required during some handover scenarios.

7.3.4 *CAMEL-BASED SERVICES IN THE CS DOMAIN*

Functional Architecture

The CAMEL Service Environment describes 'the logical entity that processes operator specific services' [7]. *Operator specific services*[6] are customised applications implemented by operators that may be unique or differentiated from similar applications offered by competitors. These applications enhance services already available in the network such as basic call and mobility control and supplementary services. Such services must be available to subscribers even when roaming in other networks. The CAMEL Service Environment, comprising the gsmSCF, gsmSSF and gsmSRF, is an overlay on the basic 3G network, similar to IN in fixed networks. Services implemented in the CSE are therefore referred to as *features*. The CAMEL features are not restricted to the CS domain but enhance GPRS-based services, Short Message Services, and are also applicable in the IP Multimedia System.

CAMEL uses key concepts from the IN Conceptual Model without the full four-plane model. Important parts of the model are the functional architecture, the basic call state models, the triggering model and the CAMEL Application Part (CAP).

---

[6]We avoid the abbreviation OSS to avoid confusion with the more frequently used Operations Support Systems.

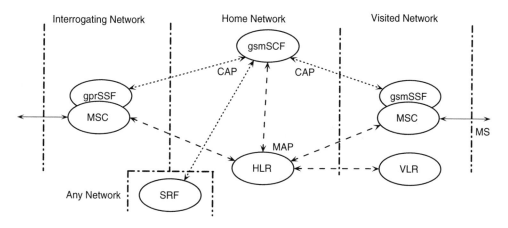

Figure 7.10. Functional architecture for accessing the CAMEL Service Environment in the circuit-switched context.

The functional architecture for CAMEL in the CS domain is expressed in Figure 7.10 using functional entities identified in [11]. The definitions of home, visiting and interrogating networks given in Section 7.2.1 are applicable here. The gsmSCF that implements the custom application is located in the home network, as is the Home Location Register. A mobile station is registered in a visiting network. The application may be invoked in both mobile-originating and mobile-terminating calls. The SSF associated with the visited MSC normally triggers home SCF applications in the case of a mobile originated call. The Gateway MSC processes requests for calls terminating on the mobile station. Triggering of application execution in the home SCF normally takes place in the G-MSC in a mobile-terminated call.

Basic Call State Models and Triggering Mechanisms

The gsmSSF triggering mechanism has important differences to the fixed line IN mechanisms. The latter defines trigger detection points (TDP) that are armed statically by a service management system. In IN, a TDP has two types: first, service logic is invoked and must return a response (type TDP-R) and, second, a notification is sent to the service logic and no response is required (type TDP-N). In CAMEL, trigger detection points are set within the call instance from data stored for the user in the HLR. Only response (TDP-R) types of trigger are allowed. Event detection points (EDP), that is detection points that are armed by the service logic within the call instance, are similar to the IN case. Both EDP-R and EDP-N types are permitted in CAMEL. Detection point data stored in the Home Location Register are called the CAMEL Subscription Information (CSI). Different types of CSI are stored, for example for originating calls and for terminating calls to the subscriber.

The basic call state models for CS-mode services in CAMEL are shown in Figure 7.11. Conventions are those introduced in Section 1.2.2. Detection points

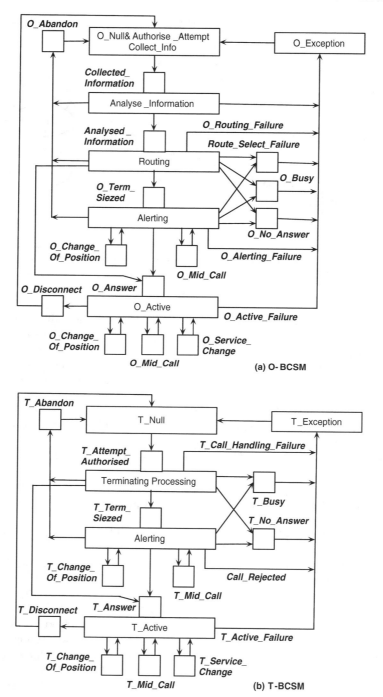

Figure 7.11. Basic Call State Models for circuit-switched 3G CAMEL.

are shown as small squares while an extended rectangle represents a point in call. Arrows show the possible execution paths.

The originating BCSM occurs in the MSC in the case of a mobile originated call or a forwarded call. Forwarding can take place at the G-MSC and the O-BCSM would occur there. The terminating BCSM occurs in the visited MSc for a mobile terminated call and in the G-MSC for a mobile originated call.

The points in call (PIC) are similar to those found in the PSTN call model. A PIC hides the detail of call processing that takes place between detection points. During call processing, call control signalling may be received and emitted by a PIC, for example ISUP messages between MSCs and Q.931 access signalling from and to a mobile station.

Detection points represent the places at which call parameters are compared with values obtained from the relevant CSI to determine whether the service logic on the gsmSCF should be activated or notified. Detection points are similar to their IN counterparts, for example *Collected_Information* and *Analysed_Information*. Three mobile network-specific DPs are *O_Change_Of_Position* (mobile originating only), *T_Change_Of_Position* (mobile terminating only) and *T_Service_Change*. The last indicates that the bearer service has changed.

In the originating BCSM, only *Collected_Information*, *Analysed_Information* and *Route_Select_Failure* may be set as trigger detection points (TDP-R). All other detection points may be set only as event detection points as allowed by data in the subscriber's CSI. Similarly, only detection points *Terminating_Attempt_Authorised*, *T_Busy* and *T_No_Answer* may be set as TDP-R type in the terminating BCSM. All other DPs may be armed as EDPs.

The CAMEL Application Part

The functional architecture in Figure 7.10 identifies the signalling protocol used for communication between the CS-domain MSCs and the gsmSCF as the CAMEL Application Part [18]. This protocol is also used between the gsmSCF and the Specialised Resource Function (gsmSRF). CAP is a development of the INAP protocol illustrated in Figure 1.11. Like its parent INAP, the transaction-oriented operations of CAP are transported using the TCAP and SCCP protocols.

Table 7.3 lists the principal CAP operations applicable in CS domain services. Operations for control of enhanced services in the CS domain are similar to the INAP Capability Set 2 used in fixed networks. CAP operations have additional parameters to pass data such as location information particular to mobile networks. Phase 4 of CAP also contains operations specialised for enhancement of Short Message Services and data services based on GPRS. The latter are discussed in Section 7.4.6.

Table 7.3 divides the CS-oriented CAP operations into five groups. The InitialDP operation carries a request for service logic execution from the gsmSSF to the gsmSCF when the first trigger detection point conditions are satisfied in a switched circuit network. The parameters of the InitialDP operation specify the service logic programme to be executed, for example to translate a short code into a full E.164 called party number in an abbreviated dialling service.

The second group of operations is concerned with the control of bearer connections by the gsmSCF. For example, after translating the B-party number in abbreviated

Table 7.3. Selected CAMEL Application Part operations for the CS domain

Initial SCF Invocation	Charging and Reporting
initialDP	furnishChargingInformation
	applyCharging
Bearer Connection Control	applyChargingReport
connect	sendChargingInformation
continue	callInformationReport
continueWithArgument	callInformationRequest
initiateCallAttempt	
releaseCall	**User Interaction**
disconnectLeg	assistRequestInstructions
moveLeg	establishTemporaryConnection
splitLeg	connectToResource
entityReleased	disconnectForwardConnection
	dFCWithArgument
Event Report Management	playAnnouncement
requestReportBCSMEvent	playTone
eventReportBCSM	promptAndCollectUserInformation
cancel	specializedResourceReport

dialling, the operation Connect instructs the MSC to connect the B-party to the originating party. Operation Continue requires the switch to complete the connection to the B-party using the original dialled number. The operation initiateCallAttempt issued by the CSE initiates a third party call. Operations moveLeg, splitLeg and disconnectLeg support call party handling, that is manipulation of multiparty connections during the active phase of the call. The role of these operations is illustrated in the Connection View States in Figure 3.31.

The third group of CAP CS-domain operations enable the setting of event detection points. The service logic issues a requestEventReportBCSM to the MSC. Specified event detection points are armed. When the requested event is detected by the detection point processing, the eventReportBCSM operation passes to the gsmSCF.

A number of charging and reporting operations are listed in the fourth group. For example, furnishChargingInformation instructs the MSC how to record charging information on the call data record or billing ticket, for example to reverse charges. Operation callInformationRequest allows the service logic to request the return of a specified set of call information from the MSC in a callInformationReport operation.

The final group of operations is concerned with user interactions by playing announcements and collecting digits at the SRF. The first five operations support setting up and clearing the bearer connection between a call party and the SRF. The remaining operations control the interaction of the SRF with the user. Three operations would be used to prompt a calling party to enter a choice by pressing a keypad digit. The gsmSCF sends and establishTemporaryConnection operation containing the E.164 number of the gsmSRF to the MSC. The MSC sets up a bearer connection between the calling party and the specialised resource. The gsmSCF then sends a promptAndCollectUserInformation to the gsmSRF. The specified announcement

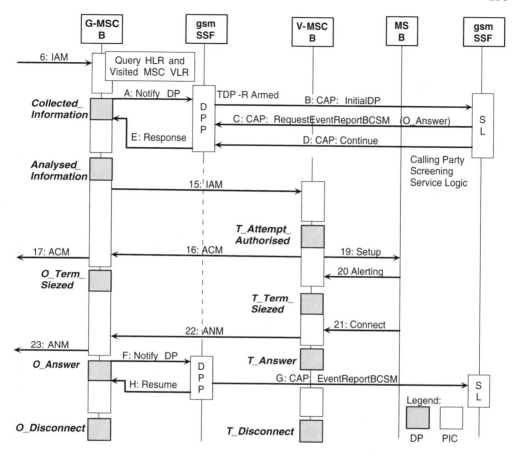

Figure 7.12. Relationship between BCSMs, call signalling and CAP. Message numbers are taken from Figure 7.9.

is played and when the user has entered the required digit(s), the result is returned to the gsmSCF. The gsmSCF then instructs the MSC to disconnect the temporary connection via a disconnectForwardConnection operation.

BCSMs, Call Signalling and CAP

Figure 7.12 is a hybrid BCSM–message sequence chart diagram illustrating the relationship between signalling, points in call and detection points. Access and inter-MSc signalling takes place during points in call. The main sequence for a successful mobile terminating call is shown. The O-BSCM is located in the Gateway MSC while the T-BSCM is in the terminating MSC.

We demonstrate the interaction of signalling with the BCSMs by mapping the normal sequence of processing through a BCSM onto a message sequence chart timeline. A sequence of detection points and points in call in the O-BCSM is superimposed on the timeline for the Gateway MSC. Detection points and points in call are shown using the convention for a processes that elapses time. The T-BCSM is similarly imposed on the Visited MSC timeline. The IAM in message 6 initiates the terminating call process. The subscriber information retrieved from the HLR as in Figure 7.9 includes trigger information.

In the example in Figure 7.12, the user has an originating call screening service. Every number dialled by the user is checked against a list of forbidden destinations. If the dialled number is on the list the call is terminated. The TDP-R at Collected_Information is set in the user's CSI to invoke the service for all A-party numbers dialled. Service logic in the gsmSSF is invoked using the CAP InitialDP operation. The gsmSCF may set an event detection point as shown at message B, for example, to be notified when the call is answered, that is at the *O_Answer* DP. The request to set the EDP is shown at message C.

The service logic compares the dialled number with the forbidden list. Assume that the A-party number passes the screening test, that is not on the list. The call control is instructed by message D to Continue processing after the *Collected_Information* DP, that is to analyse the routing number. Further detection point processing is not shown; all other DPs except *O_Answered* are unarmed in this example. Routing continues with the IAM in message 15 sent to the visited MSC. Once the terminating attempt has been authorised, the Setup and Alerting messages result in the called terminal being seized: no other caller can reach it (19 and 20). The ACM in message 16 relays the seized state to the G-MSC which forwards it to the called party in message 17. Having passed the unarmed DP, the O-BCSM enters the Alerting PIC.

The ANM in message 22 informs the G-MSC of the called party answering. The event is forwarded toward the called party in message 23. Call processing now notifies the gsmSSF that it has reached detection point *O_Answer*. This DP is armed and triggers a notification to the service logic in message G. The call is now active and continues and terminates without intervention of the service logic.

7.4 PACKET-SWITCHED DOMAIN: GPRS-BASED SYSTEMS

Initially, second generation mobile networks provided only circuit-mode connectivity. Circuit-switched data services are provided as shown in Figure 1.12. Data is carried in time slots on the radio interface and switched by the MSCs to an interworking unit to reach the destination packet network. Circuit-switched data is limited in transmission rates and invariably has an unfavourable time-based charging model. Scalability to higher bit rates is limited by available time slots and the proportionate increase in tariffs when extra time slots are allocated.

End-to-end packet-mode transfer was introduced into 2G networks by adding a second, packet-switched access network and switching domain in the core network: the GPRS. The purpose of GPRS is to provide end-to-end transfer of packets between mobile stations or between a mobile station and a node in another packet network via a gateway. Thus, one or both end-stations may be mobile stations [6].

GPRS is basically a bearer service. Teleservices based on logic in the terminal and server that may use GPRS connections are, in general, outside the GPRS specification. The GPRS packet bearer service can be used to support the Short Message Service and Multimedia Messaging. The GPRS protocol stacks, described in Section 7.4.5, are designed to give packet transport with quality of service and to allow the end-user terminals and servers to use a selected protocol, for example IP or PPP. Connections using GPRS can be enhanced using applications in the CAMEL Service Environment.

GPRS provides two basic packet transport services for the end-user. The point-to-point service is a connectionless IP service. Point-to-multipoint packet transport provides IP multicast connections. A teleservice may use either point-to-point or point-to-multipoint transport services.

The GPRS core network architecture is independent of the data rates and is unchanged from 2G to 3G. The data transfer rate available to the end user is limited largely by the speed of the radio interface. Efficient operation of GPRS networks is based assumptions about the traffic offered by a typical user: in common with most packet networks, the average bit rate offered should be well below the peak value available. Usage patterns meeting this requirement include intermittent, low duty cycle data transfer; frequent, short transactions; and infrequent transfer of long files. The GPRS standards did not originally envisage high bit rate continuous transfer.

7.4.1 GPRS-BASED PACKET-DOMAIN ARCHITECTURE

Implementations of GPRS in 2G networks requires both core and access network enhancements. A packet-switched core network is added to the existing GSM circuit-switched core. The original BSS to core network A-interface supports circuit-oriented bearers. In GPRS, packets carried on the air interface in time slots must extracted before encountering the CS core network, that is at the base station controller. A new packet-mode interface called Gb is introduced between the BSC and the PS core network. The enhanced 2G BSC is often described in terms of the dual interfaces, namely A/Gb.

Figure 7.13 shows the PS-domain based on GPRS with a 3G access network against the backdrop of the NGN Framework layers. The packet-oriented 3G RNS access network presents the IuPS interface to the core network. The Gb and IuPS interfaces differ in their protocol stacks and detail of procedures used. The RNC performs functions such as encryption that are implemented in the core in 2G networks.

The core network in GPRS consists of GPRS Support Nodes of two types shown in Figure 7.13. First, the 3G *Serving GPRS Support Node* (SGSN) has IuPS interfaces to RNCs and Gb interfaces to BSSs. The SGSN is the counterpart to the MSC, providing support to mobile stations operating in GPRS mode, but its functions differ in key aspects. For example, inter-GSN signalling via reference point Gn sets up a special form of packet connection between nodes. Second, the *Gateway GPRS Support Node* (SGSN) is located at the edge of the mobile network and controls access to or from other GPRS networks or other packet networks such as the Internet.

The extent of the GPRS system is demarcated by two reference points and an interface. The R-reference point in the terminal shown in Figure 7.7 represents the point of access for user protocol data units (PDU). The Gi reference point shown in Figure 7.13 defines the interworking between a GPRS-based network and another packet network, for example the Internet. An inter-GPRS system interface is denoted Gp.

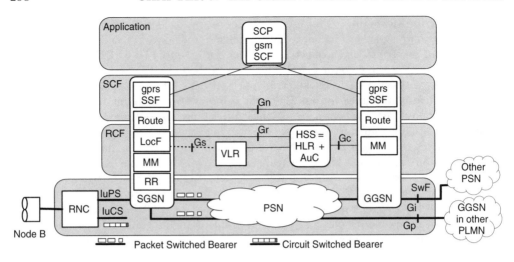

Figure 7.13. 3G architecture based on GPRS packet-switched core network.

7.4.2   GPRS SUPPORT NODE FUNCTIONS

The two types of GPRS Support Node are functional groupings in the PS-domain core network. Each GSN contains functions that are classified into different NGN Framework layers as shown in Figure 7.13.

The Serving GSN (SGSN) performs functions that fall into several groups. The *network access control* functions are authentication of users and authorisation of use of services, collecting charging information and applying barring restrictions imposed by the operator. Packet-level admission control is applied according to the applicable QoS profile.

The *packet routing and transfer* (Route in Figure 7.13) functions are address translation and address mapping. Address translation converts an address associated with the packet data protocol user to an address type used within the GPRS network. Address mapping performs address translation within the GPRS network. Packet routing is the determination of the next hop node in the low level routing protocol used to transfer packets across the GPRS network, for example using IP or ATM. The relay function is the actual forwarding process in the node. User PDUs are encapsulated with information necessary for transfer across the GPRS network in point-to-point tunnels.

Mobility management, as in the CS domain case, tracks the location of the mobile station. The SGSN accesses the HLR through the Gr interface shown in Figure 7.13. The identity of the SGSN to which a mobile station is attached is held in the HLR. The temporary location information resides in the SGSN Location function. Use of a VLR is optional. Use of location information in establishing data communications in a GPRS system is examined in Section 7.4.3.

In the case of the access network with Gb interfaces, the SGSN must in addition handle the compression and ciphering of packets to and from the MS. These functions

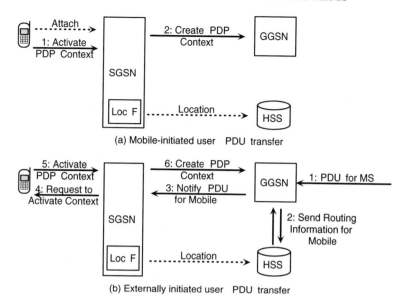

Figure 7.14. Interaction diagram showing essential steps in initiating packet data transfer in a GPRS network. Preconditions are shown dotted.

are carried out in the Iu-mode RAN. The Gb interface also requires logical link control procedures for establishment, supervision and release of communications channels across the radio interface. The 2G SGSN also manages the radio resources. These functions are performed in the 3G RNC.

The Gateway SGSN has specific network access control functions of message screening and collecting charging information. The packet routing and transfer responsibilities of the GGSN are similar to those of the SGSN, namely routing, relay, encapsulation and tunnelling. The GGSN has a mobility management relationship with the HLR described in Section 7.4.3.

7.4.3 MOBILITY MANAGEMENT IN THE PS DOMAIN

The interactions in Figure 7.14 summarise the principles of mobility management in GPRS systems. Case (a) shows the preconditions, namely the mobile station must have attached to the SGSN and the current location is stored in the HLR within the HSS. A mobile originating call is straightforward. The mobile station initiates a PDP context described in Section 7.4.4 (1) and the SGSN requests the GGSN to create the context (2).

The mobile terminating case shown in Figure 7.14(b) is initiated by the arrival of a PDU for the mobile at the GGSN (1). The GGSN retrieves information from the HSS, including the identity of the SGSN to which the mobile station is attached (2). The GGSN notifies the SGSN that a PDU has arrives for the mobile station (3). The

SGSN requests the mobile station to establish a PDP context. Interactions 5 and 6 are identical to 1 and 2 in case (a).

7.4.4 PACKET TRANSPORT IN GPRS

User data packets are transported between mobile stations or a mobile station and the interface to an external network using the Packet Data Protocol (PDP). A PDP is any protocol for the transfer of packets, for example IP. Before transferring user packets, a PDP context must be initiated. A *PDP context* is used for managing a data communications session between GSNs. A PDP context is described by a number of data fields including QoS parameters, address information, sequence numbers and an identifier for the tunnel used to convey packets associated with the context. PDP contexts exist in the mobile station, the SGSN or the GGSN. The address may be statically assigned or only when the context is activated.

Three processes are required to support the transport of end user packets across the GPRS network [10]. The mobile station must first attach itself to the SGSN. Second, a PDP context must be established. Third, packets are encapsulated using the GPRS Tunnel Protocol (GTP) for transmission between nodes.

GPRS provides unrestricted transfer of digital information, that is the end-user data is not interpreted by the network. Data transfer is asynchronous with variable bit rate and delay. Data transfer is performed according to a quality of service profile that is stored in the subscriber profile and is transferred to the PDP context. The QoS profile has four parameters:

- *Precedence*: the order in which data transfer commitments will be addressed: values are high, normal and low precedence;

- *Reliability*: probabilities for a service data unit (SDU) of loss, duplicate delivery, and undetected error (corruption);

- *Delay*: between R-interface and boundary of GPRS network, expressed as mean and 95th percentile values;

- *Throughput*: values are mean and maximum bit rate.

Figure 7.15 shows the signalling messages used to activate PDP contexts. For a terminating mobile station, the first PDU (message 1) arriving at the GGSN starts the process.

2, 3. The GGSN queries the HLR for the mobile station location.

4, 5. The GGSN notifies the arrival of a user PDU to the SGSN.

6. The SGSN requests the mobile station to activate a PDP context.

Steps 1–6 are unique to the terminating mobile station case. Steps 10–18 are required in both the mobile terminating and originating cases.

10. The mobile station requests the creation of PDP contexts.

11. The MS-SGSN context C1 is now active.

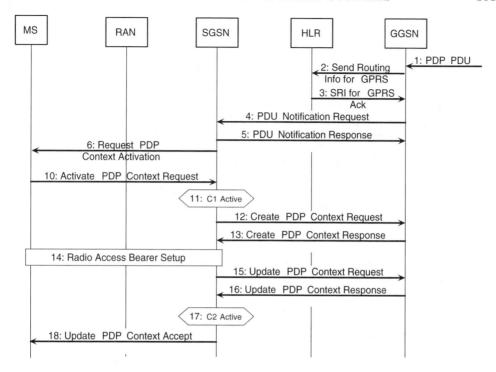

Figure 7.15. Activation of PDP context in GPRS (Iu case). Messages 1–18 show the network initiated activation case. Messages 10–18 represent MS-initiated activation.

12, 13. A PDP context for transmission between SGSN and GGSN is requested.

14–16. The radio access bearer is set up and the details of the PDP context may need updating.

17, 18. The second PDP context is active and the MS is notified.

7.4.5   PROTOCOLS USED IN GPRS ACCESS AND CORE

The protocol stacks used in the access and core networks to implement a GPRS system are based on the ISDN practice of separating signalling into a control plane and a user plane. The *user plane protocol stacks* support end-to-end signalling between applications, one of which may be outside the PLMN. *Control plane signalling* supports network functions such as GPRS Mobility Management (GMM) and Session Management (SM). In the case of UTRAN, the Short Messaging Service is supported as an application on the control plane stack. Four sets of protocol stacks shown in Figure 7.16 are defined in the 3GPP standards to serve user and control planes and UTRAN and GERAN access networks [10].

Discussion of the four sets of protocol stacks in Figure 7.16 is simplified by noting that all stacks divide into three sets of layers. First, all stacks have a set of low layers,

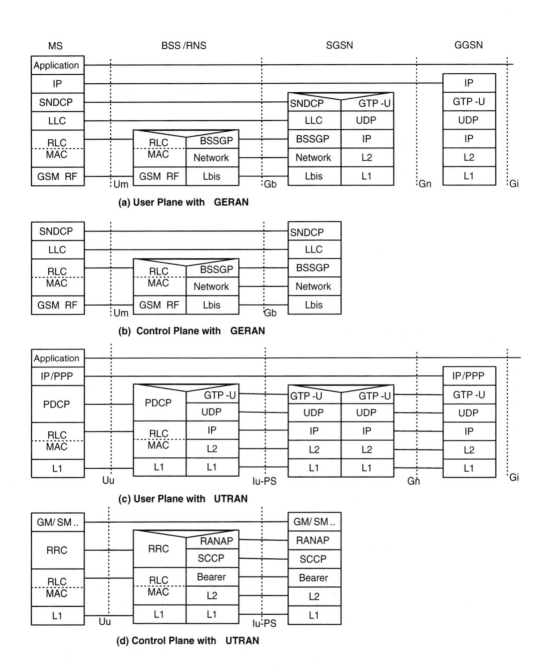

Figure 7.16. Protocol stacks for User and Control Planes in GPRS for GERAN and UTRAN cases.

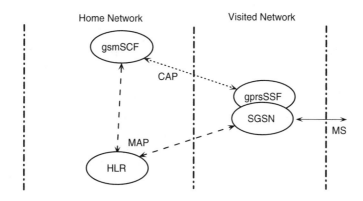

Figure 7.17. Functional architecture for accessing the CAMEL Service Environment in the GPRS context.

(LLC and below in GERAN and RLC and below in UTRAN) that are common to the control and user planes across the air interface.

Second, in the user plane, the actual application is supported by the PDP used, for example, IP in Figure 7.16(a) and (c). Control plane applications are typically mobility management and session management. In the control plane, the mobile station and the SGSN have a peer-to-peer relationship at the application layer. The control application protocol is supported by connections in a link layer protocol.

In the control plane for GERAN, the Base Station System GPRS Protocol (BSSGP) conveys routing- and QoS-related information between the BSS and the SGSN. In the UTRAN case, a tunneling protocol, the Radio Access Network Application Protocol (RANAP) conveys control information between RNS and SGSN.

Third, in the user plane cases, intermediate layers adapt the high layers to the low layers. In the GERAN case (a), the Subnetwork Dependent Convergence Protocol (SNDCP) is used in the MS and SGSN to map the user PDUs to the lower layer logical links. In the core network, user PDUs are encapsulated and tunneled between nodes using the GPRS Tunnelling Protocol for the user plane (GTP-U) [10]. In the UTRAN user plane, user PDUs are adapted to the low layers using the Packet Data Convergence Protocol (PDCP) for transmission across the radio interface [20]. The RNS, SGSN and GGSN transfer user PDUs by tunneling using the GTP-U.

The user plane structure in both GERAN and UTRAN cases decouples routing of application PDUs from routing within the GPRS network. The lower IP layer in the core supports the transfer of packets in tunnels and is local to the GPRS network. The upper IP layer performs no routing function in the GPRS network, other than identifying the endpoint.

7.4.6 GPRS AND CAMEL

The functional architecture for CAMEL-enhancement of GPRS services is shown in Figure 7.17 [11]. Only two functional entities are required in the PS domain: the

Table 7.4. Selected CAMEL Application Part operations for GPRS

Initial SCF Invocation	**Event Report Management**
initialDPGPRS	requestReportGPRSEvent
	eventReportGPRS
Bearer Connection Control	cancelGPRS
connectGPRS	
continueGPRS	**Charging and Reporting**
releaseGPRS	furnishChargingInformationGPRS
entityReleasedGPRS	applyChargingGPRS
	applyChargingReportGPRS
	sendChargingInformationGPRS

gsmSCF already introduced in the CS domain and the gprsSSF, a service switching function based on a state model appropriate to the GPRS.

In general, the mobile station is located in a visited network. The functional entities in the visited network are therefore the SGSN and the gprsSSF. The Home Location Register and the gsmCSF are located in the home network. As in the CS-domain case, the CAP protocol is used between the gprsSSF and gsmSCF. GPRS-oriented CAP operations are summarised in Table 7.4. The MAP protocol is used for updating and querying subscriber data in the HLR.

The GPRS-based system provides end-to-end data transfer by establishment of PDP contexts. GPRS has no service control functionality. Creation of enhanced data services relies on external logic, for example hosted in a CAMEL Service Environment or Open Service Architecture platform described in Chapter 8.

CAMEL-based services are invoked via the gprsSSF. Services are likely to be initiated as a result of attachment, detachment and location changes of mobile stations involved in the service. Similarly, establishment, changes to and clearing PDP contexts are potential service triggers.

Attachment and PDP context events are not necessarily correlated. Two state models shown in Figure 7.18 are therefore defined for GPRS. The state model is constructed from detection points having similar significance to those of the CS BCSM. The wrappers that hide processing details are called *points in association* (PIA) rather than points in call. Transitions connect PIA to DP. Only one form of each state model is defined because a single association rather than two half-calls is relevant in the GPRS case.

The *GPRS Attach/Detach State Model*, a form of session model, provides three detection points: *Attach*, *Detach* and *Change_of_Position_GPRS_Session*, shown in Figure 7.18(a).

The *GPRS PDP Context State Model* has the four detection points shown in Figure 7.18(b) for PDP-context related events: *PDP_Context_Establish* marks a request to establish a context and *PDP_Context_Establish_Ack* signifies success. Detection point *PDP_Context_Disconnection* allows reporting of ending a context to the service logic. Detection point *Change_Of Position_Context* supports invocation of location-related service logic.

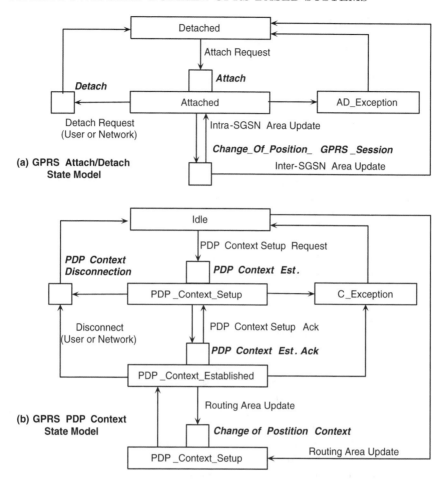

Figure 7.18. State models in GPRS for mobile station attachment and PDP context.

Arming of detection points follows the general principles described for the CS domain. A subscriber using GPRS has CSI data stored in the HLR. This data includes the detection points to be armed as trigger detection points.

CAP operations for supporting the Short Message Service and GPRS [18] are listed in Table 7.4. These are restricted to CSE invocation, control of GPRS connections and charging for GPRS services. User interaction has no significance in GPRS.

The gprsSSF sends the initialDPGPRS operation to the gsmSCF after encountering an armed TDP-R detection points in either the Attach/Detach or PDP Context state models, for example, the *Attach* DP. Operation Continue instructs the gprsSSF to continue from the detection point at which processing is suspended. Operation Connect provides address information for completion of a PDP context setup.

7.5   IP MULTIMEDIA SUBSYSTEM

7.5.1   *ORIGINS AND OBJECTIVES OF THE IMS*

The IP Multimedia Subsystem shown in Figure 7.1 is a subsystem of the packet-switched domain core network. The IMS is defined as 'all Core Network elements for the provision of IP multimedia services comprising audio, video, chat, etc and a combination of them delivered over the PS domain' [9]. The IMS core elements, together with an access network that is capable of providing IP connectivity to the mobile station are intended to support applications that rely on IP transport provisioned both in the mobile network and outside.

The IP Multimedia Subsystem reflects an important convergence between mobile networks and IP networks, particularly the Internet. This form of convergence allows mobile stations to function efficiently as terminals for accessing services provided in the Internet or an intranet as well as allowing multimedia applications based on Internet standards and services to be provisioned within a mobile network. A *multimedia application* controls or enhances basic media transfer services or is an information processing application that is enhanced by multimedia communications. Interfaces in the IM system are defined as far as possible using existing Internet standards such as SIP and Diameter.

The concept of a multimedia service has two facets. First, media streams must be negotiated, setup, modified and cleared. Some applications such as videoconferencing require the control of co-ordinated media streams. Second, the end user's view of services must hide the complexity of the media session. For example, a user should perceive and be able to handle all services in an integrated way. Assume that a user is subscribed to voice calls, video conferencing, e-mail, multimedia messaging and chat. The user should be able to set in a single application preferences such as the diversion of incoming communications for each of these media while on vacation.

The IMS is built on the existing PLMN standards and on a number of IETF protocols. For example, the IMS supports roaming users and therefore relies on the mobility management functions already embodied in the 3G network. The CS domain provides effective control over voice calls while GPRS provides transport of user PDUs across the PLNM. Alone, these two capabilities cannot support multimedia services. Two developments within the PLNM facilitate this convergence. First, softswitch-type decomposition of the MSC described in Figure 7.5 identifies the Call Session Control Function (CSCF) as a functional entity. Second, the call/session control signalling is not restricted to ISUP and Q.931 as used in the CS domain. The Session Initiation Protocol (SIP), supported by the Session Description Protocol (SDP), is adopted as the call/session control protocol for multimedia services. With the adoption of SIP as the call/session signalling protocol, the CSCF is redefined using SIP concepts. The IMS CSCF thus differs from that for the CS domain based on ISUP signalling and the IN Basic Call State Models.

The IMS has a number of objectives that guide the development of standards [1]:

- *Flexible access arrangements*: while the IMS is essentially a core network standard, it must be possible for users to access IMS services from suitable terminals via any IP-capable access network. UTRAN, with its IuPS interface, and GERAN (Gb interface) access networks are IP-capable. Alternative access networks may require interworking standards, for example those developed for wireless LANs. The access network must hide detail of mobility management from the IMS, freeing it from low-level mobility management.

- *Roaming*: IMS users must be able to roam between operators. The provisioning of roaming agreements should be automated rather than manual. Users must be able to use services provisioned in the home environment as the norm and in serving networks where these are provided.

- *Session characteristics*: a *session* in IMS has the meaning used in IP telephony in Chapter 4, namely a set of sources and receivers and the media streams they exchange. While identifying default configurations to ensure interoperability, for example codec type, a range of media types must be supported. A single multimedia session (set of media streams) may be acted on by more than one application. The quality of service associated with individual media streams may be negotiated by the user or operator both at session establishment and during the session. The baseline quality for speech is that of circuit-switched wireless systems.

- *User-oriented service requirements*: the IMS assumes that a subscription mechanism is in operation but does not prescribe it. A number of functions are required: user access control, capability negotiation and redirection of communication intended for a session member. Users may be identified using an E.164 number, a TEL URL or a SIP URL. Various modes of establishing, controlling and clearing sessions are permitted. A *conference focus* or model is defined [8].

- *Privacy, security and policies*: the baseline for security and privacy in an IMS service must not be worse than that available in a system implemented using GPRS. The IMS must allow the operator to set and implement policies and to enhance security of the network by not exposing the full network topology to another operator.

- *Standardised capability*: services in IMS must not be standardised. Rather, to allow operators to develop and deliver differentiated services, the underlying capabilities are standardised.

- *Interworking with other networks*: three cases are identified [8]. First, voice calls must be supported to users via the CS domain of a mobile network, a PSTN or ISDN. Features normally associated with voice calls must be supported, for example, call barring, call forwarding and calling line identity presentation and restriction. Second, interworking with the Internet is required. In particular, applications developed outside the mobile network must operate successfully when a party is in the mobile network. Third, the 3G mobile IMS network must interwork with future networks of various types including fixed networks that are based on IMS.

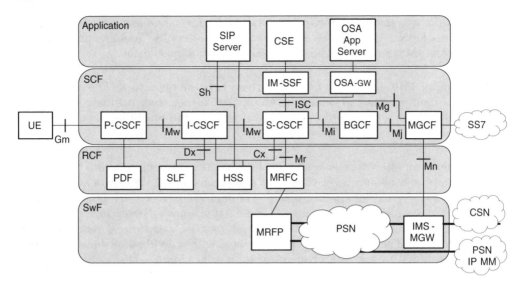

Figure 7.19. 3G IP Multimedia System architecture with CAMEL SCP shown against NGN Framework layers.

7.5.2 IMS ARCHITECTURE AND COMPONENTS

The IP Multimedia Subsystem architecture concentrates mainly on multimedia session control and applications that control or are enhanced by multimedia communications. The principal elements of the IMS are shown in Figure 7.19 against the NGN Framework layers.

The underlying transport network is an IPv6 network. Two functional entities defined in IMS are classified into the NGN Framework's Switching layer:

- The *Media Gateway* (MGW): like the media gateway described in Chapter 4, the IMS Media Gateway terminates channels from a switched circuit network and media streams from a packet network. The IMS Media Gateway performs transcoding and echo cancellation. Connections, including necessary media conversion, are made between circuit-oriented channels and packet streams. The Media Gateway is also the location for conference bridging. The media gateway presents the Mn reference point to the Media Gateway Control Function. The Mn reference point is defined by the H.248 Recommendation.

- The *Media Resource Function Processor* (MRFP): this processor terminates streams and performs functions including sourcing streams (announcements), analysing incoming streams (digit collection, voice recognition), transcoding, and mixing streams. The MRFP presents the Mp reference point to its controller, the Media Resource Function Controller. The Mp reference point must comply with the H.248 Recommendation as well as other open standard extensions.

In the Service Control Layer, the Call Session Control Function is a basic building block that is deployed in a number of modes to control multimedia sessions. The session control signalling protocol is SIP with media descriptions defined in SDP. The CSCFs therefore process SIP signalling transactions. The CSCF roles are designed to meet several requirements: roaming users must be supported; service control and applications are located in the home network; and inter-operator gateways between networks must hide the topology of the network.

Figure 7.19 shows three types of CSCF having specific roles as follows:

- *Proxy Call Session Control Function* (P-CSCF): this type of CSCF serves as the call signalling point of contact for the mobile terminal (UE). For a roaming user, the P-CSCF is in the visited network, otherwise it is in the home network. On receiving a request to start a session, the P-CSCF locates the S-SCSF in the home network or a gateway that will find the server. The P-CSCF is also responsible for allocating resources to meet quality of service requirements. A Policy Decision Function (PDF), described in Section 7.5.3, is contained in the CSCF or is available to the P-CSCF to ensure that resources are allocated according to policies.

- *Interrogating Call Session Control Function* (I-CSCF): this element is needed at the gateway between the networks of two operators. On receiving a session request from the P-CSCF, the home network I-CSCF queries the HSS for the S-CSCF to be used for the individual user. The I-CSCF acts as a proxy server with the HSS as its supporting location server. The primary purpose of the I-CSCF is to hide the topology of the network to which it belongs from other providers. For example, entities in other networks must not know the addresses of servers other than the I-CSCF.

- *Serving Call Session Control Function* (S-CSCF): the S-CSCF is the locus for decisions on actions to be taken on receipt of a session request and maintenance of the session state. The S-CSCF can interact with an application server. The IMS service model, including further detail of the S-CSCF, is described in Section 7.5.4.

The call session signalling reference point between CSCFs, denoted Mw, is based on SIP.

Two functional entities that deal with call routing to and from circuit-switched networks are shown in the SCF layer of Figure 7.19:

- *Breakout Gateway Control Function* (BGCF): for a call from an IMS network to a switched circuit network of another operator, the network in which breakout is to occur must be identified and signalling relayed toward the network. If the BGCF is in the network where breakout occurs, the MGCF to be used must be identified by the BGCF and signalling must be forwarded to the MGCF. The BGCF has call signalling interfaces Mi, Mj and Mk to the CSCF, MGCF and another BGCF respectively. Like the CSCF–CSCF interfaces, these interfaces are based on SIP.

- *Media Gateway Control Function* (MGCF): the IMS MGCF performs limited call state control for connecting media streams and channels in the media gateway.

The MGCF presents a SIP-based Mg interface to a CSCF and an ISUP interface toward a circuit-switched network.

The NGN Framework Service Control Function Layer in Figure 7.19 also contains an interface to a *service platform* that hosts value-adding logic and may be external to the home operator's domain. The service platform takes three possible forms in the 3G standards:

- The *SIP Application Server* (SIP-AS) can perform canonical SIP server functions: registrar, redirect, proxy and back-to-back user agent servers. Other unspecified logic may be hosted.

- The *CAMEL Service Environment* is defined principally for circuit-switched services but also supports SMS and GPRS. The CSE hosts applications in the CSE that enhance IMS sessions.

- The *Open Service Architecture application server* is the newest and possibly the most flexible means of providing value added services, and is discussed in Chapter 8.

The IMS standards require that all three types of service platform must communicate with the S-CSCF through a single reference point, the IP Multimedia Subsystem Service Control Interface (ISC). The ISC is essentially a SIP interface that can handle more than one transaction. For example, in the case of a third party session set up, two user agent clients exist in the application server. The CAMEL and OSA elements in the Application Layer require adaptation to the SIP-based ISC interface.

7.5.3 IETF PROTOCOLS USED IN IMS

The IMS uses IETF protocols for multimedia communication. The principal protocol is SIP, where its canonical form described in Chapter 4 is applied in the IMS architecture as a specific profile [172]. Other protocols are used in lieu of traditional telecommunications protocols. The Diameter protocol is used rather that MAP for querying the HSS from the CSCFs. The COPS protocol is used to implement a policy-based network approach to integrating QoS in GPRS and other IP networks. The RSVP protocol may used to reserve network resources for stream flows.

COPS

Quality of service in an IM network is based on the concept of *policy-based admission control* [208]. Admission refers to the usage of a resource such as a network router. A *policy* is a set of rules that determine whether is requestor is permitted to use a resource. A policy-based system has two principal architectural elements shown in Figure 7.20. The Policy Decision Point (PDP) contains logic that can apply rules to user data provided in a request. The PDP is often supported by a database. The PDP produces a decision, that is information on whether the set of data supplied satisfies the applicable rule. The resource has functionality called a Policy Enforcement Point (PEP). The PEP makes requests for decisions to the PDP and implements the decision, that is it allows or prevents use of the resource.

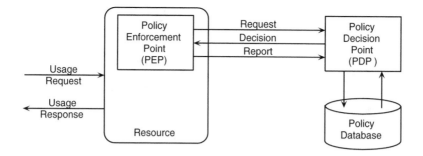

Figure 7.20. Elements in a policy-based admission control system.

Figure 7.20 illustrates the entities and relationships in a policy-based network. Assume the resource is a network router. A party requests that capacity be reserved in the router for a media stream. The PEP sends a request containing information on the proposed usage to the PDP to determine whether the capacity may be reserved in terms of applicable rules. The PDP makes the decision and informs the PEP. If the decision is affirmative, capacity is reserved for the flow.

The policy-based framework for admission is defined in general terms [208]. A specific protocol for conveying requests, decisions and state information as well as for co-ordinating queries and responses is described below.

The Common Open Policy Service (COPS) protocol [47] supports policy control by providing QoS signalling between PDP and PEP elements in a policy-based network. COPS defines ten messages. Three are of interest here. The Request (REQ) message is issued by the PEP to obtain a policy decision from the PDP. As COPS may be used to support different QoS mechanisms, the request is carried in a field provided for the client-specific request object. The Decision (DEC) message returns the policy decision to the PEP. The Report (REP) message informs the PDP whether the PEP has been successful in executing the policy decision. This exchange is included in Figure 7.26.

Diameter

The Diameter[7] protocol is an AAA protocol: it supports user authentication, access authorisation and usage accounting. A Diameter Base Protocol [31], provides the essential definition of sessions, transactions messages, security and basic messages, including support for usage accounting. Diameter is an extensible protocol using attribute-value pairs to define commands. To apply Diameter in a particular context, additional attribute value pairs must be defined. For example, the commands required to apply Diameter to the Cx and Dx interfaces in a 3G network are defined in [16].

The base protocol specifies requirements for secure, reliable transport of Diameter messages. Transport layer security is required and TCP or SCTP must be used to ensure reliable delivery. A particular application, for example the Cx and Dx interfaces, may further refine the requirement. For example, SCTP must be used

---

[7]Diameter is not an acronym. The name derives from Diameter being an extension of the protocol Radius (an acronym).

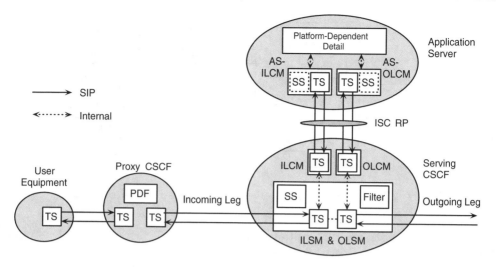

Figure 7.21. IP Multimedia System service model showing transaction state (TS) (user agent) and optional session state (SS) elements.

in the 3G context for these interfaces. Diameter identifies client and server roles. Transmission of Diameter messages is aided by agents having relay, proxy, redirection and translation capabilities. The last translates Diameter messages to and from other AAA protocols.

The base protocol defines request–answer pairs of messages. The base protocol, for example, defines messages with commands Accounting-Request and Accounting-Answer. The Cx and Dx interface application of Diameter defines six request–answer pairs of messages including User-Authorization-Request and User-Authorization-Answer and Location-Info-Request and Location-Info-Answer.

7.5.4 IMS SERVICE MODEL

Figure 7.21 identifies internal architectural detail of the P-SCSF, C-CSCF and the Application Server interface [12]. Each form of CSCF is a combination of SIP user-agent client and user-agent server. SIP signalling between user agents is structured as a series of transactions. The principles underlying SIP entities illustrated in Figure 4.26 apply. For each SIP transaction, transaction state machines (TS) are set up with user-agent client and user-agent server roles at appropriate ends of the interaction. The *Application Server* (AS) is an entity in the IMS architecture that offers value-added services and is normally located in the user's home network.

The S-CSCF model describes an interaction with an originating party as the *incoming leg* while that with a terminating party is an *outgoing leg*. These legs are terminated in SIP transaction state machines in the *incoming* and *outgoing leg state models* (ILSM and OLSM). The *session state* (SS) is also kept in this combined unit.

On receiving a session request, say an INVITE message on the incoming leg, a decision must be made to forward the request to the outgoing leg or forward to the

Application Server. The S-CSCF therefore retrieves user data from the HSS, including filter data. The *filter data* is used, together with the state of the session, to determine the next action: forward to a terminating party, invoke the Application Server or reject the message.

Communication with the Application Server takes place using SIP via the ISC reference point. The SIP transaction states are held in the *incoming* and *outgoing leg control models* (ILCM and OLCM). The Application Server interface has similar models (AS-ILCM and AS-OLCM) and may also maintain session state information. Of the three permitted service platforms, only the SIP Application Server is directly compatible with the ISC interface. A specific adaptation is required for the CAMEL and OSA types of service platform. For example, the CSE requires a gsmSSF-like element, the IP Multimedia Service Switching Function (IM-SSF).

The I-CSCF, not shown in Figure 7.21, has a simple model. It is essentially a proxy server that does not maintain call state and uses the HSS as its location server.

The Call Session Control Functions are basically SIP servers functioning mainly as proxy servers during multimedia session control. Only the S-CSCF exercises control over the service. This control is limited largely to testing session data against filter criteria retrieved from the HSS. Possible outcomes of the test are to continue session signalling, to reject a request or to forward the request to the Application Server (AS). Invocation of the Application Server extends the signalling path to include the AS. Incoming signalling, for example an INVITE from an originating mobile station, received by the S-CSCF, is forwarded to the AS if the filter conditions so dictate. Outgoing signalling initiated by the AS or forwarded by the AS toward a terminating party is forwarded by the S-CSCF.

Figure 7.22 shows the topology of a number of interactions involving the S-CSCF and the Application Server. Only the initial request (INVITE or REGISTER methods) and one related response are shown. In case (a) the INVITE is directed to the AS which functions as a redirect server. The AS responds with a response such as 302 Moved Temporarily. The UE may then direct an INVITE message to the party at its moved-to location. Case (b) occurs when the end user registers with the Application Server to be notified of an event. Case (c) shows the application as the initiating user agent, inviting the UE to a session. Case (d) shows end-to-end signalling when the AS is not invoked, that is filter conditions for invocation are not satisfied. Case (e) shows the AS acting as a SIP proxy server. This topology also applies when a service control application is invoked in the AS. Case (f) extends case (c) to two call legs with the AS acting as a back-to-back user agent (B2BUA) initiating a third party call.

7.5.5 SESSION CONTROL

The set of functional entities that take part in session control depends on a number of factors. For example, is the user in the home network or roaming; and, is the user an IMS user or a PSTN or CS-domain user? In each call, every user can be roaming or at home or be a fixed line user and may belong to different operator's networks. An IMS mobile user is served by a S-CSCF in its home network. In general, in a mobile to mobile call, there are originating and terminating home networks that could in general be distinct. A number of service scenarios therefore arise. Figure 7.23 identifies the various use cases as a function of originating party's access network

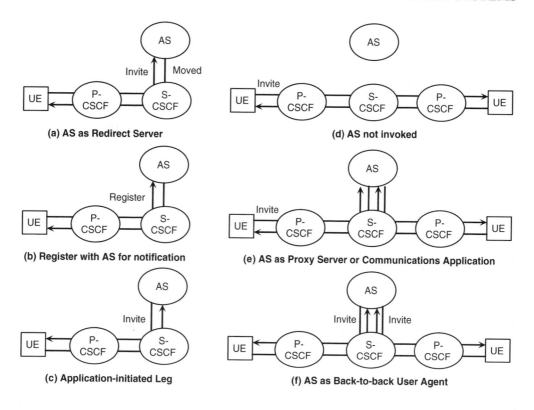

Figure 7.22. Application server roles.

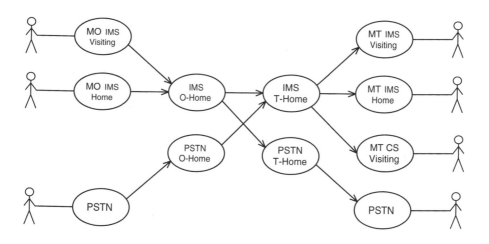

Figure 7.23. Service scenarios in the IP Multimedia System defined by the originating, serving and terminating network types.

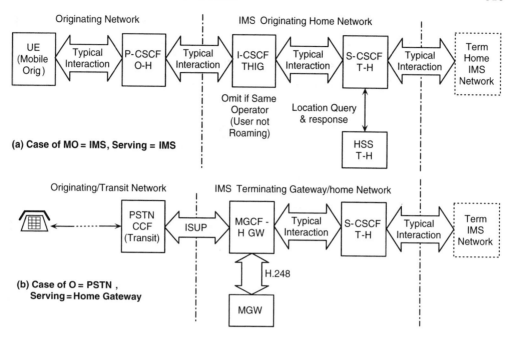

Figure 7.24. Originating use cases for IMS mobile (roaming and home) and PSTN originating users.

and serving network and the terminating party's access and serving network. Arrows linking the use cases show the expected end-to-end connection scenarios.

Use Cases for Different Originating Users

Figure 7.24 records the call scenario and signalling path for three cases of originating user and serving home network. Case (a) is for a mobile user. In the roaming case, the signalling path includes the P-CSCF of the visited network. As the serving home network is in general under the control of a different operator, the signalling path enters the home network via an I-CSCF before reaching the S-CSCF. The I-CSCF acts as a *Topology Hiding Inter-network Gateway* (THIG). The typical signalling interactions between IMS elements occur as shown in Figure 7.24(a). The typical horizontal signalling pattern is defined in Figure 7.25.

Figure 7.24(a) also covers the case of the mobile originating user in the home network if the internetwork boundary and the I-CSCF are removed.

The PSTN originating user case is shown in Figure 7.24(b). The Media Gateway Control Function receives ISUP signalling from the switched circuit network and issues the typical sequence of IMS signalling toward the S-CSCF. The MGCF issues ADD and MODIFY commands to the media gateway to set up the connection between the TDM channel on the PSTN side and the stream on the IMS side of the MGW.

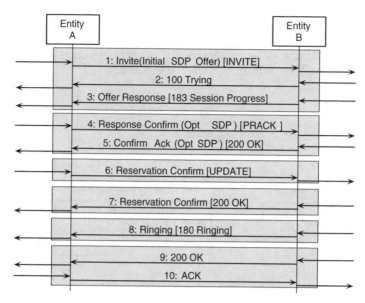

Figure 7.25.   Call signalling pattern between two adjacent entities in the IP Multimedia Subsystem.

3G-IMS-Signalling-Pattern

IMS exploits the capability of SIP to locate servers and users by the proxy or redirect server processes, supported by suitable databases. In particular, the object is to allow a roaming user to access services located in the home network. While a number of servers may be involved in the signalling path, interserver signalling is similar between pairs of servers. To simplify the description of the use cases, we identify the typical pattern of inter-server signalling.

Call signalling flows within the service control layer, that is *horizontally*, and between layers, that is *vertically*. Horizontal switching layer elements are: UE, P-CSCF, I-CSCF, S-CSCF, BGCF, MGCF and MRCF. Horizontal elements occur in typical pairs, for example UE and P-CSCF, and P-CSCF and I-CSCF. Vertical signalling messages are interspersed with the horizontal messages, for example for resource reservation. Vertical flows occur to the Application Server, Home Subscriber System, Media Gateway and Media Resource Control Function as required. While vertical signalling differs from entity to entity, the same pattern of basic SIP horizontal signalling occurs between adjacent IMS elements.

Figure 7.25 shows the typical sequence of *horizontal* signalling messages between two adjacent entities in the IP Multimedia Subsystem. Entity A is closer to the originating party than entity B or is the originating party. If entity A is the originating UE, it generates signalling messages, otherwise the entity responds to messages received from entities toward the originating and terminating parties.

Ten signalling messages in a typical exchange are shown in Figure 7.25 using descriptive names found in [13] and the corresponding SIP methods used in implementation as described in [15]. Messages are grouped into functional entity actions shown by the shaded blocks. Negotiation of the media details follows an offer–response approach.

1–3. The Invite message implemented using the SIP `INVITE` method carries the Initial SDP Offer, proposing the media details to the terminating party. The SIP 100 `Trying` message provides reassurance on a hop-by-hop basis. The response to the SDP offer is returned in the Offer Response message implemented as a SIP 183 `Session Progress` response. Entity B may invoke logic appropriate to its role and service data. For example, when an S-CSCF receives an `INVITE` message, it invokes a filtering process using data retrieved from the HSS to determine whether to direct the `INVITE` to the Application Server or the next CSCF toward a called party.

4, 5. The Response Confirmation, implemented as a SIP Provisional Acknowledge or `PRACK` method, carries the originating party's confirmation of the response to the original media offer. The distant party acknowledges the confirmation with a Confirm Acknowledge message, implemented using a SIP 200 `OK` response.

6, 7. Resource reservation is carried out in both UEs. The forward Resource Confirm message (SIP `UPDATE` request) and reverse direction Resource Confirmation (SIP 200 `OK`) convey confirmation of reservation.

8. The SIP 180 `Ringing` response indicates that the terminating UE is alerting its user.

9, 10. The terminating UE notifies its readiness to participate in the media session by sending a 200 `OK` SIP response. The originating party sends an ACK message to acknowledge the establishment of the session.

The typical SIP signalling horizontal sequence is used in Figures 7.24 and 7.27 in abstracted form.

Mobile Originating IMS Use Case

Figure 7.26 illustrates two aspects of IMS signalling. First, the detailed interaction between IMS elements is shown. Second, call session signalling must link with the control of the underlying network to authorise and commit resources to ensure the specified QoS. One network capable of meeting the QoS requirements is GPRS. Signalling to set up the necessary PDP contexts in an underlying GPRS network is shown in Figure 7.26.

The message sequence shown in Figure 7.26 may start once the originating mobile station has attached to the SGSN.

1, 2. Prior to issuing any SIP messages, the UE initiates the creation of a PDP context that will carry the IMS signalling messages. All application layer packets are routed through the GGSN to the P-CSCF.

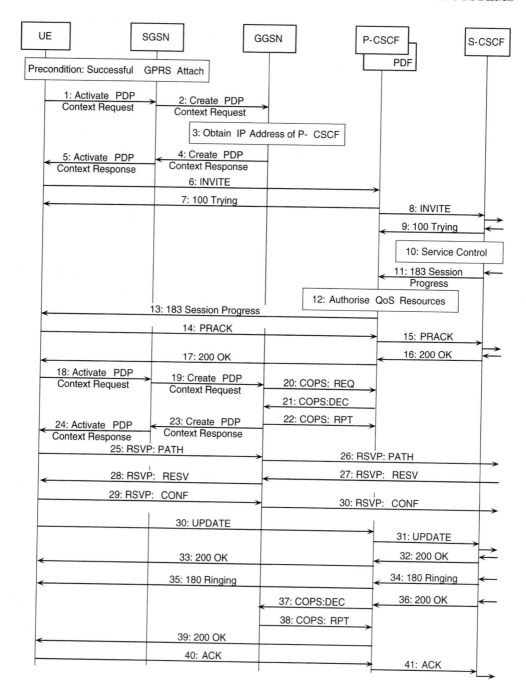

Figure 7.26. Mobile origination signalling in IMS with GPRS providing transport.

3. The GGSN obtains the IP address of the P-CSCF by a method implemented by the network, for example DNS.

4, 5. The response and acceptance of the PDP context are returned to the UE. The IP address of the P-CSCF is carried by these messages to the originating UE.

6, 7. The UE directs a SIP INVITE message to the P-CSCF. The proposed session description is carried in this message. The 100 Trying response reassures the UE that the INVITE is being processed.

8, 9. The INVITE is forwarded to the Serving CSCF. The I-CSCF is not shown.

10. The S-CSCF performs service control, including determining the next entity on the signalling path from filter conditions and call data. Retrieval of this data from the HSS is not shown. Messages and responses beyond the S-CSCF are not shown.

11. After receiving a 183 Session Progress response carrying a response to the SDP offer from an entity not shown, the response is forwarded to the P-CSCF. Media to be used are now agreed.

12. The P-CSCF, aided by the associated PDF, authorises the resources needed to fulfill the required QoS for the agreed media. A Media Authorisation Token is issued.

13. A 183 Session Progress Response is sent to the UE. The Media Authorisation Token is carried in a header. The session path is now known.

14–17. The Provisional Acknowledgement (PRACK) and the 200 OK response confirm that preconditions to the session, namely that the use of resources must be authorised at both ends, have been satisfied.

18, 19. The UE requests the activation of a PDP context for the media streams.

20–22. The GGSN is the Policy Enforcement Point. The GGSN sends the COPS protocol REQ message to the P-CSCF for the PDF. The PDF verifies the Media Authorisation Token and other data. The COPS DEC message conveys the policy decision to the GGSN. The COPS RPT message confirms receipt of the decision.

23, 24. Having checked that its resources are adequate, the GGSN responds to the request to create a PDP context. Messages 23–30 are asynchronous with the SIP messages but must start after message 13 and complete before the UPDATE is emitted in message 30.

25–30. The Resource Reservation Protocol (RSVP) is used to reserve end-to-end resources for the media in the forward direction. Not shown are the RSVP messages for media in the reverse direction sent by the terminating UE between receipt of the PRACK related to message 15 and the issuing of an OK that leads to message 32.

30–33. Using the session path established during transfer of the initial INVITE, this exchange of messages confirms the reservation of resources for the uplink and downlink directions.

34, 35. The SIP 180 Ringing response indicates the distant party's alerting state.

36. This 200 OK response relates to the message 3: INVITE.

37, 38. The COPS messages enable the use of the reserved resources, allowing the media to flow.

39. The 200 OK is sent to the UE.

40, 41. The setup is completed by the ACK sent to the terminating party via the signalling path.

Terminating Use Cases

Figure 7.27 show the signalling path for the four terminating use cases identified in Figure 7.23. Diagram (a) shows the case of both mobile originating and terminating home networks based on IMS. If the mobile terminating party is roaming, the I-CSCF hides the detail of the home network from the terminating network. The I-CSCF is not needed if the terminating MS is in its home network. The Serving CSCF is in the terminating home network. The S-CSCF uses HSS data for location information, the P-CSCF address, and filter data. All nodes are SIP-based and the typical signalling interaction occurs between pairs of nodes.

Figure 7.27(b) shows the configuration for the terminating party in a switched circuit mobile network. The mobile user is registered for CS services. The terminating network is different to the IMS network. The S-CSCF invokes the assistance of the BGCF. The BGCF determines the point of breakout into the terminating network and forwards signalling to the MGCF.

Figure 7.27(c) shows the case of a PSTN terminating party with an originating party served by an IMS network. IMS signalling terminates on the Media Gateway Control Function.

7.5.6   VALUE-ADDED IM SERVICES

Applications that are hosted by and execute on a service platform may be invoked by a request from the S-CSCF or may be initiated by another mechanism. Applications may issue session control signalling via the S-CSCF and may therefore influence the multimedia session. Applications support the requests by a UE to register for notification of events. Applications may make use of user data stored in the HSS.

The range of possible applications is large. Examples of applications that could be supported are listed in Table 7.5.

The ISC Reference Point

The IMS standards provide for three service platforms on which applications execute: a SIP Server, a CAMEL Service Environment and an Open Service Access application server, shown in Figure 7.19.

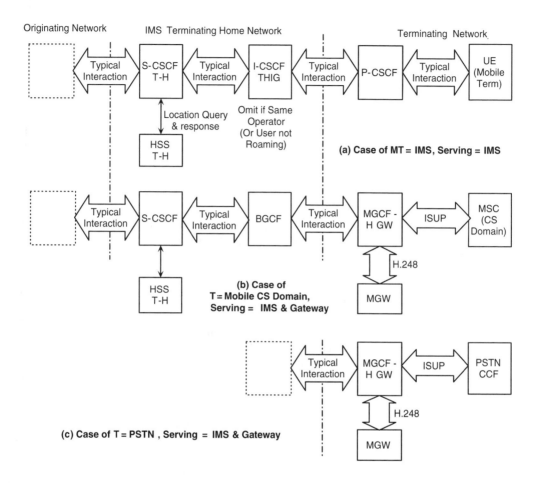

Figure 7.27. IMS originating use cases for mobile and PSTN terminating users.

Table 7.5. Benchmark applications for IMS

Selective diversion to mailbox of specified media
Change of mode in mid-session
Call completion to busy subscriber
Call completion at discretion of called party
Gaming with voice and video channel
Teleworking
Whiteboarding

The IP Multimedia Subsystem Service Control Interface (ISC) reference point is defined in [13]. The protocol used to implement the ISC reference point is SIP. Signalling is necessary for both multimedia session control assisted by the AS as well as allowing users to register with the AS to receive notifications of events. For example, in a call completion to a busy subscriber application, notification of the specified called party leaving a current session may be requested.

The ISC based on SIP must be used for all three types of service platforms. The three platforms differ in their normal interface to the network and their constraints and capabilities. The SIP Application Server is geared to receiving and emitting SIP signalling and has specific behaviours: redirect, proxy, registrar and B2BUA servers. The SIP Application Server may access user data in the HSS through the Sh reference point implemented using Diameter.

The CSE in contrast uses the CAP protocol [18]. Adaptation between SIP and CAP is therefore required. The CAP protocol supports multiparty (but not multimedia) CS services, GPRS and SMS. The CAMEL service environment expects to interact with basic service control represented by state machines at predefined detection points. The gsmSSF functional entity meets these requirements. An adaptation element, the IM-SSF described below, is therefore required.

The Open Service Access architecture defines an open API that abstracts the network detail. An application hosted on the OSA application server can access network functionality such as making connections, sending messages and interrogating network data. This API supports multimedia, multiparty calls, user interactions, mobility management and data session connections using GPRS. The API assumes that the necessary service capabilities are incorporated in an OSA *service capability server* (SCS), defined fully in Chapter 8. A set of such servers is contained in an OSA gateway implementing different functionality including call control, user interaction, messaging and mobility management. Adaptation of SIP signalling to the internal operations and data of the service capability servers is required.

CAMEL Service Environment

The relationships between the CAMEL Service Environment and the IMS functional entities are shown in the functional architecture in Figure 7.28. The gsmSCF is the functional entity that hosts service logic programmes. When CSE services are invoked from an IMS-based core network as shown in Figure 7.19 the SIP invocations across the ISC interface must be adapted to CAP operations at the gsmSCF interface. The functional grouping responsible for this adaptation is the IP Multimedia Service Switching Function (IM-SSF).

The IM-SSF is divided into two sets of functions communicating via an internal interface [5]. The lower group receives and emits SIP requests and responses across the ISC interface to and from mobile terminals. The lower group is specialised for different uses: mobile originating call, mobile terminating call and registration. The upper part contains detection point processing and a CAP interface to the gsmSSF. An internal interface is defined between the two sections. The internal interface is not unlike the interface between the MSC and the gsmSSF shown in Figure 7.12.

SIP messages are processed to determine the mapping to detection points in the IM-BCSM. For example, an INVITE received from a mobile originating terminal, examined

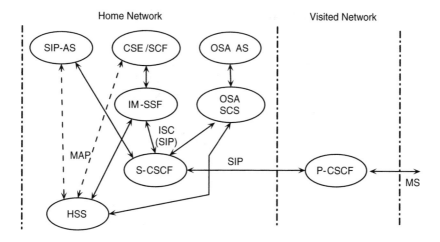

Figure 7.28. Functional architecture for accessing the application servers in the IMS context.

against the data in the CSI, corresponds to the *Collected_Information* detection point. Not all detection points correspond to a SIP message, for example arrival at the *Analysed_Information* detection point results from internal processes rather than SIP signalling. A notification is passed across the internal interface to the upper part of the IM-SSF, designated the IM CN Service Switching Function (imcnSSF).

The IM-SSF, unlike the gsmSSF, has an interface with the HSS (Si) and communicates using the MAP protocol. On starting an association with a call, the IM-SSF retrieves the IM-CSI data for the subscriber from the HSS. The detection points specified in this data are then armed.

The BCSMs are used to keep track of the flow of processing and events in the imcnSSF. The BCSMs need not be explicitly implemented but guide the specification of the logic of the imcnSSF. Detection point processing is built into the imcnSSF logic. The IM-BCSMs are simpler than those for the circuit-switched case shown in Figure 7.11. In the originating case, the Routing and Alerting PICs are merged. No mid-call or change of position detection points are available. Similarly, the terminating BCSM merges the Terminating Processing and Alerting PICs and has no mid-call or change of position detection points.

When an armed detection point is encountered, an InitialDP operation is passed to the gsmSCF. CAP operations received from the gsmSCF are converted to internal messages and passed to the lower part of the IM-SSF. Internal messages are converted to SIP requests or responses as required. For example, a CAP Connect message issued by the gsmSSF is converted to a SIP INVITE directed to the party with a specified destination address.

7.5.7   THE 3GPP2 ALL-IP APPROACH TO 3G NETWORKS

The ITU's International Mobile Telecommunications (IMT-2000) initiative seeks to develop global standards for 3G wireless communication systems integrating a variety of access networks, both terrestrial and satellite. A set of ITU Recommendations would ultimately define a single, worldwide system. Practical realisation of 3G standards has been shaped by two factors. First, a partnership project model was adopted as a speedier means of developing the standards relative to the normal ITU process. In a partnership project, a limited number of bodies having the resources to expedite the standardisation process co-operate to produce the standards. Second, regional needs and preferences potentially hamper a single global process. Thus, two partnership projects emerged. The 3GPP was started with a view to evolving from the base of GSM to standards for a third generation mobile communication system. The 3G Partnership Project 2 (3GPP2) is a parallel project aimed at evolving the ANSI second generation mobile communication systems to the third generation. The 3GPP2 set the objective of an *all-IP network* at an early stage. The description all-IP indicates the intention to base the network on both an IP network and IETF protocols including SIP and Diameter. The intention is to converge the all-IP network standards with the IMS.

The all-IP network is essentially a core network that is expected to support different access network technologies, including radio access networks. The objectives are similar to those of the 3GPP standards. The network must provide levels of reliability and quality of service at least as good as in legacy networks. A range of terminal types must be supported. The transport network must support the interoperation of IPv4 and IPv6 and provide for migration to IPv6.

Most of the functional entities in the 3GPP2 architecture are shown in Figure 7.29, arranged by the NGN Framework layers. The SCF layer, with the characteristic chain of CSCF elements, is similar to those of the IMS shown in Figure 7.19. The Application Layer also reflects the SIP Server and OSA service platforms. The all-IP architecture does not support the CAMEL Service Environment.

Entities allocated to the Resource Control Functional Layer differ from the IMS HSS by having explicit user authentication, access authorisation and accounting (AAA) functional entities. Reference points Sh and Cx are as in IMS.

The all-IP network incorporates elements necessary to implement *Mobile IP*: a protocol that enables a mobile device user identified by a permanent IP address to move from one network to another [169]. The *Home Agent*, located in the home network, is a router that maintains information about the location of the user. The Home Agent receives datagrams destined for the user and transmits these through a tunnel set up to the Foreign Agent (FA). The *Foreign Agent* is located in the network visited by the user. The FA receives the tunnelled packets from the Home Agent and delivers them to the user. Mobile IP thus supports user mobility at the network layer. Mobile IP is implemented as a Diameter application: mobile IP commands are defined for transport by Diameter.

Figure 7.29 shows a cdma2000 access network consisting of a BSC and BTSs. Other access networks are possible: digital subscriber loop, cable networks and wireless LANs. The point of access to the core network contains an Access Gateway, providing both access control through interactions with the AAA elements and allocation of

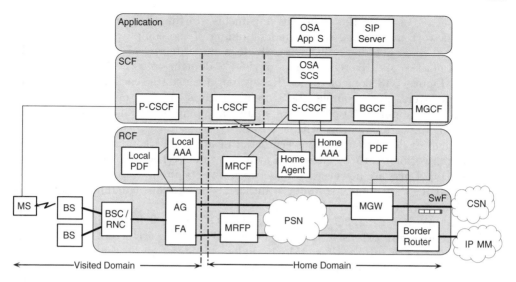

Figure 7.29. All-IP network architecture of the 3GPP2.

resources. The Access Gateway works with the PDF to manage resources to achieve QoS.

7.6 CONCLUSION

The third generation mobile communications system represents a stage in the evolution of mobile systems. Third generation systems exhibit the leading characteristics of a next generation network: they interwork with switched circuit access and core networks; the multiservice core transport network is based on packet switching and provides carrier-grade QoS; the softswitch principle is applied with call and session control decoupled from the transport network; value adding applications are accommodated on a separate service platform; and, while the current options are limited, the principle of user access via a variety of access networks has been adopted.

The treatment of third generation networks is based mainly on the 3GPP standards. The circuit-switched domain supports interworking with legacy switched circuit access networks, for example GSM BSS with A-interface and PSTN core networks. GPRS, already added in the second generation provides packet bearer communication that is secure and quality assured. The IP Multimedia System brings the benefits of protocols such as SIP, COPS, Diameter and RSVP to mobile systems, supporting the move from voice to multimedia communications. The IMS represents a significant convergence between GSM-based mobile networks and multimedia communications supported by IETF protocols. In the process, the canonical protocols such as SIP are applied in a defined architectural environment with profiles appropriate to a carrier-grade network.

The 3GPP standards espouse layering [1], identifying access and connectivity as a layer, service enablers as another and applications/services as a third. This layering is not however pervasive in the standards documents. In this chapter, the layers of the NGN framework are applied to aid understanding of the complex 3G architectures. Four layered diagrams depict 3G networks in a uniform way. Figure 7.6 maps the functional entities and reference points of the CS domain onto the NGN framework. Similarly, Figure 7.13 depicts the GPRS system against the backdrop of the NGN Framework layers. The complex IMS architecture is mapped onto the framework in Figure 7.19. Figure 7.29 performs a similar mapping of the 3GPP2 architecture, bringing out the commonalities with the IMS.

This chapter does not explore every detail of 3G mobile communication systems such as interworking with wireless local area networks (WLAN) as IP-capable access networks [14]. Links are however made between aspects often treated in isolation. For example, GPRS PDP contexts, resource authorisation, session signalling, and resource reservations are integrated in Figure 7.26.

At the time of writing, 3G networks based on GPRS are well established and IMS-based systems are in extensive trials. As ever, ICT system architects and standards writers look to the future. The IMS is under consideration as the basis for fixed networks, a logical development as interworking between fixed and mobile networks using a single protocol is an important enabler of multimedia services across both types of networks. Similarly, the as yet undefined concept of the fourth generation communication system must, in due course, unfold from the base of 3G networks. These topics are taken up in Chapter 10.

Chapter 8

Opening the Network using Application Programming Interfaces

The NGN Framework focusses attention on three principal aspects of communication systems: a multiservice transport network, generic service control functionality and on applications that interact with communications services. The third generation communication systems depicted in Figures 7.19 and 7.29 illustrate these aspects as system layers. The separation of applications from generic functionality and the provision of an open interface was argued in Section 5.4 to provide an opportunity for convergence between enterprise applications and enterprise networks. In this chapter, we describe the extension of this concept to enable convergence between information technology (IT) applications and public network services. Convergence results from moving from a closed to an open network.

8.1 CLOSED NETWORK EVOLUTION

Convergence with computer applications emerges from a long evolution of telecommunication services. A *basic bearer service* underlies all enhanced or value-added services. The basic bearer service transports user data in allocated channels in switched circuit networks, as shown in Figure 8.1(a), or as frames in a packet-switched network. Call control signalling is exchanged between terminals and call control entities in PSTN switches or softswiches.

The narrowband ISDN introduced two concepts for enhancing basic bearer services, namely teleservices and supplementary services. Figure 8.1(b) illustrates the *teleservice* paradigm. The network provides only basic bearer services. Logic is hosted in intelligent terminals that are able to signal to each other at an application level. In the case of *supplementary services* in switched circuit networks shown in Figure 8.1(c), enhanced logic is hosted in PSTN switches. Supplementary services are accessed by messages that are part of the call signalling protocol, for example the Facility message in Q.931. The supplementary services concept also exists in H.323 packet multimedia communications. Service logic may be hosted in a terminal or network node.

The *classical Intelligent Network* (IN) introduced the principle of hosting service logic on a computing node separate from the switches, the Service Control Point (SCP). The IN-SCP communicates with the switches through a service-oriented protocol, INAP or CAP, as shown in Figure 8.2(a).

Several objectives of service creation were formulated during the development of the Intelligent Network. Telcos wished to minimise the time-to-market for a new service: the time from conceiving a new service to revenue-earning deployment. They wanted

---

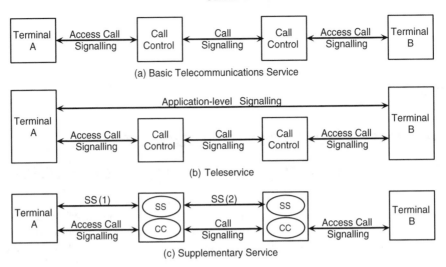

Figure 8.1. Interaction diagrams for switch and softswitch-based service paradigms.

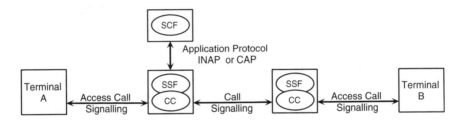

Figure 8.2. Interaction diagrams for IN service paradigms.

to gain competitive advantage by differentiating their service offerings from others in the marketplace. The objective of the IN was therefore to standardise the underlying capability in a way that permits a range of innovative, differentiated services to be offered.

The emphasis of IN falls on adding value to basic bearer communications in the PSTN [171]. The user terminal is, in general, a plain telephone and interactions with users take place through subscriber loop signalling. Most services are initiated by triggering on a network event such as a particular number being dialled or a switch being unable to complete a call. This mode of operation is encapsulated in the terms *network event-triggered service* or *application-enhanced communications*.

A second property of IN is its closed business model. The service control point is generally part of the telco's infrastructure and the telco owns the service logic programmes and data. The INAP interface to the SCP is inside the telco domain and relies on the SS7 network. This model does not support independent service providers. The service model is described as *closed*: only the telco can offer value-added services. This mode of offering services is also termed *local*, that is provisioned within the secure domain of the telco. In addition, service creation the IN is based on an arcane software reuse method, namely service creation using Service Independent Building-blocks (SIB). Consequently, only a limited, specialised band of programmers is available for development of IN-services.

Convergence between IT applications and public telecommunications services requires the network to be opened, allowing applications in an external domain to invoke network functionality in the secure telco domain. This chapter deals with the principles of open networks and particular standards for achieving this goal. Section 8.2 describes the properties of an open network.

8.2 OPENING THE NETWORK

The purpose of opening the network is to allow applications, for example in an enterprise domain, to invoke communications capabilities in a public network. Capabilities include voice and multimedia calls, messaging, accessing content, mailbox usage, and access to network data such as user status, location and account information. Access to network capabilities must be secure, disciplined and billable. An open interface is the enabler of such an open service creation and provisioning environment.

An *open interface* is defined in a generally accepted standard. The interface abstracts the detail of the underlying network service capabilities and data. An open interface is preferably defined in an implementation-independent form and is implementable in a distributed computing environment. The interface occurs at the boundary between administrative domains and definition should take security into account. These requirements indicate that the interface definition must be an applications programming interface rather than an applications protocol.

Several motivations for opening the network are advanced. First, telcos seek to increase traffic on their networks, and hence revenue, by offering attractive services. Rather than develop these services in the closed IN-environment, a strategy is to encourage independent software developers to develop and deploy applications that enhance communications. Second, the developers of information services, seek to use network connections and messaging to enhance their IT applications, for example transaction notification messages in a banking application. These *communication-enhanced applications* are new sources of revenue for the telco in addition to network-event triggered services.

Several initiatives sought to open the network. In the private domain, the CSTA and TAPI interfaces described in Chapter 5 allow programmers working in an IT paradigm to control telephony services in an enterprise environment, including initiating calls and sending messages. The TINA architecture described in Chapter 6 enables service providers in different administrative domains to work together to provide services. In particular, a third party service provider business role is supported.

The PINT and SPIRITS initiatives are specific initiatives to open PSTN Service Control Points to Internet clients and servers. Definitions are provided for interfaces between hosts in an Internet domain and the SCP in a PSTN as shown in Figure 8.3. The PINT standard [170] defines methods that allow the initiation of third party calls in a PSTN by an Internet client, for example to initiate a PSTN call from an Internet

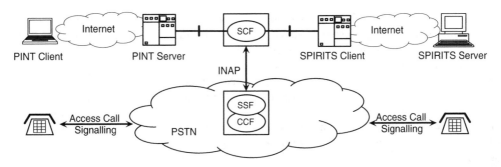

Figure 8.3. Internet–PSTN convergence through PINT and SPIRITS protocols.

application. SPIRITS [103] has methods that allow a PSTN SCP to notify an Internet party of an event, for example a call not completed because the called party is busy. Both PINT and SPIRITS require interfaces to be opened on the IN-SCP.

The Internet is held up as an open environment for service creation. All application logic is hosted on end stations and the network provides basic bearer services only. The number of application developers is large and the range of software paradigms is also large. Inspired by the Internet model, interest grew in opening the telecommunications network to applications 'at the edge of the network'. The Internet however does not have services such as call control or messaging, nor does it keep user data; these are embodied in end-stations. While the Internet provides the inspiration for an open network, the model for opening the telco networks must take account of the service capabilities in the network. A new type of secure network edge is opened to allow service providers to access those capabilities in the network provider domain.

Several approaches to achieving programmability in networks exist [141]. In this chapter, we concentrate on a particular approach to opening the network by providing an interface between the generic service and application layers, namely *Open Service Access* (OSA) [159, 211]. The 3GPP standards define an Open Service Access architecture [17] that enables application developers outside the mobile telco domain to make use of network functionality and to receive information from the network via a standard interface. An open applications programming interface defines the interface. In this context, network functionality includes call and session control, messaging as well as access to network databases. The attributes of such an API are:

- *Openness:* the interface is defined in a generally accepted standard.

- *Security:* access to use the service interfaces is restricted to authenticated, authorised parties and external service providers may require similar assurance of the authenticity of the network provider.

- *Integrity:* network functionality must not be compromised by the application making excessive numbers of requests and, conversely, an application server must not be overloaded by excessive numbers of network-initiated requests.

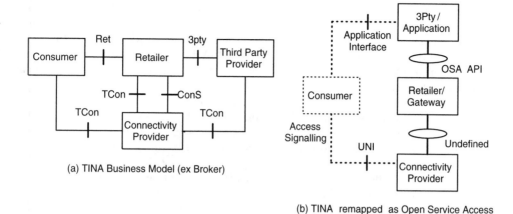

(a) TINA Business Model (ex Broker)

(b) TINA remapped as Open Service Access

Figure 8.4.   (a) TINA business model without broker and (b) remapped to application, SCF and network layers as basis for Parlay.

- *Flexibility:* the interface allows a variety of applications to be supported and the interface implementation must allow a range of bearer networks to be used.

- *Abstraction:* details of the underlying network and its heterogeneity must be hidden from the application programmer: the programmer is not expected to understand the network types and signalling protocols.

- *Technology neutrality:* the API definition is not locked to specific language and distribution technology implementations.

- *Service Discovery:* an application can establish the services available in a serving node.

The Open Service Access API has the seven properties listed. The 3GPP standards define *Open Service Access* as a 'concept for introducing a vendor-independent means for introduction of new services' [2]. A system presenting an interface conforming to these requirement can, in addition, be described as *vendor independent*.

8.2.1 BUSINESS MODELS

Opening the network allows new business models that depart from the historical vertically integrated telco.   The TINA initiative introduced both the concept of business modelling and the specific business model shown in Figure 8.4(a) without an explicit broker function. The TINA business roles are defined in detail in Section 6.4. The third party provider (3Pty) supplies services requested by the consumer via the retailer. TINA predated the NGN layered model. No position was taken in the TINA standards on the permitted or advised division of generic and specific service logic between the retailer and the third party provider. Also the third party reference point was not fully defined.

Figure 8.4(b) transforms the TINA business model (ex broker) to conform with the open service access model and NGN layering. The retailer domain is constrained to contain generic service logic: it retails network capabilities. Network functionality is made available to the application through an OSA interface. The retailer domain is decoupled from the transport network by an interface that is not at present defined but is relegated to an implementation detail.

The retailer reference point in the TINA business model contains powerful multimedia service control methods, implying that part of the service logic is in the consumer domain. In the remapped business model, two possible methods are shown for the user to signal to the network. The first, favoured in the 3G OSA standards, is to signal through a user-to-network interface (UNI) using a call/session protocol such as SIP. Signalling is deflected toward the application if trigger conditions are satisfied. Second, an adaptation of the TINA approach would be for the terminal application to signal to the provider's application. The IMS standards identify an interface, Ut, between the terminal (UE) and an application server for operations such as subscriber group and list management [13].

8.3   THE OSA/PARLAY ARCHITECTURE

The open API described in this section arises from the activities of three bodies. The Parlay Group is a consortium of telcos and vendors with the goal of developing open application programming interfaces to support the development of applications that operate across various networks. The 3GPP, concerned with all aspects of third generation mobile communication systems, adopted the Open Service Access principle for the 3G mobile communication system.   The third body is the ETSI Services and Protocols for Advanced Networks (SPAN) initiative. The three initiatives have converged and a single set of APIs now exist. The specifications are published as ETSI standards for Open Service Access (OSA); Application Programming Interface (API) [66]. The APIs are commonly referred to as Parlay/OSA or OSA/Parlay.[8]

8.3.1 ARCHITECTURAL CONCEPTS

The OSA standard defines an applications programming interface allowing applications to interact with network functionality. An architecture is defined to identify the roleplayers and functional groupings and to locate the constituent interfaces [66]. Figure 8.5 shows the OSA/Parlay architecture.

The OSA/Parlay architecture has two main domains with an open API defining the interdomain boundary. The underlying network and the functionality that implements the open API is in the *network operator* or *telecommunication service provider* domain, as indicated in Figure 8.5. The application that uses the open API is in the enterprise operator domain.   The *enterprise operator* is, in general, an independent service provider that has a business relationship with the network operator. The enterprise operator subscribes to network level services offered by the network operator.

The enterprise operator domain may host one or more application described as *client applications* in relation to the gateway services.

---

[8]The latter form is preferred in this text.

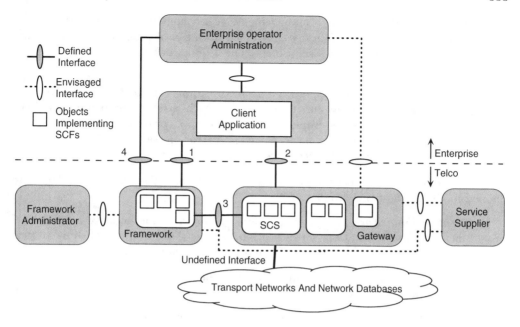

Figure 8.5. OSA/Parlay architecture.

A number of concepts are used to define the OSA/Parlay architecture:

- A *Service Capability Feature* (SCF) is an individual network level function or *capability*, such as setting up a connection, handling a message or interrogating a database, that is accessible by an application via an OSA-standardized interface. A SCF has a number of methods collected in one or more interface. The term service used without qualification normally has the same meaning as SCF in this context. The user of services is termed an application. A SCF is a logical grouping of associated interfaces that provides access to network connectivity, messaging or data.

- A *Service Capability Server* (SCS) is a functional entity that makes OSA-standard interfaces accessible by application. A SCS may host one or more SCF. A SCS may have implied network-facing functionality that is assumed to be present but is undefined in the OSA/Parlay standard. For example, the call control SCFs assume a network-event triggering mechanism while the mobility management SCF relies on mobile network HLR or HSS databases.

- The term *Gateway* is used in OSA/Parlay to express the notion that a Service Capability Server provides the application with access to network functionality. A Gateway is one or more SCS.

- The *Framework* is a form of SCS that provides interfaces to functionality required to support the use of SCFs. Support functionality includes registration of

SCFs designated as service management, access control and authentication of applications, discovery of service control features, establishing service agreements and managing service subscriptions. The Framework also supports integrity management.

- A *client application* or *application logic* or simply *application* is the user of Framework and Gateway functionality. For each SCF, a number of interfaces are defined that must be implemented in the application domain to allow the SCF to return results, send notifications and error messages in a consistent way. These interfaces are termed *callback interfaces*.

A number of interface locations are shown in Figure 8.5. Four sets of interfaces are defined in version 5 of the standard [74]:

1. *Client to Framework Interface*: this interface provides methods for a number of functions: authentication, authorisation, service discovery, establishing service agreements, access control to SCFs and integrity management.

2. *Client to Gateway Interface*: up to 15 different SCFs that may be used by the client application to access network functionality.

3. *SCS to Framework Interface*: registration of SCF managers with framework, operational management (load management, monitoring, and fault management).

4. *Enterprise to Framework interface*: service subscription.

Undefined interfaces shown in Figure 8.5 acknowledge that management functions are required for both the Framework and the Gateway.

8.3.2 API DEFINITION METHOD AND TECHNOLOGY INDEPENDENCE

The OSA/Parlay standard is essentially an API definition. The 15-part standard consists of an architectural overview [66], definitions of common data types [73], the Framework definition [74] and individual SCF definitions. At the heart of the standard are the interfaces belonging to the Framework and various SCFs, the methods available on these interfaces and supporting data definitions. Interfaces are designed in a technology-neutral, object-oriented form using the Unified Modelling Language (UML).

Class diagrams for the interfaces, with detailed definitions of the methods contained in each interface, provide the formal definitions of interfaces. A number of use cases expressed as sequence diagrams describe typical use of the Framework and SCFs by applications. State transition diagrams are given using UML state concepts.

Several distribution and language technology realisations of the interfaces are possible. Three routes to implementation of the interfaces are shown in Figure 8.6. The OSA standards provide OMG Interface Definition Language (IDL) definitions of each interface and data types that map exactly from the UML definitions. The IDL definition is well suited to a distributed implementation using CORBA and languages having IDL mappings.

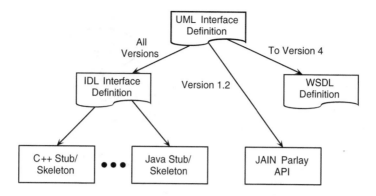

Figure 8.6. Hierarchy of definitions in OSA/Parlay and JAIN Parlay.

The SCF definitions also provide as annexures mapping of the interfaces and data types into Web Services Definition Language (WSDL). These mappings have informative status [66].

Java language mappings are possible but rely on a constraint applied on the UML method definition. The method call parameters are restricted to in-parameters: out-parameters are not compatible with Java syntax. Two mapping routes exist. First, a distributed Java realisation may be obtained using IDL mapping in a CORBA environment. Second, a direct mapping to a Java language realisation shown in Figure 8.6 is the JAIN Service Provider API (SPA), described in Section 8.9.1 for use in a Java technology environment. This mapping is also regarded as informative in the OSA standards. It is described as a local technology realisation, that is not inherently distributed.

8.4 FRAMEWORK INTERFACES AND USE CASES

The interface used for accessing network service features must be protected from unauthorised access and from misuse. The OSA/Parlay Framework [74] provides the mechanism that allows application providers to access network services securely and efficiently. The Framework requires knowledge of the SCFs available in the gateway and therefore also interacts with the Gateway. We describe the functions of the Framework through a number of use cases identified in Figure 8.7. The actors shown are defined in the architecture in Figure 8.5.

8.4.1 USE CASE: PROVISIONING A GATEWAY SERVICE

The first use case shows the installation of a service in the Gateway. To provide a context for this use case, assume that an interactive voice response (IVR) unit has been provisioned for use under the control of client applications. Access to control the IVR by an application is supported by a SCF with the **IpUIManager** interface described in Section 8.6.6. There must one such manager for each application using the service. Each manager can control a number of simultaneous connections and user interactions associated with different calls. An essential step is the installation

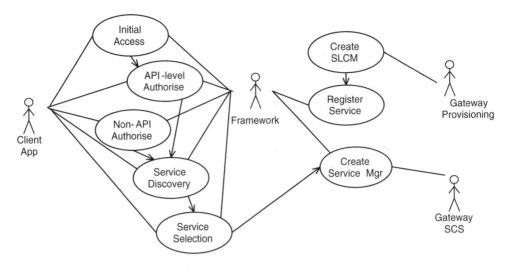

Figure 8.7. Framework use cases.

of a service lifecycle manager (SLCM) object in the Gateway that can create objects implementing the **IpUIManager** interface. Other than one interface, the details of the service lifecycle manager are implementation dependent.

Figure 8.8 shows the *Create Service Lifecycle Manager* and *Register Service* use cases.

1. A management system, the Service Supplier in Figure 8.5, represented by the SCS Manager entity instantiates the service lifecycle manager. The **new()** is an implementation-dependent method for creating an object and returning a reference to the created object.

2. The SCS Manager uses a Framework interface to authenticate itself using a method agreed between the two entities. Details of authentication are not shown.

3. The service lifecycle manager now registers the service with the Framework. The **registerService** method has two parameters: **serviceTypeName** identifying the service by a standardised name and the service attributes in parameter **servicePropertyList**.

4. An identifier for this service is generated by the Framework and is returned as **serviceID**. This identifier is significant between the Framework and the Gateway.

5. The service lifecycle manager invokes the **announceServiceAvailability** method, identifying the service by **serviceID** and giving the Framework a reference to itself.

The actual service manager is not instantiated at this stage. This use case occurs solely during provisioning, for example, initial installation of or software upgrade to a SCF. The service is now registered with the Framework and can be discovered by an authorised application.

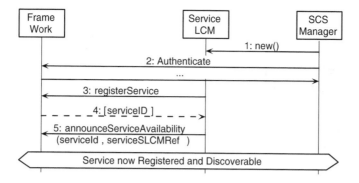

Figure 8.8. Creation of a service lifecycle manager, and registration with Framework.

8.4.2 USE CASES: AN APPLICATION GAINS ACCESS TO A SERVICE MANAGER

Figure 8.9 shows a series of interactions that take place after an application is started and needs to access a service in the Gateway. The use cases are *Initial Access* and *API-level Authentication* shown in Figure 8.7.

1. The application initiates the authentication process by invocation on a Framework interface that is already known by the application.

2. An agreed authentication method (not shown in detail) is used. Authentication may be two-way: the application may require the Framework to authenticate itself. Authentication may be at API level, that is, carried out using API interfaces or may be by another method, for example, using authentication within CORBA.

The application is now able to access other Framework interfaces.

3. The application uses a requestAccess method that provides a reference to itself to the Framework. The framework returns a reference to the interface to be used by the application.

4. The application and Framework agree on an algorithm to be used for signed exchanges.

The *Service Discovery* use case follows. Only step 8 is required if the application already knows the serviceTypeName and servicePropertyList.

5. The application uses the obtainInterface method to request a reference to an interface on the Framework that can be used to discover and gain access to Gateway services.

6. The application may not know the services available and may request a list of service types by invoking listServiceTypes.

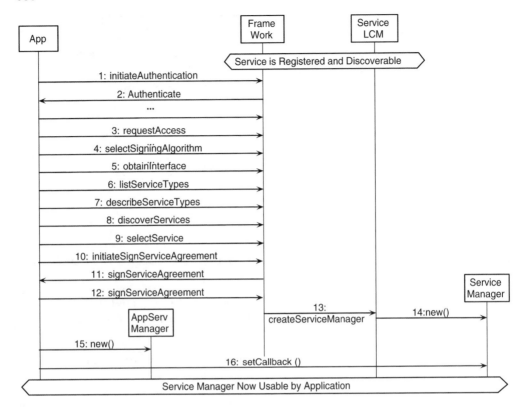

Figure 8.9. Registration of a service manager, discovery and usage by an application.

7. The application may need to examine the leading properties of selected services using describeServiceTypes.

8. The core operation, discoverServices informs the Framework of the required service using parameters serviceTypeName and servicePropertyList as well as the maximum number of matches the application wishes to receive. The Framework returns a list of SCFs meeting the requirements and their service properties.

The *Service Selection* use case follows. The application selects a particular service manager SCF to be used.

9. Using the selectService method, the application informs the Framework of the ServiceID of the SCF it wishes to select. The Framework provides a token that is private to the application.

10–12. The application and Framework sign a service agreement electronically. This agreement is made in terms of a prior agreement entered into offline between the network operator and the enterprise operator.

The final use case is the *Creation of a Service Manager* that serves this application.

13. On successful completion of the service agreement, the Framework uses the
    method createServiceManager to request the service lifecycle manager to create
    an instance of the service manager, identified by serviceID, for the requested
    service.

14. The Service Manager is created in the SCS by the service lifecycle manager.
    A reference to the Service Manager is returned to the Framework.

15. The application creates an object implementing the service manager's callback
    interface.

16. The method setCallback conveys the reference to the callback interface to the
    service manager.

The application and the service manager can now communicate. The application is
now able to invoke methods on the service manager in the gateway. The service
manager communicates with the application by invoking methods on the service
manager's callback object created in step 15.

8.5 THE OSA/PARLAY GATEWAY

8.5.1 *STANDARD SERVICE CAPABILITY FEATURES*

The SCFs defined in OSA/Parlay version 5 fall into broad groups. The first group is
concerned with voice, multimedia, conference and data calls, messaging and charging:

- *Generic Call Control SCF*: allows the application to control simple two-party
  calls [76].

- *Multiparty Call Control, Multimedia Call Control*, and *Conference Call Control*
  SCFs: a set of related SCFs that support control of multiparty, multimedia
  and conference calls. Calls and conferences may be created, manipulated or
  released. Call legs may be controlled, for example be attached or detached as
  required [77, 78, 79]. Individual media flows may be controlled.

- *User Interaction SCF*: allows interactive voice response (IVR) functions to be
  used within a simple or multiparty call already in existence [80]. This SCF also
  supports *audiocalls*: an application requests an announcement to be played to
  an end user outside an existing call.

- *Data Session Control SCF*: controls a data connection initiated at network level,
  for example a GPRS connection [83].

- *Charging SCF*: allows an application to perform charging operations for one or
  more users: levy a charge, reserve an amount from user's balance, charge against
  a reserved amount, credit and debit and query credits [69].

Two messaging-oriented SCFs are defined in the OSA/Parlay Standard:

- *Generic Messaging SCF*: is essentially a mailbox service. The application can
  manipulate mailboxes, folders and send and retrieve messages [84]. Messages
  may have various formats: text, binary or audio.

- *Multimedia Messaging SCF*: allows an application to send, receive and store various voicemail or electronic mail messages, within or outside a mailbox context [72].

A third group of SCFs focusses mainly on allowing an application to interact with data about users such as status, location and accounts:

- *Mobility SCF*: allows the application to obtain information about a user's status and location. Three query modes are supported: on demand, when the information changes (triggered) or periodically updated [81].

- *Terminal Capabilities SCF*: allows the application to determine the attributes of a user's terminal [82].

- *Account Management SCF*: allows the application to query account information [68].

- *Presence and Availability Management SCF*: allows a application to manage, retrieve and publish user-related information including identities, communication capabilities, content delivery capabilities, state information, presence and availability of entities for different contexts and communication methods [71].

Two further SCFs provide operations support like functions:

- *Policy Management SCF*: anticipates a need to support policy-enabled services that are served by applications. The SCF interfaces allow policies to be provisioned and compliance of service usage with policies to be evaluated [70].

- *Connectivity Manager SCF*: assuming that the underlying packet network can be configured as a virtual private network, this SCF provides methods that allow the inter-site virtual connections to be configured [67].

8.5.2 *INTERFACE DEFINITION DESIGN PATTERN*

Service Capability Features in the first and second groups of SCFs conform to a common design pattern. Each Service Capability Feature has a service manager interface and may have one or more additional interface. For example, the service manager for the Generic Call Control SCF, has interface IpCallControlManager with methods for controlling notifications from the network and initiating a call using the method createCall. When an application invokes createCall, the manager instantiates an object implementing the interface IpCall to represent the call. This object has methods that allow call leg objects to be created and controlled. Table 8.1 lists the SCFs that use the design pattern. Each has a manager object and one or more objects specific to the service instance.

In general an SCF object in the Gateway must have an object implementing the corresponding *callback interface* that receives responses, error notifications and network event notifications on behalf of the application. Callback interfaces serve to provide a uniform way of returning responses and forwarding network generated notifications to the application. Typically, an SCF interface has request methods, for example routeReq connects a call leg. The related responses are defined as methods on the callback interface, for example routeRes and routeErr.

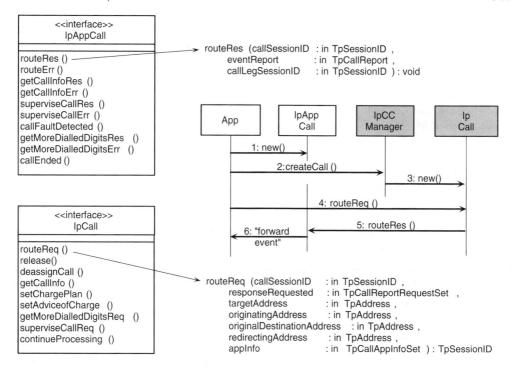

Figure 8.10. Example of Gateway and application callback interface definitions with selected UML request and response method definitions.

Figure 8.10 illustrates the form of the SCF and callback interfaces and the method of definition used. This example is taken from the Generic Call Control SCF. A call control service manager object exists in the gateway as a result of the process describe in the Framework use cases. Each call is represented by an object with interface IpCall.[9] The methods available on this interface are shown in the class diagram for IpCall. A callback interface, IpAppCall, as defined in the class diagram, must be implemented in the application domain.

The UML definition of one of the methods of IpCall is listed in Figure 8.10. Method routeReq allows the application to request the network to connect a party specified by an address parameter to a call. *Connect* means that the required call/session signalling is sent. For example, to connect a B-party in a PSTN, an INAP Connect message is sent to the switch to request completing the connection. In an IMS system, an INVITE message is sent to the call party. Successful connection is reported to the application by the routeRes method on the callback interface. Some methods return results through a return parameter, for example routeReq returns a leg identifier of type TpSessionId.

---

[9]Interfaces in OSA/Parlay have names starting with Ip.... Callback interface names start with IpApp.... Data types have names starting with Tp....

Table 8.1. SCFs using manager–instance design pattern with hierarchy of instances

SCF using pattern	Instance-1	Instance-2	Instance-3
Generic Call Control	Call		
MP/MM Call Control	MPCall	Call Leg	
Conference Call Control	Conference	Subconference	Call Leg
User Interaction	UICall/UI		
Data Session	Data Session		
Generic Messaging	Mailbox	Mailbox Folder	Message
Charging	Charging Session		
Multimedia Messaging	Mailbox		

The message sequence chart in Figure 8.10 shows part the sequence required when an application sets up a call.

1. The application creates the callback object implementing the IpAppCall callback interface. An implementation-specific communication mechanism is used between IpAppCall and the application.

2. The application invokes a createCall method on the service manager implementing the IpCallManager interface.

3. The service manager creates an object implementing the IpCall interface. The reference to this object is returned to the application.

4. The application requests the first party to the call to be connected by invoking the routeReq method on the IpCall object.

5. If the connection is successful, the IpCall object invokes a routeRes method on the IpAppcall callback interface.

6. An implementation-specific mechanism is used to forward the information to the application. A notation such as "forward event" is used when a message is not defined in the OSA/Parlay standard.

Steps 4–6 could then be repeated to connect the second party to the call. Use of other methods is illustrated in a set of use cases identified in Figure 8.14.

8.5.3 INTERACTION IN COMMUNICATION-ORIENTED SCFS

Figure 8.11 shows the interaction within the design pattern used in the group of OSA/Parlay SCFs listed in Table 8.1. The structure of the application logic is not defined by the OSA/Parlay standards other than the callback interface mechanism.

SCFs concerned with call control, user interaction, data sessions, charging and mailbox management have one or more service instance interface classes, defined in a hierarchy shown in Table 8.1. For example a multiparty call has a service manager with interface belonging to class IpMultiPartyCallControlManager. An object with interface defined by class IpMultiPartyCall is instantiated for each call. Each leg of the call is represented by an object with interface defined by class IpCallLeg.

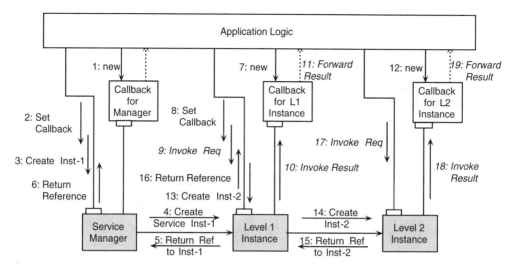

Figure 8.11. Computational viewpoint and principal interactions showing design pattern used in OSA/Parlay.

Figure 8.11 shows the pattern of interactions in setting up an instance of a service, for example a call or a mailbox query. The precondition is the existence of a service manager that has been selected by the application and created by the Framework as described in Framework use case *Select Service*. The application must also create a callback object (1) for the service manager and supply its reference to the Service Manager (2). The objects required for the service instance are created as follows.

3. The application invokes a method on the Service Manager that creates an object for the first-level service instance.

4–6. The Service Manager creates the first-level service instance object and returns the reference to this object to the application. Steps 4 and 5 are implementation dependent.

7. The application creates a callback object for the first-level service instance.

8. The application passes the reference to the callback object to the service instance using the setCallback method.

The application may now invoke a method on the first level service instance (9). The result is returned to the callback interface (10) and forwarded to the application (11).

A similar sequence is used to create a second-level service instance by invoking a method on the first-level service instance. This sequence shows the following steps.

12. The callback object for the second-level instance is created.

13. The application requests the first-level instance object to create the second-level instance object.

14. The first level instance creates the second level instance object.

15, 16. A reference to the new object is returned.

In this case, the callback object is created first and its reference is passed in the method that requests the creation of the gateway object, obviating the need to use setCallback in message 8. This pattern occurs in the multiparty call control use cases described subsequently. A method may now be invoked on the second-level instance (17), with results returned via the callback interface (18–19). A similar sequence allow the creation and use of a third level instance, if one exists, for example a call leg.

8.6   COMMUNICATION-ORIENTATED USE CASES

8.6.1   CALL CONTROL AND USER INTERACTION INTERFACES

OSA/Parlay provides two forms of call control interface. The Generic Call Control SCF supports simple two-party voice calls. Generic Call Control dates from 3GPP Release 99 and is not being developed further by the standards bodies. The three sets of interfaces are defined for control of multiparty (MP), multimedia (MM) and conference (Conf) calls. The MP, MM and Conf interfaces are 'to be considered as the future base call control family' [75]. Closely allied to call control SCFs is the User Interaction (UI) SCF that allows the application to use IVR capabilities.

The three sets of interfaces for control of multiparty, multimedia and conference calls are, in reality a single set of SCFs of increasing capability. The multiparty interfaces provide methods required to set up, modify and clear multiparty calls. The multimedia interfaces inherit all the methods of the multiparty interfaces and therefore forms an enhanced interface that allows the control and manipulation of media within a multiparty call. The conference call control interfaces inherit the methods of the other two interfaces while adding specialised capabilities, for example the ability to split conferences into sub-conferences.

In referring to interfaces in the following use case descriptions, the notation MP refers to the multiparty interface, MM refers to multimedia (including MP capabilities) while Conf indicates conference control (including MP and MM capabilities).

8.6.2   CONCEPT OF A CALL IN OSA PARLAY

Figure 8.12 shows the concepts embodied in the object model for the MP/MM call. A *call* in the OSA/Parlay context is a relationship among zero or more parties to the call. Each call is represented by a *call object*. The call object presents an interface to the application and can invoke methods on and supply notifications to a callback interface.

Each party is represented by an *address object*. A comprehensive set of address types is defined in OSA/Parlay, including E.164 addresses, IP unicast and multicast, URLs and SIP address types. A *call leg* object represents the association between an address and the call object. This relationship develops through the stages of a call. The call object can *create* a call leg object. The associated bearer or media connections are not connected until the leg is *routed*. A call leg that has a connected bearer or media flow is described as *attached to the call*. A *detached leg* has no bearer or media flow. In a multimedia call, a media flow can be allowed or barred. The call

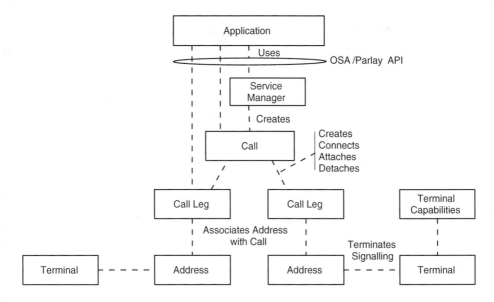

Figure 8.12. Conceptual base for the objects used in call control in OSA/Parlay.

leg object presents an interface to the application. This model is similar to the JTAPI object model shown in Figure 5.14. The SCF objects have an unspecified mechanism that exercises control over network resources, For example, if the application requests that a leg be routed, signalling to initiate a physical connection to the party flows in the underlying network.

Call objects are defined representing different levels of capability. The Generic Call Control SCF, while it uses the concepts of address and call leg, only defines and instantiates a call object described by class lpCall. As the number of legs cannot exceed two, they are represented and controlled within the lpCall object. Multiparty, multimedia and conference control requires objects implementing the lpCallLeg interface to be created to represent call legs.

The structure of the classes comprising the Multiparty Call Control, Multimedia Call Control and Conference Call Control SCFs is outlined in Figure 8.13. Methods available on each interface are listed in Table 8.2. Horizontal lines demarcate methods added by the inheritance relationships. Each set of interfaces conforms to the design pattern described in Figure 8.11. An object implementing the lpMultiPartyCallControlManager service manger interface exists in the Gateway for every application with a signed service agreement. The callback interface is defined by interface class lpAppMultiPartyCallControlManager. Each multiparty call is represented by an object implementing the lpMultiPartyCall interface. The related callback interface class is lpAppMultiPartyCall. Objects representing call legs in the gateway are defined by interface class lpCall, with callback interface class IPAppCall. As shown in Figure 8.13, one call manager can handle more than one call and each call can have a number of legs.

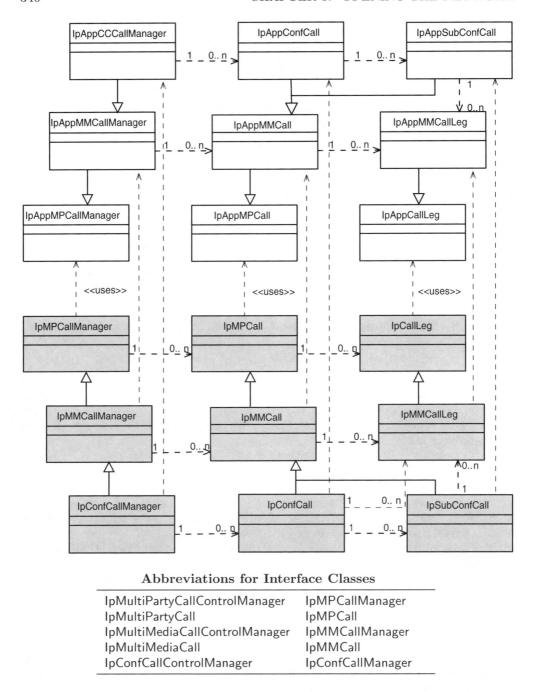

Figure 8.13. Multiparty, Multimedia and Conference Call Control SCF interface class diagram. Shaded classes are in the Gateway.

Table 8.2. Multiparty, Multimedia and Conference Call Control SCF interface methods

Manager Callback	Call/Conf Callback	Leg Callback
reportNotification	getInfoRes	eventReportRes
callAborted	getInfoErr	eventReportErr
managerInterrupted	superviseRes	attachMediaRes
managerResumed	superviseErr	attachMediaErr
callOverloadEncountered	callEnded	detachMediaRes
callOverloadCeased	createAndRouteCallLegErr	detachMediaErr
abortMultipleCalls	superviseVolumeRes	getInfoRes
reportMediaNotification	superviseVolumeRes	routeErr
conferenceCreated	partyJoined	callLegEnded
		superviseRes
		superviseErr
		mediaStreamMonitorRes
Manager SCF	**Call SCF**	**Call Leg SCF**
createCall	getCallLegs	routeReq
createNotification	createCallLeg	eventReportReq
destroyNotification	createAndRouteCallLegReq	release
changeNotification	release	getInfoReq
setCallLoadControl	deassignCall	getCall
enableNotifications	getInfoReq	attachMediaReq
disableNotifications	setChargePlan	detachMediaReq
getNextNotification	setAdviceOfCharge	getCurrentDestinationAddress
createMediaNotification	superviseReq	continueProcessing
destroyMediaNotification	superviseVolumeReq	setChargePlan
changeMediaNotification	getSubConferences	setAdviceOfCharge
getMediaNotification	createSubConference	superviseReq
createConference	leaveMonitorReq	deassign
checkResources	getConferenceAddress	getProperties
reserveResources	splitSubConference	setProperties
freeResources	mergeSubConference	mediaStreamAllow
	moveCallLeg	mediaStreamMonitorReq
	inspectVideo	getMediaStreams
	inspectVideoCancel	
	appointSpeaker	
	chairSelection	
	changeConferencePolicy	

The multimedia call control manager, call and call leg classes inherit from their multiparty counterparts. The conference manager and call interfaces are defined similarly, inheriting from the multimedia counterparts. The Conference Control SCF introduces a new class, the subconference, to support the ability to divide a conference into subconferences. Call legs are associated with a conference or subconference and are as defined for the multiparty, multimedia cases.

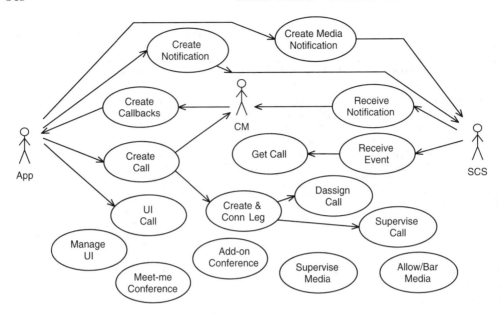

Figure 8.14. Multiparty/multimedia call control use cases.

8.6.3 INTERFACES AND INTERACTIONS

The Multimedia, Multiparty and Conference Call Control SCFs are defined in parts 4-3 to 4-5 of the OSA/Parlay standard [77, 78, 79]. Interfaces are comprehensive and are explained through a number of high-level use cases at the level of self-contained applications, for example an application initiated call, prepaid service, call barring, number translation, media flow enabling or barring, volume-based metering, meet-me conferencing and add-on conferencing. These high level examples illustrate the use of most methods available on these interfaces. We adopt a different approach to understanding the capabilities of the three API interfaces. Simple use cases are extracted from the comprehensive service description in the standards. The use cases are identified in Figure 8.14. Most full services can be synthesised from the elementary use cases. A number of elementary use cases are described in the following sections.

8.6.4 NETWORK EVENT DETECTION AND REPORTING

The OSA/Parlay standard defines *event types* rather detection points as in IN. The set of event types defined in [77] is depicted in Figure 8.15. Pathways between detection points are illustrative and not defined as in the IN BCSMs. While the set of events is evocative of the IN CS-2 trigger detection points, the OSA/Parlay standard cautions against attempting a one-to-one mapping [77]. The standard is silent on how individual signalling protocols interact with the set of events. Guidance given by 3GPP documents is reviewed in Section 8.8.1.

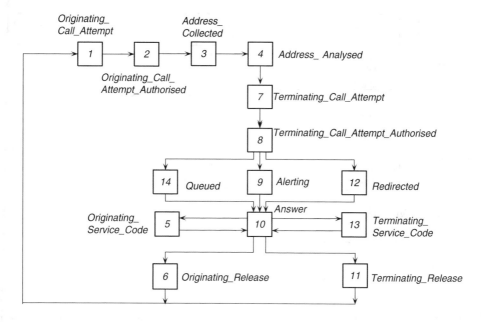

Figure 8.15. OSA/Parlay events depicted as detection points.

The Service Capability Server that accommodates call control SCFs interacts with network signalling by an unspecified mechanism. The SCS is assumed to have an event detection point mechanism that is not unlike triggering in the IN CS-2 Basic Call State model. Events can be enabled for detection in two ways. First, a mechanism in the network can arm an event for detection, for example using data contained in a user profile. These events are provisioned by the network operator.[10] The OSA/Parlay application must decide whether it wants to be notified of such events. Second, in several SCFs, the application is able to provision network signalling events that must be detected and reported to the application.

Two modes of *event monitoring* are available. The network-level call process may be *interrupted* and wait for an instruction from the application analogous to trigger detection points in IN. Alternatively, the network-level process may continue and simply *notify* the application, similar to the event detection point in IN.

Use case *Create Notification* in Figure 8.16(a) shows the creation of an event notification in the network by the application. The entity identified as SCS contains the unspecified mechanism for receiving and sending network signalling and setting and notifying trigger events.

1. The application creates a callback interface for the MP call manager to receive subsequent notifications.

---

[10]This class of event is referred to in the OSA standards as Type "B"; no reference is made to a Type "A".

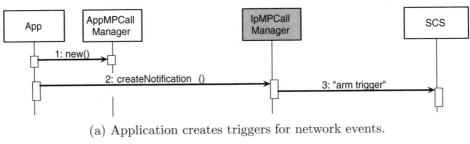

(a) Application creates triggers for network events.

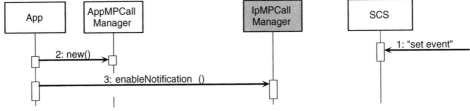

(b) Application enables notification of trigger events set in the network.

Figure 8.16.   Multiparty call control use case: two ways of initiating trigger conditions.

2. The application uses the method createNotification on the MP call manager interface. The parameter list contains a list of events to be armed.

3. The MM call manager uses an unspecified method to request the SCS to enable detection of the event(s).

Figure 8.16(b) describes the corresponding use case in which a trigger is set by a mechanism in the network. The objective is to inform the MP call manager that the occurrence of such an event must be notified to the application.

1. An action in the network arms a particular trigger.

2. The application creates a callback object for the MP call manager.

3. The method enableNotification informs the MP call manager that it wishes to receive notifications of a particular event that has been provisioned in the network.

This use case is typically invoked after an application is initiated.

Use case *Receive Notification* in Figure 8.20 shows how such events are reported to the call manager and typical consequent actions.

8.6.5   CALL, CALL LEG AND CALLBACK OBJECT CREATION

When a call is to be controlled by an application, an object implementing the IpMultiPartyCall interface must be instantiated, together with a callback object.

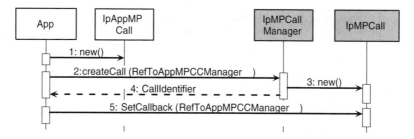

Figure 8.17. Multiparty call control use case: create call object and callback object.

Figure 8.17 describes the *Create Call* use case. A call manager callback interface (not shown in the figure) must exist and the call manager must have its reference.

1. The application creates the callback object implementing the IpMultiPartyCall interface by an implementation-dependent means. The application obtains the reference to the callback object.

2. The application invokes the createCall method on the IpMultiPartyCallControlManager interface. The callback interface reference may be passed as a parameter.

3. The call object is created.

4. The createCall method returns a multiparty call identifier as a result. The identifier has two fields: a reference to the call object interface and a session identifier.

Message 5 shows an alternative method to passing the reference to the object implementing the IPAppCall callback interface to the call object, had this not been done in message 2.

Once a multiparty call object has been created, call legs may be created and connected. Figure 8.18 shows the message sequence in the *Create and Connect Call Leg* use case. Prerequisite to this use case is the existence of a call object and its callback object as in the *Create Call* use case.

1. A call leg callback object is created.

2. The application invokes the createCallLeg method on the call object's interface. Two parameters are passed: the call session identifier and the reference to the call leg callback interface.

3. The call object creates the call leg object, say to be the A-leg in a new call.

4. Method createCallLeg returns a result containing the call leg identifier: the interface reference to the call leg object and a session identifier for the leg.

At this stage, the call leg object exists but the physical connection associated with the call leg has not been attempted. The leg must be *routed* to make the connection.

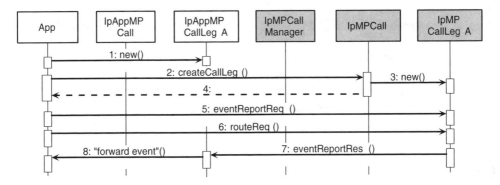

Figure 8.18. Multiparty call control use case: create and connect call leg.

5. Before making the connection in step 6 below, the application uses the eventReportReq method to provide a list of events, from those shown in Figure 8.15, that it wishes to be informed of.

6. The application invokes the routeReq method to initiate the physical connection. Parameters include the target address, originating address and the desired connection properties. The last parameter indicates whether the connection should be made explicitly, that is, the media flow starts immediately, or whether a subsequent attachMedia operation must be invoked to initiate the media flow.

7. When one of the listed events occurs, for example *Alerting*, the call leg object uses the eventReportRes method on the callback interface to notify the occurrence of the event.

8. The notification is forwarded to the application.

To connect a second party to the call, this use case is repeated. A leg object and its callback object must be created and the leg routed.

An alternative to this sequence is the single method createAndRouteCallLegReq available on the multiparty call object. This method has a parameter that specifies a set of events to be notified. If successful, the call leg is implicitly attached to the call.

8.6.6   *USER INTERACTION*

While the User Interaction SCF supports stand-alone user interaction, it is also used within calls controlled via the Generic Call Control or Multiparty Call Control. The UI SCF definition distinguishes between two levels of user interaction. The generic user interaction supports sending information or sending and collecting information from a user using an interface class IpUI. The IpUiCall interface inherits the methods of the IPUi interface and provides methods that support recording messages and retrieving and deleting recorded messages. Figure 8.19 shows the message sequence required in the generic user interaction use case. This use case requires a call to be in existence having a call leg established as in the *Create and Route Call* leg use case. A user interaction manager object must exist.

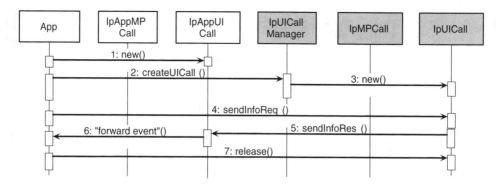

Figure 8.19. User interaction use case: create user interaction call object and callback object.

1. The application creates a callback object for the UICall object.

2. The application invokes createUICall on the UI manager object, passing the reference to the callback object created in 1.

3. The UI manager creates a UICall object. The reference to this object is returned to the application as a result for message 2.

4. The application invokes a user interaction operation on the UICall object, for example sendInfoReq requests a specified announcement to be played.

5, 6. The result of the user interaction, for example successful completion is returned.

7. The connection to the UI is released.

8.6.7 *HANDLING A TRIGGERED NOTIFICATION*

Use case *Create Notification* shows in Figure 8.16 how notification of a specified event is requested as a prerequisite to the *Receive Notification* use case discussed here. This use case assumes that a call is being processed using network signalling, for example in an SS7 or IMS network, and no application is yet in control of the call. A multiparty call manager exists: that used originally to create the notification. A trigger event, for example indicating the need to translate the destination address or to forward the call, takes place.

1. The trigger event, for example Address Analysed, is reported.

2. The call manager checks the whether the event is relevant to the application with which the manager is associated.

3. In anticipation of the application taking control of the call, a multiparty call object is created.

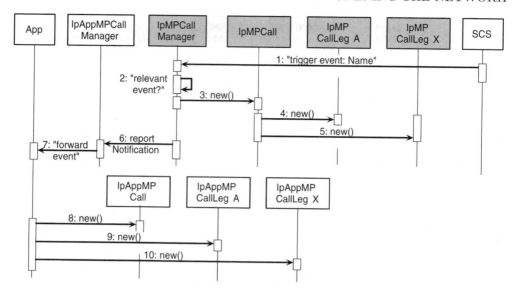

Figure 8.20. Multiparty call control use case: receive and handling a triggered notification from the network.

4, 5. Call leg objects are created for each of the legs in the call already existing at network level (two in this example but may be only one with originating leg triggering).

6, 7. The notification is reported to the application via the manager callback interface. The reportNotification method transfers a call identifier and identifiers for each leg. These identifiers include a reference to each object.

The application now executes and may need to issue method calls on the call and call leg objects created in steps 3–5. The application therefore creates callback objects for the call (8) and call legs (9, 10). The setCallback method may be used to transfer the references to the objects to the corresponding gateway objects. The application logic now continues execution.

8.6.8 CALL SUPERVISION

Calls and legs may be monitored and interrogated during a call. The Multiparty Call Control SCF allows a call to be *supervised* by the application. A period is specified in the superviseReq method call. The application is notified by the superviseRes method when the period expires, the call or user interaction ends or the quality of service is renegotiated at the network level. The action on expiry of the period may be specified: release the call, respond to the application or apply a tone.

The method getInfoReq enables the application to request information from a call or leg object, for example information required to calculate charges within the application.

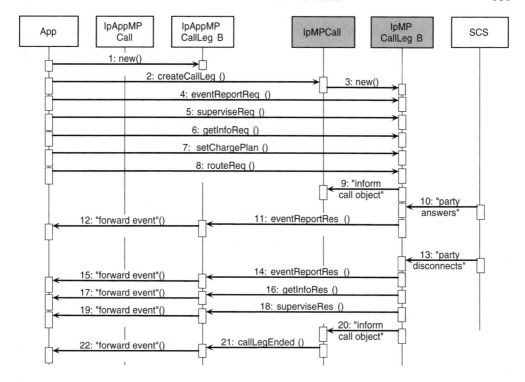

Figure 8.21. Multiparty call control use case: supervise a call.

The application is able to initiate a charging scheme that is specific to the operator by invoking setChargePlan on a call or leg object. Where terminals are able to receive charging information, the application uses the setAdviceOfCharge method to instruct that charging information must be sent to the terminal.

The *Supervise Call* use case shown in Figure 8.21 illustrates requests for supervision, information and charging and the related responses within a call setup. The multiparty call and callback objects are already in existence as is an A-leg (not shown). A supervised leg, B, is now added to the call by the following actions.

1–3. The call leg B is created as in the *Create and Connect Leg* use case.

4. Prior to routing the leg, the leg events to be reported are specified in the eventReportReq method call, including answering and disconnection by this call party.

5. The superviseReq method initiates supervision of the call and specifies the treatment of the call, for example action to be taken at the end of the supervision period.

6. The getInfoReq method specifies the information to be returned about the call when it ends.

7. The charging mode is set. Modes are specific to operators.

8. Having specified the supervision, reporting and charging requirements, the party is now connected to the call by the routeReq method.

9. The state information in the call object is updated.

10. The B-party answers the call. At this stage, supervision and charging come into operation.

11, 12. The answer event, requested in message 4, is reported to the application.

13. The call continues for some time and the B-party is the first to disconnect. The A-party has not yet disconnected.

14, 15. The B-party disconnect event is reported to the application.

16–19. These actions return information requested in methods superviseReq (5) and getInfoReq (6) to the application.

20. The B call leg reports its imminent self-destruction to the call object.

21, 22. callLegEnded reports the ending of the call leg by an event in the network to the application. The call and A-Leg still exist. For example the A-party may be allowed to make a continuation call.

8.6.9 MULTIMEDIA MEDIA STREAM CONTROL

The multimedia call interface IpMultiMediaCall adds one method to those inherited from the multiparty call interface, namely superviseVolumeReq. This method allows the application to set a maximum amount of data that can be transferred before a specified action takes place, similar to those in call supervision: release the call, respond to the application or play a tone. Figure 8.22(a) shows the actions in supervising the volume of data transferred in a call. The call has been set up and the legs are connected.

1. The application initiates the supervision of the volume of data to be transferred, for example, the basic increment allowed in volume-based charging.

2, 3. When the specified amount has been transferred, the call leg reports the result to the application.

4. The application initiates some action, for example, it initiates a user interaction to inform the user that the volume allocation is exhausted. In prepaid charging, sequence 1–3 repeats.

The IpMultiMediaCallLeg interface adds three methods to the multiparty interface: mediaStreamAllow, mediaStreamMonitorReq and getMediaStreams. When a call leg is routed, the media stream is either connected implicitly or must be completed using a separate attachment.

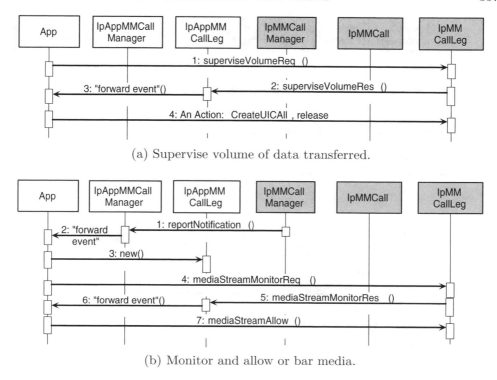

(a) Supervise volume of data transferred.

(b) Monitor and allow or bar media.

Figure 8.22. Multimedia call control use case: supervision and barring of media.

8.6.10 *CONFERENCE CONTROL*

A *conference* is an association of call parties, usually three or more, and the media connections between parties. OSA/Parlay embodies the concept of subconferences. A *subconference* is a grouping of parties within a conference that may be connected to each other. A conference always contains at least one subconference.

The Conference Call Control SCF is an extension of the multiparty and multimedia call control SCFs according to the inheritance relationships shown in Figure 8.13. The manager interface, IpConfCallControlManager, allows conferences to be created and can reserve and release resources for the conference. Conference call objects implementing the IpConfCall interfaces have methods to create subconferences. The subconference call interface IpSubConfCall inherits the basic call control methods to create, route and release call legs within the subconference, to supervise the conference and the volume of data transferred, to set charging modes and to get information about the call. The application may split and merge existing subconferences by calls on IpSubConfCall. The call leg object within a subconference is simply the multimedia call leg.

The conferencing SCFs support the common modes of conferencing. In a prearranged *meet-me conference*, resources are reserved for the conference, including the address or number to be used. Parties call in to the specified address and, after authentication, if required, are connected to the conference. In an *add-on conference*,

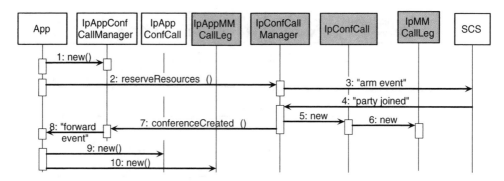

(a) Starting a meet-me conference

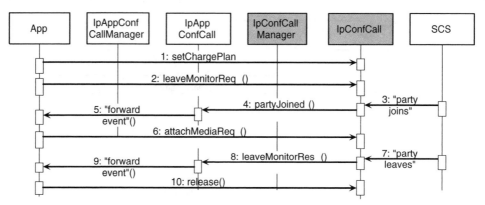

(b) A party joins and leaves a meet-me conference.

Figure 8.23. Conference call control: meet-me conference use cases.

parties are added to the conference by the application following some control action by the party arranging the conference, for example via a graphical user interface to the conference call application. The controlling party can in general add, drop, suspend or reinstate a party to a subconference.

Two conferencing use cases are shown in Figure 8.23. In (a) the method of starting a prearranged meet-me conference is shown. Here the number of subconferences defaults to one and the subconference object is not explicitly created.

1. The application creates a callback object for the conference call manager.

2. The conference is *prearranged* by the application reserving resources for the conference. This method specifies the date and time at which the conference starts, the duration, the maximum number of call parties, the address that they must call, and the conference policy: what the parties may and may not do.

3. An event is armed in the network to trigger on calls to the conference address.

4. When the first party dials-in, the event is reported to the conference call manager.

5, 6. The conference call manager creates a conference call object and a leg object for the first party.

7, 8. The creation of the conference is reported to the application.

9, 10. Callback objects must be created for the conference call and call leg objects.

The second use case in Figure 8.23(b) shows the sequence when a party joins a meet-me conference by calling the specified address.

1. The application may invoke methods such as setChargePlan to determine how the conference must be charged.

2. The application uses leaveMonitorReq to ask for reports of parties leaving the conference. A signalling-induced event must be set such as a *Terminating Release* occurring as a result of a network message.

3–5. An event indicating that a party has dialled into the conference is reported to the application.

6. The application allows the new party to take part in the conference by allowing the media stream(s) to flow. When other parties join the conference, the sequence 3–6 is repeated.

7–9. Some time later, a party leaves the conference and the event is reported to the application.

10. When the second to last party leaves the conference, the conference is released. Other rules may apply for terminating the conference.

The meet-me conference is simple and the application's role is largely reactive when network events are reported. An add-on conference is more complex as the application actively controls the connection of parties, creation of subconferences and assigning control of the conference to a party. The sequence to start an add-on conference with subconferences is shown in Figure 8.24(a). An add-on conference call is similar to an application-initiated call: the application joins the parties to the call.

1. The application creates a callback object for the conference call.

2–3. The application creates a conference call object by invoking the createConference method on the conference call manager. A parameter specifies the number of subconferences to be created.

4. The required subconference objects are created. Only the first is shown.

5. The application retrieves the details of the sub-conferences that have been created.

6. Other requests, for example to set the charging plan for the conference, may be invoked.

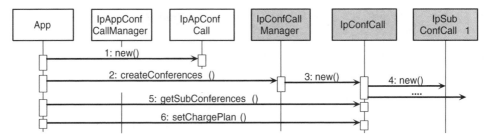

(a) Start an add-on conference with subconferences.

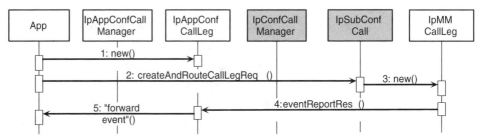

(b) Adding a party to a conference with subconferences.

Figure 8.24. Conference call control: add-on conference use cases.

A party is added to the conference or a subconference using the sequence shown in Figure 8.24(b). We illustrate the use of the **createAndRouteCallLegReq** method as a simplification of the separate creation and routing shown in Figure 8.17. A party can be suspended and resume participation by stopping and restarting the media flows. A party's participation may be ended by the application invoking a release on the call leg or a disconnection event from the network.

8.6.11 *DATA SESSION CONTROL*

The Data Session Control SCF allows an application to control a data connection in the network, for example a GPRS PDP context. The Data Session Control SCF is similar to its Generic Call Control counterpart in having only two parties. The manger interface allows notifications to be set in the network and notifications to be received from triggers provisioned in the network. The first party enters the data session through an action in the network, for example a request to set up a PDP context. The application becomes involved in the data session due to a trigger and may perform functions such as translate an address, connect the second party, supervise the session or instruct the network to continue processing the connection itself.

A single object implementing the **IpDataSession** interface is created by a method call on the manager interface. This SCF has no leg objects. The data session object allows sessions to be created, supervised and released. As in call control, a charge plan can be initiated and advice of charge requested.

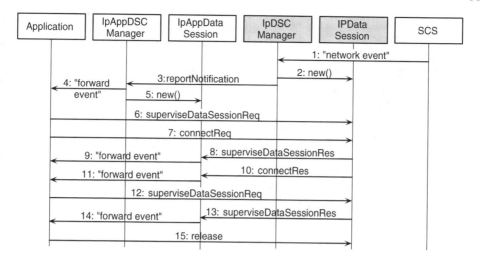

Figure 8.25. Data session control use case: supervise a session and end session on a supervision result, for example time elapsed.

An illustrative data session control use case is shown in Figure 8.25. A data session requested by a party is to last for a predetermined period, for example to play a video clip. The prerequisite is the existence of a data session control manager in the gateway and its required callback object. The use case commences with a notification of the data connection request from the network reported to the application (messages 1–5), following the pattern established in the call control use cases.

6. The application indicates that it wants to supervise the data session.

7. The application instructs the data session object to complete the connection to the second party, for example complete the establishment of a PDP context.

8, 9. A result relevant to the supervision of the data session is returned.

10, 11. The connectRes message indicates that the connection is made.

12–14. If the data session is to last a predetermined period or volume of data, the application requests to supervise the session. When the period expires, as reported in message 13, the session must be ended.

15. The application releases the data connection.

8.6.12 GENERIC MESSAGING

The Generic Messaging SCF provides a means for an application to access messages in a mailbox and be alerted of the arrival of such messages. The application may initiate transmission of a message by passing it to the mailbox. The SCF is generic in that messages with different voice and data encodings can be sent and received by the application.

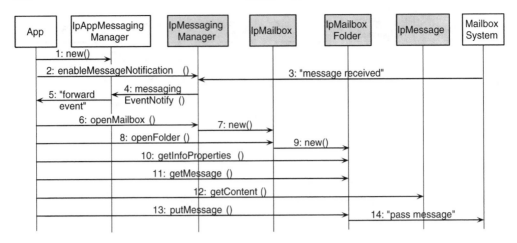

Figure 8.26. Composite example of using the Generic Messaging SCFs: accessing a mailbox service.

The entire set of objects in a mailbox service is shown in Figure 8.26. The entity Mailbox System represents the implementation of the mailbox. A generic messaging manager exists after the application has selected the service and a callback object must be created (1). Three other gateway interfaces are required to retrieve a message. The IpMailbox interface allows the application to access the mailbox implementation. The mailbox is a system of folders containing messages. The mailbox interface has implementation independent methods to create new folders and open existing folders. An object implementing the IpMailboxFolder interface must be created to access messages in the folder. This object has methods for the application to place a message in the mailbox folder for processing by the mailbox system. Message-level operations are implemented in an object implementing the IpMessage interface.

Message 2 in Figure 8.26 shows the application enabling the notification of message receipts by the mailbox system. Messages 3–5 show the receipt of a notification that the mailbox system has received a message. The steps to retrieve the message and to reply are shown in Figure 8.26. No callback interfaces are defined for the mailbox, folder and message objects: the returned result is used in each method call.

6, 7. The application requests the messaging manager to create a mailbox object.

8, 9. The application requests the mailbox object to create a folder object.

10. The application may then query the folder contents, for example the number of entries in the mailbox (not shown) and the properties of the entries.

11, 12. The application retrieves a message from the mailbox system by obtaining a reference to the message (getMessage) from the folder and requesting that the content be transferred (getContent).

13, 14. The application then composes a reply and passes it to the mailbox system via the mailbox object. The mailbox system then transmits the message.

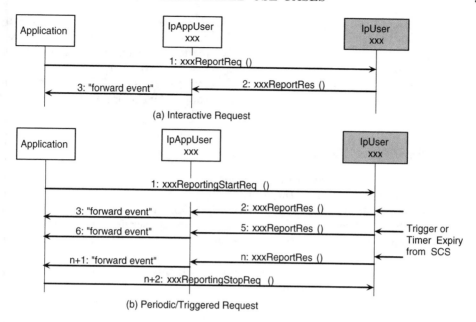

Figure 8.27. Modes of operation of data-oriented SCFs. (a) Interactive query. (b) Periodic or triggered queries. xxx takes on one of the following forms: location, periodicLocation, triggeredLocation in the Mobility SCF.

Table 8.3. Data-oriented SCFs and interface classes

SCF	Gateway Interface	Query Modes		
Mobility	IpTriggeredUserLocation	I	T	P
	IpUserLocationCamel	I	T	P
	IpUserLocationEmergency	I	T	P
Account Management	IpAppAccountManager	I	T	
Terminal Capabilities	IpAppExtendedTerminalCapabilities	I	T	

8.6.13 *MOBILITY, ACCOUNT MANAGEMENT AND TERMINAL CAPABILITIES*

Four OSA/Parlay SCFs enable an application to obtain various types of information held in the network: the Mobility, Account Management and Presence and Availability Management SCFs. A fourth, the Terminal Capabilities SCF, allows the application to determine the properties of a user's terminal. The four data-oriented SCFs follow a simple design pattern illustrated in Figure 8.27. Table 8.3 list the gateway and callback class names for SCFs conforming to the pattern shown in Figure 8.27.

Three modes of querying network or terminal information are possible. First, an *interactive query* follows a simple request–response model. Second, a sequence of *periodic responses* is started and repeats at a defined interval until stopped by the application. Third, *triggered responses* occur whenever a defined event takes place. Figure 8.27(a) generalises the interactive request message, with the exception of Terminal Capabilities, where the returned result is used rather than a response method. Figure 8.27(b) shows the application starting a series of triggered or periodic responses, the responses and the sequence being stopped by the application. Table 8.3 shows the modes allowed for each data-oriented SCF.

8.6.14 PRESENCE AND AVAILABILITY

Next generation networks strive to provide services that meet the needs, preferences and work or personal situation of the user.

This user-centric approach is supported within OSA/Parlay by the Mobility and Terminal Capability SCFs. Full support for user-centric service, however, requires more information about the user. User information is both static and dynamic. The former is held in network databases and directories. Dynamic information must be captured over time. User location is an example of dynamic information that is kept current by mechanisms in the mobile network.

Dynamic information about a user includes the available communication means, for example voice, e-mail and instant messaging, as well as the user's preferences and state. The objective is to better enable another users to communicate with the user, while respecting the privacy of the user. An extensive set of dynamic data about the user is therefore required. The OSA/Parlay Presence and Availability Management (PAM) SCF allows an application to provision user data, gain access to the data and be notified of changes to data. As with all OSA/Parlay SCFs, no implementation detail is specified. The PAM SCF is, however, designed to be compatible with the IETF Model for Presence and Instant Messaging [44].

User-centric applications depend on knowledge of the *user context*: a set of user information, derived from various sources that characterizes a users environmental, personal, resource and service state. The concepts of presence and availability are used in OSA/Parlay. *Presence* is a set of characteristics describing the context in which the user, or an agent representing the user, exists. In general, presence includes simple attributes such as the user being online, the communication modes that can reach the user, as well as the user's disposition and activity. In reality, presence data is limited: OSA/Parlay presence data supports voice, SMS, MMS and instant messaging. *Availability* in contrast expresses the user's ability and inclination to share information and communicate in a specified context. An availability profile has two parts: first, the restrictions that apply to distribution of the user's availability profile, and second, the availability attributes.

The OSA/Parlay standard does not define a set of availability attributes. The PAM provisioning SCF allows an application to name and define attributes as simple variables, structures or XML documents.

The third concept used in the definition of the PAM SCF is *identity*. Individuals or organisations are termed *entities* in PAM nomenclature. An identity is a representation of one or a group of entities. An identity has attributes and may be typed. One type is the *presentity* defined in the IETF presence and instant messaging model [44]. This type has attributes including subscriber status, network status, communication means, contact address, location and priority. An identity is therefore a name or names and a set of attributes. A rich set of methods for managing identities is provided in the PAM SCF.

8.6.15   CHARGING AND ACCOUNT MANAGEMENT

The Charging and Account Management SCFs fall into different groups: the former provides methods within service control while the latter supports service management functions.

The Charging SCF is structured with an interface class IpChargingManager that creates IpChargingSession instances when a charging session is required within a service, for example in a call or data session. Methods are available on the charging session object that support crediting or debiting an account by an amount specified in a currency or a number of units, for example kilobytes. The credit or debit may be made directly on the user's account or on an amount reserved for the session. Reservations may be extended.

8.6.16   CONNECTION MANAGEMENT

The OSA/Parlay Connection Management SCF is unlike other types: it is more of an operations support nature than real-time service control. The underlying network is assumed to be an IP network that is capable of supporting virtual private networks (VPN). The interfaces provide capabilities that would be required by a network management application for configuring VPNs.

The Connection Management SCF may be used by an enterprise operator that has entered a relationship with a network provider to provide VPN services to customers within a policy-based agreement. Applications using the CM SCF may also be hosted within a network operator's operations support system.

A possible business case is the provisioning of VPNs by an enterprise operator, the VPN provider, using a telco's IP network infrastructure. The VPN would support enterprise applications for an entity that may be the same or different enterprise operator. For example, an IT service provider may offer the VPN configuration service to its enterprise customers rather than the customer sourcing VPN services directly from the telco. The VPN service provider has a business relationship with the telco involving a payment model for use of network resources and policies on use of those resources. The telco gives the VPN provider access to the Connection Management SCFs in a Gateway subject to the safeguards provided by the Framework.

The network is provisioned using a virtual leased line concept termed a Virtual Provisioned Pipe in the Connection Management SCF. The various customer sites are interconnected by pipes having the required capacity and quality of service defined by a chosen template.

8.6.17   OSA/PARLAY DATA STRUCTURES

Some of the use cases described above refer to parameters and returned values of some method calls. This is a small sample of the rich set of data types, both common across the SCFs and that support the individual SCFs. We illustrate the richness of the data types in Parlay by examining the data associated with a selected method call, sendInfoReq in the User Interaction SCF.

```
sendInfoReq
    (userInteractionSessionID    :    in TpSessionID,
    info                         :    in TpUIInfo,
    language                     :    in TpLanguage,
    variableInfo                 :    in TpUIVariableInfoSet,
    repeatIndicator              :    in TpInt32,
    responseRequested            :    in TpUIResponseRequest)
                                 :    TpAssignmentID
```

Several data types occur in most of the SCFs. For example TpInt32 is a simple 32-bit integer while TpSessionId is a unique identifier for a call or leg session within a given instance of a SCF of type TpInt32. The info field of type TpUIInfo specifies one of nine types of data through a choice structure: the identifier of a script or data stream; free format data; a URL pointing to text; a script or a stream; UUencoded data; MIME data; Wav data, AU data; a Voice XML script; or speech synthesis data. Each of these types in turn has a definition with choice elements names:

InfoID   :   the (operator specific) ID of the user information script
                            or stream to send to an end-user.
InfoData   :   the free format data to be sent to an end-user's terminal.
InfoAddress   :   URL of the text, voice application script or stream to be
                            used.
InfoBinData   :   free format binary data to be sent unchanged to an end-
                            user's terminal.
InfoUUEncData   :   UUEncoded data to be sent to an end-user's terminal.
InfoMimeData   :   MIME data to be sent to an end-user's terminal.
InfoWaveData   :   WAVE data to be sent to an end-user's terminal.
InfoAuData   :   AU data to be sent to an end-user's terminal.
InfoVXMLData   :   string description of the VXML (Voice XML) page to be
                            executed.
InfoSynthData   :   describes the content and how speech is to be synthesised.

The type for InfoSynthData contains the text data and speech synthesis parameters such as speaker gender, age, rate, range and pronunciation. The the remaining choice elements are of basic OSA/Parlay types.

This simple example illustrates the method of defining data types in OSA/Parlay. Common data types are defined in [73]. Other SCF definitions expand the data types, for example for call control [75] and mobility management [81].

8.6.18 *SAMPLE SERVICE*

An application implementing a complex service orchestrates a sequence of elementary use cases of the types listed above. The OSA/Parlay Generic and Multimedia, Multiparty and Conference Call Control standards present a number of service-level use cases [76, 77, 78, 79] that illustrate how the elementary use cases make up the full message flow. In this section we present a *collect call* or reverse charging service implemented as a OSA/Parlay application as an example of how the elementary use cases make up the message sequence of a complex call.

The collect call service is offered under an agreement between telcos that may be in different countries. A caller dials a freephone number that is specific to the country in which the called party is located. In the originating network, the call is treated as a freephone call. The dialled number is translated to the E.164 number that ensures that call signalling reaches the entity that triggers the service in the called party's network. This entity is a switch in a switched-circuit network or a S-CSCF in an IMS network. An agreement between the telcos determines the way that revenue is shared using the off-line billing system. The terminating network must pay the originating network a settlement fee for the call.

An event is provisioned in the terminating network to detect all calls to the translated number. The message sequence proceeds through the following phases:

- Messages 1–8 show the SCS detecting the receipt of a call request with the translated number. The event is reported to the application, a call and A-party call leg objects are created in the gateway, the application is notified and call and call leg callback objects are created.

- Messages 9–14 shows a user interaction that requests the caller to give the number of the desired called party.

- A further user interaction, details not shown, requires the caller to record a message to the called party.

- When the user interaction is complete, the IVR is released and the calling party is put on hold at message 15.

- The application must now initiate a call to the intended called party, denoted B, to enquire whether the call is to be accepted. Messages 16–19 create the B-party call leg and its callback object and initiates a connection to this party.

- Having created a UI call, the application plays an announcement, including the recorded message, to the B-party and asks whether the B-Party is willing to accept the call and the charges. The B-party accepts or declines the call. (Messages are not shown.) In the case of the called party accepting the call, noticing that the B-party is a prepaid subscriber, the application asks the B-party whether a limit should be placed on the call (cost or duration).

- The UI call is released at message 22.

- In messages 23–25, the application requests to supervise the call and collect information at the end of the call. The call is connected only at message 30.

- The application now uses the Charging SCF to open a charging session (messages 26 and 27).

- In messages 28 and 29, the application reserves the amount for the call. The remaining sequence assumes that party B's balance is sufficient.

- With message 30, the application connects the parties.

- Via messages 31–33, the application receives notification that call is active, that is, the connection is successful.

- The application supervises the call, allowing it to continue for the time specified by the B-party.

- The call may end in several ways. Messages 35–40 show the case of the A-party ending the call before the time limit is reached. The call and remaining party are released at message 40.

- Only charging actions remain. The application calculates the actual call charge (41) and debits the prepaid user B's account against the reserved amount (42 and 43). The charging session is released (44). Data relating to the settlement with the originating network is recorded offline (45).

8.7   PARLAY X WEB SERVICES

8.7.1   THE CASE FOR A SIMPLER API

The OSA/Parlay API abstracts network functionality to the extent that the programmer need not understand underlying network protocols such as MAP, SIP, INAP and ISUP. The programmer must, however, work at a level indicated by the SCF use cases reviewed above, the Collect Call example and sequence diagrams given in the OSA standards. Callback objects must be created, notifications must be enabled, and calls, call legs and media streams, and the way that these are supervised and charged, must be controlled. Similarly, a programmer using mailbox facilities must work with folders and messages. Many method are asynchronous and the programmer must understand the consequences of this mode of invocation. OSA/Parlay relies on a large, rich set of data types, that must be understood by application programmers. In summary, the OSA/Parlay API is very rich in methods that operate at a level of abstraction that hides network detail but requires detailed control of calls, messaging and database operations.

A strong argument exists for a second Open Service Access interface that provides a greater level of abstraction that hides the fine grained call, messaging and data operations that are required when using the OSA/Parlay API. At the same time, the power of the OSA/Parlay API set should be made available to the programmer. For example, it should be possible to place a call or start an add-on conference with a single method call, to send or receive a batch of messages simply and to perform service management. Such an interface must provide methods that have simple semantics and are simple in operation, for example using synchronous invocations and the invoking process not having to keep track of the service state.

OSA/Parlay seeks to create convergence between information technology and telecommunications. Section 5.4 observed that a common API technology to provide supporting service for IT applications and telecommunications service features would promote the interests of enterprises. The Service Oriented Architecture described in Section 5.2.4 is such a common approach that has found widespread acceptance. The Parlay Group has therefore defined *Parlay X Web services* as a simplified, highly abstracted means of accessing network functionality [101]. Parlay X is described as a *Web services specification for Open Service Access*. The Parlay X interfaces are

published as ETSI standards [85]. For simplicity of use, the Parlay X interface is based
on a synchronous request–response model. The interfaces do not support notifications
from the Web service implementation to the application.

An argument advanced in [102] for a simplified, Web services API for accessing
network capability is based on the number of potential programmers capable of working
at different levels of abstraction relative to the underlying network. An estimate is
given of some 10 000 service developers worldwide who are capable of programming at
a level that requires knowledge of network protocols such as ISUP, MAP and INAP.
These protocols express the underlying network capability at a fine grain. The number
of programmers competent in the C or Java languages in a CORBA environment who
are capable of programming at the level required for using the OSA/Parlay API is
estimated at one-quarter of a million. By contrast, the number of programmers capable
of working in a Web services environment is described as 'into the millions' [102]. Web
services, described in Section 3.8.3, are inherently simple and cannot express detailed
network capability. The number of potential programmers for service development has
an inverse relationship to the expressive power of the interface definition.

8.7.2   THE PARLAY X WEB SERVICES ARCHITECTURE

Parlay X Web services are intended to provide open service access with a simpler
interface and greater abstraction of the underlying network and data resources. The
OSA/Parlay architecture is extended as shown in Figure 8.29 to support Parlay X Web
services. The Framework, Gateway and Parlay applications domain are unchanged.
The Parlay X Gateway is added. The additional gateway is essentially a Parlay
application that presents Web service interfaces to a different class of application,
the Parlay X application. For example, the Make Call Web service requires an
implementation of a third party call to interact with the SCFs in the full Parlay
Gateway. It is recognised that some Web service implementations may not use the
full Parlay gateway and may use protocols directly, for example for sending e-mails.
Implementation differences are not visible to the Web services programmer.

Parlay X Web services defined in the 2005 standards fall into two groups:
call/message related and service management.

- Call and conference related Web services:

    - make a two party call, query call status and end call;
    - control a multimedia conference;
    - make an audiocall, that is call a user and play content;
    - send a short message;
    - send a multimedia message;
    - pay for content;
    - query and optionally notify status or location of terminal or terminal group.

- Service management:

    - manage accounts (recharging account);
    - manage address lists: white and black lists, forwarded-to addresses;

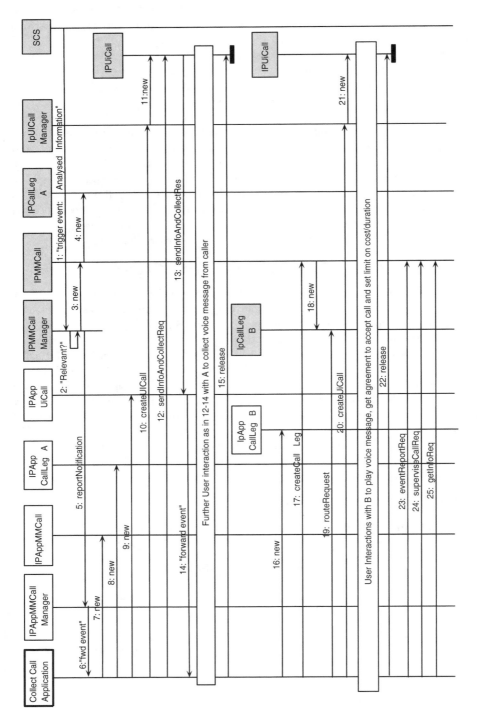

Figure 8.28. Collect call application: initial interaction and call setup.

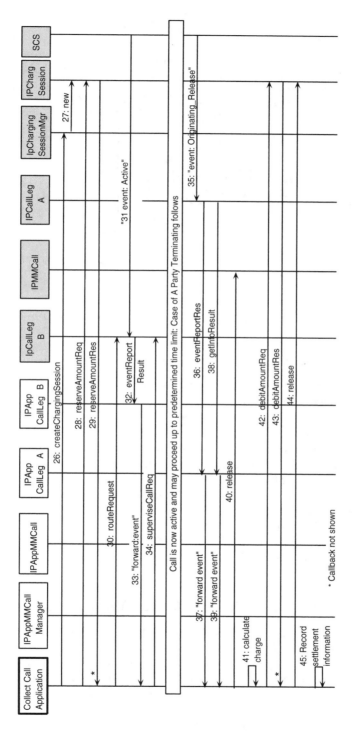

Figure 8.28. Continued.

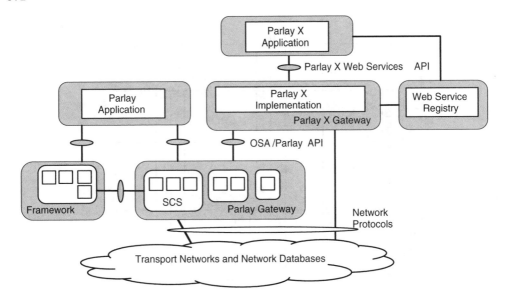

Figure 8.29. Parlay X architecture, showing its relationship to the OSA/Parlay API.

  - specify call handling treatment: conditions, announcements;
  - obtain and register presence information about a user: activity, place, privacy, environment and communication means.

The method of definition and implementation of the Parlay X Web service interface is prescribed in [85].   The interface definition must be performed using XML conforming to the Web Services Interoperability Organisation's (WS-I) Basic Profile [28].   This WS-I profile describes how the W3C's Web Services Description Language (WSDL) 1.1 is to be used. Web service messages must conform to the OASIS SOAP message security specification [161].   Use of HTTP to transport messages must also conform to the WS-I profile.

The Parlay X standards do not define a framework-like security mechanism. Rather, Web services security prescriptions must be followed. Encryption must follow either VPN or Transport Layer Security practice.

8.7.3   PARLAY X WEB SERVICE EXAMPLE: AUDIOCALL

An example of an application making use of a Parlay X Web service is given in Figure 8.30.   The Audiocall Web service interface is used [86].   This service has operations PlayTextMessage, PlayAudioMessage, PlayVoiceXmlMessage, GetMessageStatus and EndMessage.   Each operation has an input or request message and an output or response message.   In the example, a [human] user interacts with the application to request that a text message be played to a phone user via a text-to-voice converter. The message sequence shown in Figure 8.30 performs functions as follows.

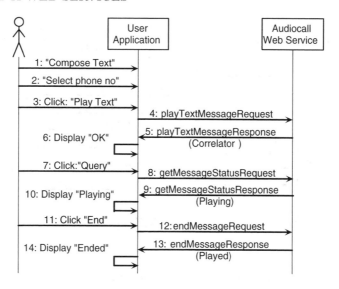

Figure 8.30. Parlay X Audiocall, playing a text message to a phone user.

1, 2. The user composes a text message in a graphical user interface (GUI) and selects the phone number of the party to be called.

3. The user clicks on a button to request that the message be played.

4. The application invokes the Audiocall Web service using the PlayTextMessageRequest message. The language to be used and the charging mode instructions are also contained in the message.

5. The PlayTextMessageResponse message is returned immediately. An identifier Correlator is returned for later references to the same call. Processing the call by the Web service implementation commences, for example using method calls on the GCC and UI SCFs in the OSA/Parlay Gateway.

6. A reassurance message appears in the GUI.

7. Time elapses and the user wishes to check the progress of the call and clicks on a button in the GUI.

8. The application invokes the GetMessageStatusRequest, identifying the call by the correlator.

9, 10. The status of the call is returned as "playing". Other allowed results are "played", "pending" and "error". The playing status is shown in the GUI.

11. The user decides that the time has come to stop the call.

12–14. The application invokes the EndMessage operation. The implementation interacts with the gateway to clear the audio call. The ending of the call is shown in the GUI.

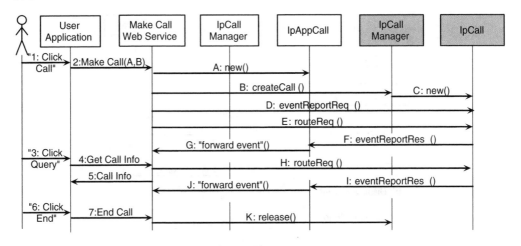

Figure 8.31. Parlay X Third Party Call, showing action at the Parlay Gateway.

8.7.4   PARLAY X IMPLEMENTATION VIA OSA/PARLAY

The Parlay X architecture shown in Figure 8.29 locates the Parlay X Gateway as an OSA/Parlay application that presents Web services interfaces to external applications. The Parlay X Gateway must therefore implement the detailed logic that allows simple Web services invocations such as Audiocall to invoke a more complex sequence of asynchronous methods on the OSA/Parlay gateway. The example in Figure 8.31 illustrates the relationship between Web services calls on the Parlay X Gateway and resulting message flows at the OSA/Parlay gateway interfaces.

Figure 8.31 shows the message flows in a *click-to-dial* service using the Third Party Call Web service [87]. Three operations are defined in this service. Make Call, with the two party addresses and the charging mode as parameters initiates a call. Operation Get Call Information allows the invoking application to query the status of the call. End Call terminates the call. The sequence of operations 1–7 indicates the interactions between the application and the Parlay X Web services interface. The implementation uses method calls A–K on the OSA/Parlay Generic Call Control interface to implement the service.

8.8   OSA/PARLAY API IMPLEMENTATION ISSUES

8.8.1   GATEWAY-TO-NETWORK INTERFACE

The application-facing or *northbound* interface is defined for each SCF. The *southbound* interfaces between SCFs and the underlying networks or resources are regarded as an implementation issue in OAS/Parlay. Application logic and therefore the sequence of method invocations and responses on the gateway SCFs is independent of the underlying network. Similarly, when setting triggers and receiving event notifications, the application does not need to know the type of underlying network.

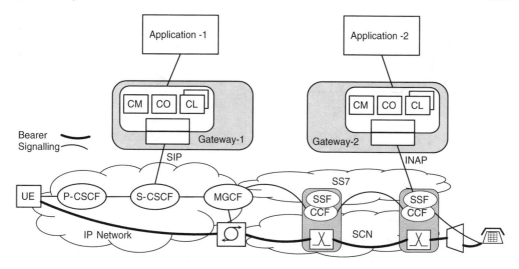

Figure 8.32. Network interface adaptation for two cases: Gateway-1 with IMS network; Gateway-2 with a switched-circuit bearer network.

No standards exist for the network interface, regarded by the OSA/Parlay standards as an implementation detail. An inherent constraint exists on any implementation. The Gateway does not co-ordinate connections, for example by orchestrating a media gateway and the circuit- and packet-mode signalling in the adjacent networks. Rather, a tacit principle, visible in some implementations, governs the interface to the transport network, namely that there is a *single point of contact* at which vertical signalling is received from the network. Two such cases illustrated in the composite example in Figure 8.32 are described in 3GPP Technical Reports for mobile systems using CAP signalling [3] and IMS systems [4]. First, mapping between a switched circuit CCF/SSF using INAP or CAP and API methods and responses allows the OSA/Parlay application to be involved in both first and third party services. The single point of contact for switched-circuit networks in the SSF of an exchange or MSC. Second, the ISC interface in IMS identifies the Serving CSCF as the single point of contact between the gateway and call/session signalling.

Bearer and media sessions therefore rely on connections made by the call/session signalling. For example, a connection between an IMS and circuit-switched network is made by call/session signalling to the media gateway controller as described in Section 7.5.5.

Figure 8.33 outlines the relationship between call/session signalling and API requests and responses. Assume that an event is enabled in the network that detects all called numbers in a stated range, for example starting with 0800. The call/session signalling protocol is SIP. The server for the ISC interface is assumed to be implemented in the SCS. The SIP server operates as a proxy server. Other SIP servers (P-CSCF and I-CSCF) are not shown as they merely relay the SIP messages. An illustrative signalling and method call sequence follows:

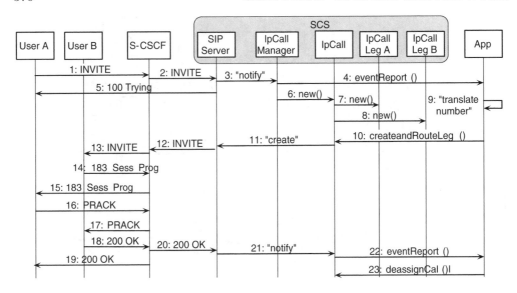

Figure 8.33. Illustrating the relationship between Gateway method calls and SIP signalling.

1. User A starts a call by issuing an INVITE message containing its proposed session description and the freephone address of the called party.

2–4. The Serving CSCF detects that the called party number requires processing by the application server. The event is notified to the application via an INVITE message.

5. The 100 Trying message reassures the called terminal that the request is being processed.

6–8. Call and call leg objects are created.

9. The application translates the freephone number into a routable number.

10, 11. The application uses method createAndRouteLeg to establish leg B and to request that the call be connected.

12, 13. The routing request is translated into a SIP INVITE directed to the B-party. The session description is forwarded to the B-Party

14, 15. Using a 183 Session Progress message, the B-party accepts or modifies the session description.

16, 17. The A-party provisionally acknowledges the session description.

18, 19. The B-party acknowledges the session setup.

Table 8.4. Degree of abstraction for signalling protocols, OSA/Parlay and Parlay X

Signalling Protocol	OSA/Parlay API	Parlay X API
Messages, Methods and Operations		
Call/session signalling messages	Rich, fine grain requests, responses, notifications	Rich functionality, coarse grain operations, no notifications
Calls, procedures and states		
Follows protocol state machine and procedures	Numerous elementary use cases, asynchronous, stateful, events	Synchronous, simple, stateless, status by polling
Data structures		
Complex, specific to protocol	Complex, typed with inheritance	Simple XML schema
Invocation of value added services		
Relies on triggering on message data fields	First or third party invocation	Third party invocation
Knowledge required		
Protocol details, call processing, media descriptions, ...	Concepts of call model, conference, message, mailbox, ...	Service: call made, message sent, preferences set, ...

20–22. Various events may be monitored. Here we show the 200 OK message from B triggering a notification to the application.

   23. The application does not require further notifications and therefore deassigns the call. The call continues under the control of SIP signalling.

8.8.2 ABSTRACTION REVISITED

An important objective of the OSA/Parlay and Parlay X interfaces is to provide access to network capability while hiding its detail from the application programmer or Web service user at different levels. Figure 8.30 illustrates the simplification the application's interactions with the Parlay X interface relative to the full OSA/Parlay API. Similarly, Figure 8.31 shows an example of the expansion of OSA/Parlay API call control method calls into call/session signalling.

Working with a network protocol such as SIP is likened to programming in assembly language [140]. Detailed knowledge of the protocols is required and the various fields in a message must be crafted. Extending the metaphor, using the OSA/Parlay API is analogous to high level language programming. Using Parlay X to create services is at a script programming level. Table 8.4 compares the visibility of a number of underlying elements at the three levels of abstraction.

Associated with each level of abstraction in Table 8.4 is an level of abstraction of the transport network. When working with signalling protocols, knowledge of the

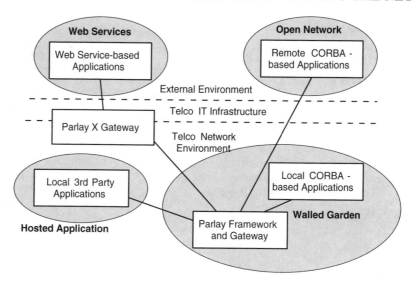

Figure 8.34. Four possible deployment scenarios for applications based on OSA/
Parlay and Parlay X Web services.

transport mechanism is required. For example, composing a SIP message requires the
transport layer protocol to be specified (UDP/TCP/SCTP). Similarly, working with
a PSTN protocol such as INAP requires knowledge of TCAP since INAP messages
are sent within TCAP dialogues. The OSA/Parlay API, defined in UML, allows
implementations with distribution transparencies, particularly implementation and
location transparency. Details of the transport and lower layers are hidden from
the application programmer. The Parlay X API requires SOAP and HTTP as the
transport mechanism, with the programmer having appropriate knowledge.

8.8.3 SERVICE DEPLOYMENT SCENARIOS

The OSA/Parlay architecture with its extension via Parlay X Web services enables
a number of models for providers to deliver services. Four illustrative scenarios are
shown in Figure 8.34. Three environments are identified. The secure domain of the
telco network environment hosts the OSA/Parlay Framework and Gateway and may
have application servers. The *IT infrastructure* is a packet network for supporting the
IT services and operations support of the telco. While the past focus of the telco's IT
infrastructure fell on the telco's internal operations, this infrastructure is also suited to
providing access to Web services. The *external environment* encompasses independent
application providers and hosts.

The *open network* scenario arises when an external service provider hosts applications that access network capabilities using the API with implementation on
the OSA/Parlay Gateway, under the control of the Framework. A CORBA-based
distribution plane is likely to be used. Such applications are described as *remote*, that
is outside the telco domain.

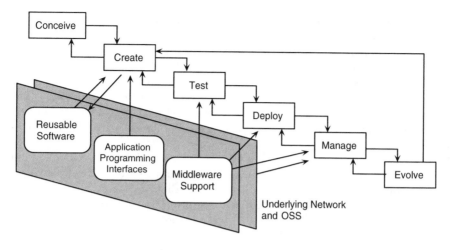

Figure 8.35. Service and application lifecycle with essential support.

The *Web services* scenario is similar. However, the application invokes Parlay X Web service operations on the Parlay X Gateway. A SOAP/HTTP distribution plane is used with WS-I security mechanisms [28].

Two scenarios locate the application within the telco environment. A *walled garden* is a situation in which a network operator implements applications on an application server located within its own secure domain and accessing network capabilities using an open API on its own gateway. Walled garden has wider meaning than that just described, namely a proprietary domain that cannot interoperate with other domains. This situation could arise if proprietary features are used at the API interface.

A network operator may also host applications developed and owned by third party providers. These applications have access to the telco's OSA/Parlay Gateway. This is the *hosted applications* scenario. These two scenarios are usually CORBA-based and are termed *local*.

These four deployment scenarios in Figure 8.34 illustrate the versatility of the OSA/Parlay architecture. Further deployment issues are discussed in [205].

8.8.4 SERVICE CREATION

Service creation is a set of methods for building software for services and applications for a particular context. Context is introduced into the definition of service creation because of its influence on the realisability of services and the approaches taken. Examples of contexts are the PSTN/IN, the Internet and a Parlay X environment. Service creation methods in the PSTN/IN are based on the use of reusable software modules called Service-independent Building Blocks (SIB) that enable the service programmer to use network capabilities, for example making connections and playing announcements, while giving a good degree of abstraction of network resources. Service creation environments allow service designers to chain SIBs together and populate parameters to provide a service. Each network or service architectural context has its own approach and support for service creation.

A desirable service creation environment is described in [27] as having two features. First, the environment must provide *service mediation*, that is, provide a unified approach and architecture that allows access to network capabilities. Second, the environment should attract the largest possible *developer community*. The first objective is served by the Open Service Access architecture. The second objective is promoted by using widely known and accepted software paradigms such as Web services.

The service creation process is generalised as shown in Figure 8.35. The lifecycle of a service or application is shown using a waterfall model. The process requires three forms of generic support. First, service mediation, that is access to network capabilities, is facilitated through application programming interfaces. Second, a software reuse methodology is required. Third, the environment must support the inherent distributed nature of the resulting system. The OSA/Parlay and Parlay X provide the first form of support, each to its own level. Parlay X, as a Web services architecture, allows reused functionality to be captured as a Web services interface and underlying implementation. OSA/Parlay does not define the structure of the application layer other than the use of callback interfaces. Software reuse for OSA/Parlay applications, and the implementation of Web services, remains an open question.

8.9 OTHER APPROACHES TO OPEN NETWORKS

8.9.1 JAIN

JAIN, a name derived from Java APIs for Integrated Networks, is an initiative that supports the development of next generation communication services and products related to those services through the application of Java technology. Java technology is more than the programming language but includes development and execution environments, Java Web services and products.

JAIN provides a series of application programming interfaces defined in the Java language for accessing and controlling network capabilities [201, 26]. Some APIs are network-technology independent, for example a Java realisation of the Parlay 1.2 API, known as the Service Provider Access (SPA) interface [200]. JAIN is more than a technology-specific Parlay realisation. Other APIs provide access to specific network protocols by providing Java methods for accessing and modifying fields in protocol messages. For example the JAIN SIP API allows signalling messages to be sent, received, synthesised and analysed. Java accessor methods are provided for the various constituents of SIP messages, for example for extracting data from SIP headers. JAIN therefore defines the northbound interfaces of the network adapters.

The architecture of a communications system based on JAIN is shown in Figure 8.36 using the three principal layers of the NGN Framework. Network and data resources are isolated from the service control functions by Resource Adapters (RA), defined for specific protocols or interfaces, for example JSIP for SIP signalling and JCC for INAP access to PSTN switches. The JAIN Service Logic Execution Environment (JSLEE) provides a secure, reliable environment for service implementations to

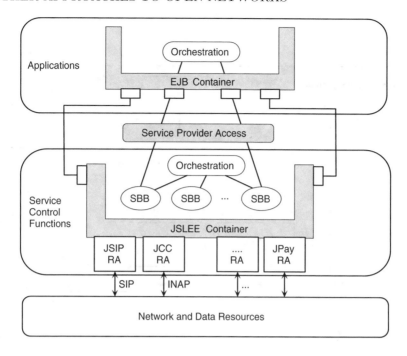

Figure 8.36. Architecture of a system based on JAIN.

execute, receive notifications and access to the resource adapters. Service Building Blocks (SBB), for example implementing service features, execute within the JSLEE. More complex services may be created by higher level logic invoking SBBs. This process is called *orchestration*. The structure of the application layer also has an execution environment, the Enterprise Java Beans (EJB) Container. The Service Provider Access (SPA) interface presents a Java-implemented Parlay API.

8.9.2 OPEN MOBILE ALLIANCE SERVICE ENVIRONMENT

The Open Mobile Alliance (OMA) is a group of telecommunications operators and vendors founded with the objective of promoting data services in mobile networks. Envisaged services include browsing, e-mail notification, games, location services and multimedia messaging. The OMA publishes a service environment architecture, the OMA Service Environment (OSE) shown in Figure 8.37 [165]. The key component in this architecture is the *enabler*: a component that has defined public interfaces with particular functionality that can be used in developing or deploying services. Some enablers perform network protocol adaptation. Enablers are specified by the OMA to ensure interoperability.

The business-oriented objectives of the OMA are similar to those of IN and OSA/Parlay. Competition should be promoted by providing an environment in which service providers, both traditional and third party, can innovate and differentiate their offerings. The pool of service developers should be expanded. Varied business models

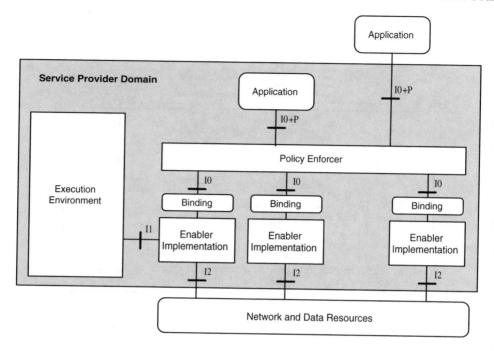

Figure 8.37. The Open Mobile Alliance Service Environment.

should be possible. Service development should be rapid and be supported by software reuse.

Enablers have three categories of interface. The northbound interface is of type I0. The enablers' characteristic or intrinsic functions are accessed through this interface. The OSE contains an Execution Environment that manages the lifecycle of enablers through a horizontal interface belonging to category I1. Interfaces to underlying resources fall into category I2. Like OSA/Parlay, the OMA does not define such interfaces.

Enablers can communicate with each other through I0 interfaces. Thus, a number of enablers may be stacked between an application and a network resource.

The OMA architecture recognises that access to enablers by an application must be carried out in a secure, orderly way. The OSE therefore interposes a layer called the *Policy Enforcer* between the application and enablers. A policy-enforced northbound interface is designated I0+P.

We illustrate the nature of enablers in the OSE by considering the IMS described in Chapter 7. SIP signalling flows from terminals (UE) via P-CSCFs to the Serving CSCF. The ISC interface directs SIP signalling to a service platform [166]. An ISC enabler has an I2 interface implementing the SIP protocol to interwork with the ISC and an I0 interface for use by an application or another enabler. Similarly, the service platform must communicate with the HSS to obtain subscriber data. A further enabler must therefore exist that communicates with the HSS via an I2-type interface

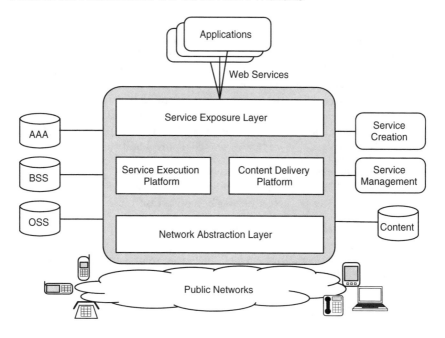

Figure 8.38. A generalised Service Delivery Platform.

implementing the Sh Diameter interface. An I0 interface with suitable methods is presented to applications.

This brief review shows the OMA Service Environment as a further instance of the three-layer subset of the NGN Framework: application, SCF and underlying resources. Applications are served by capabilities encapsulated in enablers. Network resources are accessed through and abstracted by the service capability type I0 interfaces.

8.9.3   THE SERVICE DELIVERY PLATFORM CONCEPT

The convergence of information technology and telecommunications requires a means of delivering telecommunications services at the request of IT applications. OSA/ Parlay, JAIN and the OSE are specific architectures that meet this need. A Web services interface between the application and the network capability functions or enablers is pervasive. While these standards-based architectures are maturing, many vendors have developed proprietary equipment serving similar purposes that they describe as service delivery platforms.

The concept of a service delivery platform (SDP) varies from vendor to vendor and there is little consensus on how standardised interfaces relate to SDP architectures. A contribution toward a uniform definition given in [160]. A *Service Delivery Platform* is defined as 'an IT solution which can be deployed by fixed and mobile service providers to deliver next generation value-added voice and data services to consumers and enterprises'.

This definition has several implications. The SDP is seen as part of the IT infrastructure of the telco or service provider, along with functions such as operations and business support systems (OSS/BSS). Services are value-added and do not include basic bearer services. The IT infrastructure must interwork with the underlying networks. While not part of the definition, it is clear that access to network capabilities must be secure and not compromise the integrity of the platform. Also, the technology used is likely to be characteristic of IT rather telecommunication systems.

The basic architecture and essential elements of a SDP proposed in [160] are shown in Figure 8.38. Four essential elements of the SDP are identified:

- The *Network Abstraction Layer* provides access to network capabilities such as voice and multimedia calls and messaging via standardised interfaces. The OSA/Parlay gateway is such a network abstraction layer.

- The *Service Execution Platform* provides the service deployment and execution environment. Like the JSLEE and the OSE, the service execution platform must provide lifecycle support for services, integrity management, access to network abstraction interfaces and event handling.

- The *Content Delivery Platform* is described in [160] as an optional element for delivering multimedia content in mobile networks.

- The *Service Exposure Layer* is a means of providing access to network (and content delivery) capabilities with a high degree of abstraction, greater than that of the network abstraction layer. An instance of a service exposure layer is Parlay X, where the external application accesses services implemented in the service execution layer that in turn rely on fine-grained capabilities accessed through the network abstraction layer. The service abstraction layer is described as an optional element but is likely to be part of any SDP that provides convergence between enterprise applications and telecommunications.

The generic, skeletal architecture of the SDP in Figure 8.38 recognises the need for integration of the service delivery mechanism with operations and business support functions, service creation and management as well as AAA functions.

8.10 CONCLUSION

The OSA/Parlay architecture provides a means of opening the service of a telco network for use by applications in other domains under secure, controlled conditions. A new business model has emerged with telcos and a particular type application provider as roleplayers.

This development represents a move away from intelligence in the network. It is not however a simple move of intelligence to the edge of the network. In the Internet context, intelligence at the edge simply means that all logic and data is in terminals or servers. The real issue is who owns the intelligence, who extracts value from it and what the barriers to entry are. In the IN context, a vertically integrated telco owns the service logic, extracts value and keeps other players out by working in a secure, closed domain, with technology that is not well known. With open service access, other parties can extract value.

Chapter 9

Operations Support Systems

The next generation networks described in preceding chapters support a range of information and communications technology (ICT) applications and services. The NGN Framework developed in Chapter 2 enables a complex ICT system to be analysed, described and designed using a process of abstraction. Figure 2.13 shows the set of layers of the NGN Framework together with crosscutting concerns captured as planes. One of these is the *management plane* that encompasses all concerns about the operation of facilities and services and business relationships with customers, partners and suppliers. The layers of the NGN Framework are concerned with the systems that provide communication between users or enhance applications: transmission, switching, resource and service control, content hosting and distribution and value-adding applications. These are control and user plane operations. The management plane captures the numerous behind-the-scenes operations that are required to enable service to be delivered. These operations are captured in two terms: operations support systems (OSS) and business support systems (BSS).

Operations support systems encompass the set of processes that a network operator requires to provision, monitor, control and analyse the network; to manage and control faults; and to perform functions that involve interactions with customers. Operations support includes the historical term *network management*: the control and management of network elements.

A *business support system* encompasses the processes a service provider requires to conduct relationships with external stakeholders including customers, partners and suppliers. The boundary between operations support and business support is indistinct: business support functions are the customer-oriented subset of operations support. Business support processes, for example, taking an order from a customer for a new service must flow into the operations support processes to configure the resources necessary to deliver the service. Support systems are therefore often described as OSS/BSS systems or simply OS/BS.

Operations and business support systems are complex, critical and costly parts of a service provider's functions. Much attention has been given to OSS in standards bodies and a degree of uniformity of approach has been achieved. A full treatment of OS/BS systems is beyond our scope. This chapter therefore has three objectives. First, we demonstrate the relationships between the ICT systems that deliver services to end users and the OS/BS systems required to support the technical and commercial operation of such systems. Second, we review the leading standards that guide the development of OS/BS systems and their evolution and convergence. Third, given the complexity of OS/BS systems, we illustrate methods for managing complexity and in

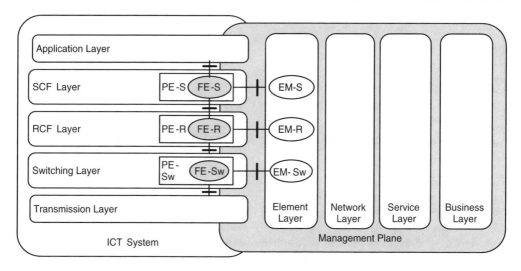

Figure 9.1. Expanding the management plane in the NGN Framework. TMN layering is used for illustration.

particular the specification and description methods described in Chapter 3 that are used in the OS/BS context.

9.1 RELATIONSHIP OF OSS/BSS TO ICT SYSTEMS

An ICT system comprises networks, service capabilities, applications and content resources that actually deliver services to end users. OSS/BSS systems provide a complex set of management functions to support the ICT system. Figure 9.1 shows the relationship between the management system and the ICT system.

The ICT system is represented by the NGN Framework layered model. Each layer contains functional entities that generally have client–server relationships between layers. An entity such as FE-R in a layer presents a northbound serving interface to entities in the layer above. The entity may in turn access the services of and receive notifications from an entity, for example FE-Sw in the layer below via a southbound interface. For example, a functional entity FE-R in the RCF layer, hosted in physical entity PE-R in the resource control layer, serves functional entity FE-S hosted on physical entity PE-S in the SCF layer.

An entity in a layer generally presents an interface to a management system. Each physical entity is managed by an element manager, for example EM-R manages FE-R. Network-wide, service-specific and customer-oriented management functions are performed by entities in other layers of the management plane, for example the TMN layers described in Section 9.2.3.

Management systems may range from simple forms, managing one or a limited number of ICT system entities, to large complex systems. Figure 9.2 illustrates OSS concepts and the reasons for the complexity of OSS/BSS. The infrastructure to provide telecommunications and value-added services consists of network and service elements.

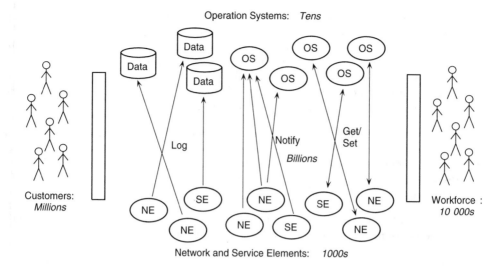

Figure 9.2. Why is operations support complex?

A *network element* (NE) is an individually manageable equipment, including software, forming part of the network infrastructure. Example of network elements in the PSTN are switches, concentrators and SDH transmission systems. Internet elements include IP switches and routers.

A *service element* (SE) is any software entity or network element that is essential to the delivery of a particular service. An IN SCP is an example of a service element while an ATM switch would normally be termed a network element. Service elements may thus be network elements incorporating software processes.

Operations support functions are generally divided into particular areas such as fault management, provisioning and billing. Each is supported by a software system, termed an *operations system* (OS). Orders of magnitude of various entities are informative. A large ICT system such as a telco network may have tens of OS, controlling thousands of service elements or network elements. Operations systems concerned with business processes must interact with customers who are typically numbered in millions. Customers may be individuals or corporate users. A service provider such as a telco has a workforce to carry out business, operational and maintenance functions. This workforce typically numbers tens of thousands.

Operations systems require the support of databases. Large amounts of data is exchanged between NE and SE and the operations system or is logged to databases. Data may be exchanged by polling or may be sent as automatic notifications from NE/SE to OS. The appropriate OS must initiate action in the form of a command or data transfer to an NE/SE or a work order to telco personnel. The number of transactions or notifications flowing in the OSS/BSS is large, potentially numbered in billions.

Management systems interact with every element that makes up the ICT system as well as customers and the workforce. Management systems are, in general, complex because the systems they manage are complex and the scale of operations is large. Because of their complexity, management systems are in general arranged in layers, for example using the Telecommunications Management Network layers shown in Figure 9.1 [122]. In a topic as complex as operations support, organising principles are essential. Various standards bodies use different approaches to the concepts and principles for defining their management standards. In general, the following issues arise in management system architectures and standards.

- What are the elements, systems or services that must be managed?

- What information is required to manage given elements, systems or services. How is this information defined to allow management systems to be structured as open systems?

- How does a managing element interact with the managed element? What management-specific application layer protocols or APIs are required?

- What communications protocols or distribution transparency support is required to support management applications?

- What management functionality is required? How does this functionality support the processes a service provider needs to support services?

Early management systems address these questions in a bottom-up sequence, focussing on protocols and management information. Evolving standards have moved to a top-down approach described in Section 9.5.

9.2   EVOLUTION OF OSS/BSS

The evolution from network management to complex OSS and BSS systems can be traced through various standards: the Open Systems Interconnection (OSI) network management model, the Internet management model, the Telecommunications Management Network (TMN), and the Telemanagement Forum initiatives. The objectives and nature of management systems change during this evolution. Figure 9.3 shows three typical stages in the evolution of OSS/BSS.

Initially, the OSI and IETF standards utilised a simple manager-agent model, together with protocol-based communication between manager and managing entity, termed an *agent*, shown in Figure 9.3(a). The element under management is represented by a defined set of information forming part of a larger structure called the Management Information Base (MIB).

Next, operators sought to manage sub-systems within their networks, for example SDH transmission systems, a set of TDM switches or a Signalling System No. 7 network. The management systems focussed on elements and how they function as a system. The SDH standards, for example, developed an architecture and information models that can represent end-to-end connections and their constituents. With a network-wide management view, services offered on the network require management. Managing systems therefore require extra layers, for example a network layer to co-ordinate the control of elements and a service layer to manage specific services. Provisioning a leased line service would, for example, require allocation of

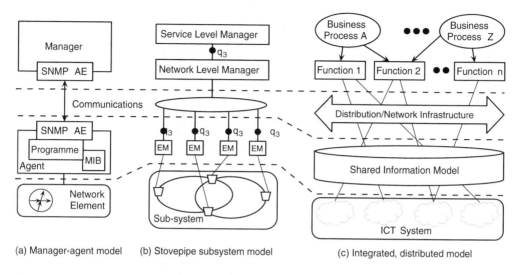

Figure 9.3. Evolution of OSS from simple manager–agent to complex distributed forms.

SDH resources to support the required bandwidth. Such management systems are essentially stovepipe or silo systems with little co-ordination between systems.

The third stage shown in Figure 9.3(c) represents several developments. To support integrated management, management information is sharable across management applications. Managing systems have a structure that separates generic functions from the business processes they support. Functionality is modular and higher level processes orchestrate its use. The system has become large and inherently distributed and proper distribution support exists.

9.2.1 THE OSI NETWORK MANAGEMENT MODEL

The need for a standards-based approach to network management was recognised in the Open Systems Interconnection initiative. The OSI network management standards, developed in the 1980s, are based on a manager–agent model and define application-level operations and a comprehensive information structure [209].

The elements to be managed are identified in abstract terms as *managed objects* (MO). A physical element such as a network switch has a characteristic set of data that describes its configuration, state and statistics. The managed object is a formal representation of this information that is part of a larger consistent set of definitions for all managed objects. The *Structure of Management Information* (SMI) is an extended, object-oriented model that for defining data, events and management operations for managed objects. The object definitions are located in a *management information tree* (MIT) that attaches to the object identifier tree shown in Figure 3.18. Because state changes in one managed object may impact on the ability of another to perform its function, relationships between managed objects may also be defined.

Table 9.1. Comparison of OSI and SNMP management protocols

Aspect	Internet Management	OSI Management
Information model	IETF SMI	GDMO
	Management Information Base	Management Information Tree
Association	Connectionless	Connection-oriented
	Uses UDP	Uses ACSE, ROSE
Connection services		INITIALISE
		TERMINATE
Information tree services		CREATE
		DELETE
Information manipulation services	GET-REQUEST	GET
	CANCEL-GET	SET
	GET-BULK	ACTION
	GET-RESPONSE	EVENT REPORT
	SET	
	TRAP	
	INFORM	

The Guidelines for Definition of Managed Objects (GDMO) extends ASN.1 to provide a method for defining managed objects. Special class templates are defined, principally the **MANAGED OBJECT CLASS**. This class definition allows an MO class to be named, to specify the superclasses that it inherits, to specify its data attributes, operations and notifications and to give the registration information for the object on the ISO object registration tree. The object characteristics, that is the data attributes, operations and notifications, are specified using **PACKAGE**s. The applicability of a package may be conditional on particular characteristics. The operations that may be performed on a data field are specified, effectively defining read-only or read/write data.

The managing and managed entities, referred to as *managers* and *agents* respectively, are viewed as peers. Entities communicate using an application layer protocol that supports management operations. A number of Common Management Information Service Elements (CMISE) are defined. Figure 9.1 names seven service elements defined in CMISE. Other standard protocols are used to support CMISE operations. The Application Context Service Element protocol (ACSE) is used to set up and end associations between entities that communicate. The Remote Operations Service Element (ROSE) is a generic protocol that transfers requests and responses between peers. These protocols have co-ordination mechanisms that allow multiple associations to exist at an entity such as a manager.

A service element such as **GET** has request and response message forms defined in ASN.1. The CMIP protocol provides the actual PDUs, for example **GET-REQUEST** and **GET-RESPONSE**. These PDUs are communicated using lower layer OSI protocols.

9.2.2 IETF NETWORK MANAGEMENT STANDARDS

The IETF in 1988 formulated recommendations for the development of Internet Network Management Standards [34]. The recommendations had two thrusts. First work on further development of the Management Information Base, the information models for managed objects, was planned. Second, the Simple Network Management Protocol (SNMP), reviewed in this section, was to be extended as short term measure with a view to using the ISO CMISE/CMIP service elements and protocols as the long term basis for network management in the Internet. This last goal became difficult to achieve and the SNMP was adopted rapidly to make 'IP and TCP implementations ... network manageable' [33]. *Network manageable* means that three requirements are met. Management information must be defined using the common structures and naming scheme defined in the Structure of Management Information (SMI). Second, the actual managed objects are defined in the Management Information Base. Third, the SNMP is used for communication. These three standards are closely coupled and are formally referred to as the Internet standard SNMP Management Framework [107] or simply by the protocol, SNMP.

The SNMP and SMI standards have evolved through a number of versions to the present version 3 [107]. Shortcoming in earlier versions, for example lack of a security mechanism, access control and a message format, have been addressed. Version 3 is, however, regarded as an extensible framework.

The management framework is intended for application in a range of contexts from a simple command-line manager for remotely configuring network elements to a complex network under the control of an advanced management system. The emphasis in the protocol and method of defining management information has always been on simplicity. The association controls of the OSI management standards were rejected in favour of simple connectionless communication using UDP and IP at the network layer. A simple template was adopted for the **OBJECT-TYPE** used to define managed objects and their attributes.

In version 3, the architectural principles of the manager and agent entities are defined in terms of a group of functions that constitute the *SNMP Engine* and a set of communications functions termed *applications*, in the sense that they provide services to managing applications or interface with the MIB implementation in the managed element. The constituents of the conventional SNMP manager and agent entities are shown in Figure 9.4 using defined applications and SNMP engine components.

The agent entity manages a network element. The network element is described by part of the Management Information Base. All data elements that can be read, changed or notified are defined. The functions of setting, interrogating and notifying management data are termed *MIB instrumentation*. The application functions available in the agent entity allow notifications to be emitted and responses to commands from a manager entity to be issued. Notifications and responses make management data available to the managing application. The *access control* function allows policies to be applied to limit access to the managed element's MIB data. The SNMP architecture allows messages to be forwarded between entities using the *proxy forwarder* application.

An important group of functions is the *dispatcher*. A response or notification from the managed element is serviced by the PDU dispatcher. The message processing

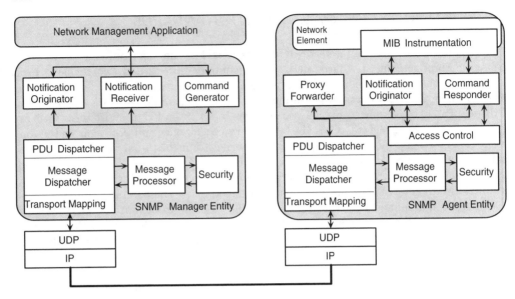

Figure 9.4. Architectural view of the Simple Network Management Protocol.

system processes the message according to the applicable version of the standard and invokes the selected security mechanism. The final step before transmission of the message is mapping onto the transport PDU. While several transport protocols may be used, UDP is normally used in IP networks. Similarly, a command received via the network is serviced by the dispatcher. The dispatcher, message processing, security and access control subsystems form the *SNMP engine*.

The SNMP engine functions, other than access control, are also implemented in the manager entity, as shown in Figure 9.4. The application functions allow commands to be issued on behalf of the management application. Notifications may be received or emitted by the management application.

Formal definitions of the primitives and sequence charts specify the interchanges between the components of the manager and agent entities.

The services available in SNMP are listed in Table 9.1. Because SNMP operates in a connectionless mode, no association control services are required. Three forms of **GET** service elements are available to retrieve a single management information value, the next value in sequence or the values of elements specified in a list. An information element may be assigned a value using **SET**. Two types of notifications are possible. **TRAP** issues a notification without expecting an acknowledgment while **INFORM** expects an acknowledgment. The set of messages is defined in [175].

The Management Information Base for SNMP is defined in [174]. The MIB is a tree of objects of type **OBJECT-TYPE**. The **OBJECT-TYPE** class [184] used throughout the definition of the MIB contains the following fields. The object has a name declared to be **OBJECT-TYPE**. Each object has a **SYNTAX** that is typically a basic type such as an **INTEGER** or **STRING** or a type defined elsewhere in the MIB. The object has an **ACCESS** declaration that indicates whether the data may be read or written. Data may

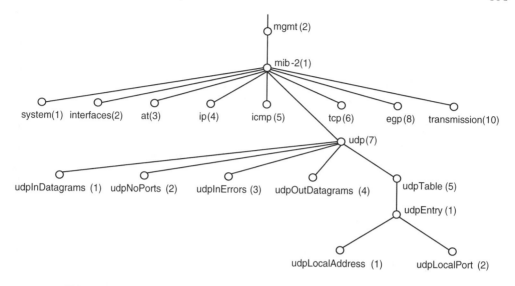

Figure 9.5. Portion of the MIB for management of UDP devices.

be not-accessible, that is, similar to a virtual class, the object is not instantiated in the MIB implementation. The STATUS element shows whether the object is mandatory, optional, obsolete or deprecated. The DESCRIPTION contains a textual explanation of the nature and purpose of the object. An optional REFERENCE field allows explanatory notes to be inserted, for example how the object was developed. The optional INDEX field allows objects to be identified that can be used for locating data in the MIB implementation. Finally, the definition value locates the object in the object tree, by naming the parent class and giving the numerical object identifier at its level in the tree.

The MIB is structured according to the object identifier tree shown in Figure 3.18. The mib-2(1) definitions attach to the node mgmt(2), with object identifier mgmt(1). Eleven groups are defined within the management area; nine are shown in Figure 9.5, one historic form and one for SNMP entities. Each of the groups has a comprehensive definition that extends the tree. The UDP group, attached to node udp(7), is shown.

Four objects from the UDP module of the MIB are illustrated in Figure 9.6 [174]. A structure udpTable is made up of objects of type udpEntry. The entries have two fields, an IP address udpLocalAddress and a UDP Port number udpLocalPort, a 16-bit integer. The definition makes the address and port fields readable by a GET operation. Types udpTable and udpEntry are not-accessible because they serve to define the structure at a higher level. An index consisting of a pair of values udPLocalAddress, udpLocalPort serves to locate a particular part of the MIB implementation.

SNMP has developed to version 3 and has defined architectures for the manager and agent entities, a communications protocol, and a well developed Management Information Base with definitions for most types of elements found in IP networks. The SNMP standards do not, however, guide the design of complex management systems: the designers must adopt a suitable methodology for structuring the system.

```
udpTable OBJECT-TYPE
        SYNTAX      SEQUENCE OF UdpEntry
        ACCESS      not-accessible
        STATUS      mandatory
        DESCRIPTION
                    "A table containing UDP listener information"
        ::= { udp 5 }

udpEntry OBJECT-TYPE
        SYNTAX      SEQUENCE OF UdpEntry
        ACCESS      not-accessible
        STATUS      mandatory
        DESCRIPTION
                    "Information about a particular current UDP listener"
        INDEX       { udPLocalAddress, udpLocalPort }
        ::= { udpTable 1 }

UdpEntry ::=
        SEQUENCE {
                    udpLocalAddress IpAddress,
                    udpLocalPort INTEGER (0..65535) }

udpLocalAddress OBJECT-TYPE
        SYNTAX      IpAddress
        ACCESS      read-only
        STATUS      mandatory
        DESCRIPTION
                    "The local IP address for this UDP Listener. ..."
        ::= { udpEntry 1 }

udpLocalPort OBJECT-TYPE
        SYNTAX      INTEGER (0..65535)
        ACCESS      read-only
        STATUS      mandatory
        DESCRIPTION
                    "The local port number for this UDP Listener. ..."
        ::= { udpEntry 2 }
```

Figure 9.6. Example of MIB definition: UDP listener table.

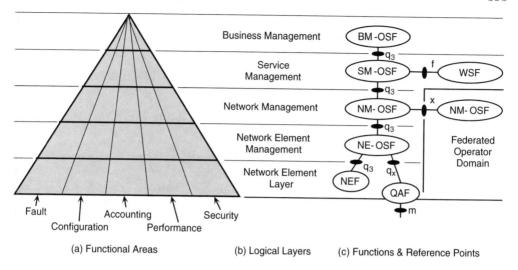

Figure 9.7. Concepts used in defining the Telecommunications Management Network.

This approach is favoured in the Internet community. The telecommunications community sought a greater degree of structuring of management systems in the Telecommunications Management Network described next.

9.2.3 THE TELECOMMUNICATIONS MANAGEMENT NETWORK

While the IETF developed a simplified approach to management of IP networks, telcos sought to manage large, complex networks with special requirements. Networks must meet five-nines availability requirements, operators federate to offer services, new services are introduced and customers demand a high level of care by the service provider. The ITU-T therefore developed an approach to management standards that was geared to the specific requirements of telcos. The approach built on the OSI network management standards, for example by adopting CMISE and CMIP and structuring management information using GDMO object definitions. The standards were to support management of services and business level activities in addition to network level management. The resulting suite of standards specified the Telecommunications Management Network (TMN) [122].

Three concepts used in specifying the TMN are summarised in Figure 9.7. The TMN standard divides management concerns into five *logical layers* [122]. The TMN system is depicted as a pyramid with five layers:

1. *Network Element Layer:* contains the agent portion of the element manager.

2. *Element Management Layer:* contains the management entity for a single managed element.

3. *Network Management Layer:* is concerned with the management of an entire network consisting of a number of elements.

4. *Service Management Layer:* consisting of the customer interface, account management, service provisioning and complaint handling. No management of physical entities takes place.

5. *Business Management layer:* planning, agreements between operators, setting, implementing and tracking goals.

The second organising principle shown in Figure 9.7 is the TMN *management functional areas*, encapsulated in the FCAPS acronym:

1. *Fault management* is concerned with detecting, isolating and correcting abnormal conditions in networks and elements, embracing operations such as: alarm surveillance; fault detection, isolation and control, logging; circuit testing; correction of abnormal behaviour; and problem detection, diagnosis, tracking and trouble ticketing.

2. *Configuration management* provides data to and controls network elements to elicit desired behaviour. Functions include: view management; topology management; software management; inventory management; provisioning and status and control.

3. *Accounting management* enables the measurement of service usage and determination of costs, including: usage of resources; collection of data and charging and billing.

4. *Performance management* evaluates and reports on the behaviour and effectiveness of the network, specific equipment or network elements. Functions include: data collection and analysis; problem reports, displaying, formatting; evaluation of effectiveness of resources; traffic management and performance monitoring.

5. *Security management* includes: controlling customer access to facilities and data; detecting, tracking, reporting security violations and maintaining security services such as encryption.

The third TMN concept shown in Figure 9.7 is the *functional architecture* and reference points. Management functions are performed in the element, network, service and business layers. Management functions are termed *operations system functions* (OSF). In general, OSFs process information to monitor, coordinate or control telecommunications functions. Four types of OSF exist, element, network, service and business, corresponding to the four layers in the layered model.

The Network Element Layer hosts two types of functions. The *network element functions* (NEF) are functional blocks associated with network elements that must be monitored or controlled by the TMN system. A NEF communicates according to TMN standards and the managed element has a TMN-compliant information model. Typical elements that have TMN-compliant management interfaces are PSTN switches and Intelligent Network building physical entities. A second functional block, the Q-adapter function, provides adaptation for managed elements that are not TMN-compliant. Interface m between the adapter and the element to be managed is an implementation detail – the element may for example have an IETF standard or proprietary MIB definition.

The *workstation function* provides the human user interface to the TMN. This function is partly inside and partly outside the TMN. The former presents TMN reference points to other functional blocks, the latter provides the human interface which is not specified in the TMN recommendations.

Not shown is the *mediation function* that ensures that information passing between OSF and the network conforms to the expectations of the communicating parties. Mediation function blocks may store, adapt, filter, threshold and condense information.

The TMN recommendation defines four reference points between functional blocks. The q_3 reference point is the standard TMN reference point for communication between OSFs or an OSF and a NEF. The q_x reference point is defined for communication with non-TMN compliant blocks. Effective management of services offered across telco administrative domains requires management systems to interact. The x-reference point is defined as an interdomain reference point. The f-interface is located between an OSF and a workstation function.

Communication at reference points is realised by a standardised interface, denoted Q_3, Q_x, X and F respectively. The Q_3 interface uses OSI protocols at layers 4–7. CMISE, ACSE and ROSE are used as in the OSI management model. The Q_x interface may use any communications protocol recommended by the ITU-T. The X interface is substantially the same as Q_3, with appropriate security for inter-administration communication.

The TMN's *information architecture* is an-object oriented approach which allows the Open Systems Interconnection systems management principles to be applied in the TMN context. A network information model is a uniform, consistent and rigorous method for describing the resources in a network, including their attribute types, events, actions and behaviours. The network information model is generic to ensure that a wide range of network resources can be modelled. ITU-T Recommendation M.3100 [119] defines a generic network information model for TMN, following the approach of the OSI management model. Physical resources are represented by managed objects, registered on appropriate branches of the object identifier tree. Definitions are inherited from the OSI management information definitions.

The definition of managed objects differs from that used with SNMP in two respects. First, the GDMO is used, resulting in more comprehensive and complex object definitions. PACKAGES and CONDITIONAL PACKAGES are used to define behaviour, a prose definition of the object and what it does, and attributes. The methods that may be used with each attribute are defined. An object definition may have ACTIONS, defining methods that can be invoked on the managed element. For example, a fabric object controls connections in a crossconnect. ACTIONS connect and disconnect are defined within fabric. The CMISE ACTION service element listed in Figure 9.1 defines PDUs that carry action requests and responses to managed elements.

The second difference relative to SNMP is the definition of elements that allow a network to be described, not just the elements. Telecommunications networks, particularly transmission systems, are connection-oriented. Management information is used to control connections, for example in SDH crossconnects. SNMP by contrast originated in the connectionless IP network where a definition of elements such as routers could be self-contained with fields containing the address of the elements to which each interface is connected. These fields are populated by routing protocols rather than centralised management systems. Managed objects are therefore defined in TMN that allow end-to-end connections to described and managed.

TMN has guided the design of management systems for telecommunications networks. The FCAPS organisation of management functions is widely applied. The emphasis in TMN falls mainly on the element and network level with service management being less developed. The business layer is relatively undeveloped. TMN identifies but does not specify functions to be performed in the FCAPS areas. The functional entities that perform these functions are located in layers and interfaces are defined. Communication across reference points is specified using standard interfaces. TMN is therefore communications-oriented. The identification of functions that support processes in the business layer is useful but not explored within TMN. A fuller conceptualisation of functions to support business processes had to wait for the Telecommunications Operation Map described in the next section.

9.3   THE TELECOMMUNICATIONS OPERATIONS MAP

The Telecommunications Operation Map (TOM) is both a response to limitations of TMN but also a precursor to the Enterprise Telecommunications Operation Map (eTOM) described in Section 9.4. TMN embodies a traditional approach: standards are developed and managed and managing systems are then built to these standards. In developing TOM and eTOM, the Telemanagement Forum focusses strongly on the business processes that a service provider needs, the functionality and information needed to support these processes and how a management system can be integrated using commercial off-the-shelf (COTS) components. The model attempts to serve the business needs of all types of service providers: traditional telcos, Internet service providers, enterprises and application service providers.

The Telecommunications Operations Map identifies the functions that are required for three groups of processes that a service provider needs to take care of customers [202]. The service provider must *fulfill* a request for service by provisioning the necessary facilities, initiating billing and activating the service. Once the service is active, its continued operation and quality must be *assured* by appropriate responses to faults or degradation of quality. During use of the service, usage information must be gathered and the customer *billed* at agreed rates. Processes in the three areas, fulfillment, assurance and billing (FAB), require a number of generic or recurring functions. The task of the TOM is to identify these functions, to group them suitably into processes and to map the interactions between processes to create end-to-end management processes.

The functions required for typical FAB processes involve interactions with the customer, services, network resources and at the network element level. Functions at different layers must interact to execute a process. The TMN layers are therefore adopted in the TOM as shown in Figure 9.8 with names that better reflect their functions. The business layer is not shown explicitly. Business processes call on the services of the layers and could be thought of as spanning the layers. The FAB areas cut across the layers by grouping functions as shown.

- The *Customer Care Processes* layer is split as a separate layer from the TMN Service Management Layer. Functions that support the purchase of a service and handling of the order are in the fulfilment area, problem handling and QoS

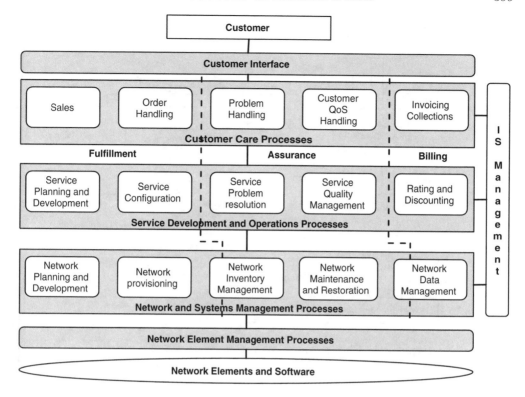

Figure 9.8. The Telecommunications Operation Map.

handling fall within assurance, while invoicing and collection is in the billing area. These functions all involve interactions with the customer through a customer interface process.

- The *Service Development and Operations Process* layer is also a development of the TMN Service Management Layer. This layer also contains the fulfillment, assurance and billing-related groups of functions shown: Service Planning and Development, Service Configuration, Service Problem Resolution, Service Quality Management and Rating and Discounting.

- The *Network and Systems Management Process* layer is a development of the TMN Network Management layer with fulfillment, assurance and billing-related groups: Network Planning and Development, Network Provisioning, Network Inventory Management, Maintenance and Restoration and Network Data Management.

- The *Network Element Management Processes* layer has no identified functions but represents the element management layer.

The TOM seeks to fulfill the promise of TMN in which the functionality of a lower management layer is made available to a higher layer and, in particular,

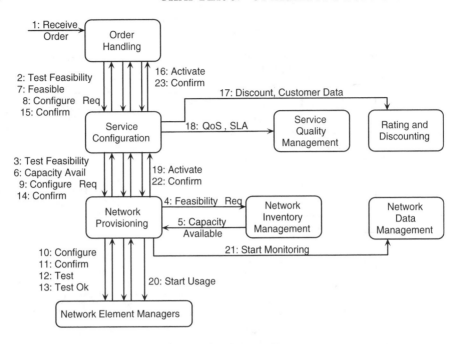

Figure 9.9. Hypothetical interaction illustrating processing an order and activating a service.

comprehensively serving the needs of the business processes. To meet this objective, each serving layer must have adequate capabilities.   These capabilities must be identified in a way that allows different processes to be synthesised using available functions.   Also, the way that functions interact should not be constrained by limited reference points and protocols. Rather, the top-down approach of the RM-ODP or Model Driven Architecture is be employed. The TOM can be regarded as part enterprise viewpoint and part information viewpoint for the suite of functions needed to support customer-related operations in a variety of service providers. The requirements are also suggestive of a service-oriented approach.   The functions in individual processes are be made available as services that can be invoked by process logic as required.

Figure 9.8 presents a static model, identifying only the basic processes, interfaces and external entities. Specification of dynamic processes are supported by *FAB process diagrams*, showing the processes involved, activities and interactions among processes and with external entities.

Each process is defined as having a number of functions. For example, the Service Configuration process includes functions that configure network and CPE, initiate installation work, manage numbers, activate/deactivate service and request end-to-end service testing.   The Service Configuration process is located in the Service Development and Management layer and in general receives requests from processes in the Customer Care layer. For example it may receive a Configure Service request from the Order Handling process. A process generally makes requests on other processes

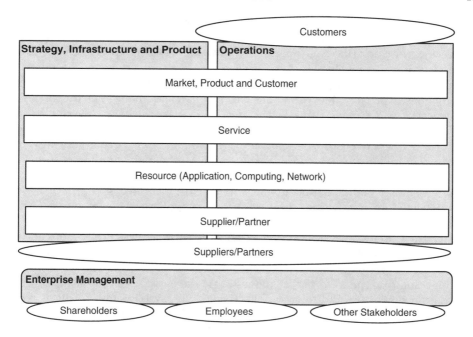

Figure 9.10. eTOM with level-0 processes.

in its layer or in the layer below. The Service Configuration process may, for example request Service Quality Management to configure QoS data for a customer or request Network Provisioning to assign resources.

Figure 9.9 gives a hypothetical example of the type of interactions that take place between processes in an end-to-end process. In this example, an order for a service, say for VPN interconnection, is received at 1 by Order Handling. An automated process is triggered, going through phases of testing whether the service can be offered as requested, configuring and activating the service, including the billing arrangements.

9.4 ENHANCEMENT OF THE TOM: ETOM

The Telecommunications Operations Map serves to unify business support processes, that is, those processes that interface with the customer, and operations support, that is processes that interface with the infrastructure relating to specific services. The framework allows end-to-end processes to be designed to meet service fulfillment, assurance and billing requirements.

The FAB processes are only some of those required by a service provider. The TOM has been expanded into the Enterprise Telecommunications Operations Map (eTOM) to cover all processes generally required by a service provider [203]. The eTOM uses a similar structure to TOM. TOM has two main levels of features, the functional layers and the individual processes. Because of eTOM's increased complexity, three main levels of features are identified.

Level 0, shown in Figure 9.10, provides a broad classification of processes and identifies the external stakeholders. The fulfillment, assurance and billing types of processes are described as operational. Other essential processes required by the service provider involve strategic decision making on services to be offered and infrastructure required to support the services. Both services and infrastructure go through phases in their lifecycles that need systematic management. A *Strategy, Infrastructure and Product* area is therefore added to the map covering processes that support decisions on what services to offer, how to offer the services and how to present the services to customers. An *Enterprise Management* area is added to contain processes required in any enterprise. As in the TOM, the map identifies customers as a stakeholder group. Because service providers often enter into alliances or partnerships with other similar bodies, suppliers and partners are identified as a group. Processes must be defined for interacting with partners and suppliers. An augmented TMN-like layering is superimposed on the operational and strategic areas.

Level 1 features shown in Figure 9.11 add the vertical classification of end-to-end processes into areas such as fulfillment, assurance and billing already defined in the TOM. In the operational area processes are needed to ensure that the service provider is in a position to respond to service requests. At the customer relationship level, sales channels must be open and function effectively. The systems and facilities required to offer each service must operate cost-effectively, within quality specifications and be properly supported. Similar requirements exist at the network level: resources must be available and operational to meet new service provisioning demands. These processes form an *Operational Support and Readiness* (OSR) group. The three FAB processes and OSR meet the operational requirements for the service provider. The horizontal layers are specialised at level 1 for the operational and strategic areas.

Processes such as Order Handling and Service Configuration shown in the TOM form the level 2 view of the eTOM. Processes may be further decomposed: level 2 process elements into level 3 elements that may in turn be decomposed to level 4.

The TOM is thus subsumed into eTOM, giving a much expanded enterprise and information viewpoints. The eTOM does not, however, exist in isolation but forms part of a strategy to define the New Generation OSS, described in Section 9.5. In the New Generation OSS, eTOM provides a top-down, enterprise-level view of a system that supports all the end-to-end processes required by a service provider.

9.5 NEW GENERATION OSS

Operations and business support systems exist on different scales. At a large scale, an OS/BS system is a complex, distributed software system performing critical functions for the service provider. The OS/BS system must support several management processes and interact, on the one hand, with a large number of customers and, on the other, with a number of service and network elements. The analysis, design and implementation of a large OS/BS system should advisedly follow a proven methodology such as RM-ODP or the Model Driven Architecture. These generic approaches must be adapted to meet the objectives of the system.

The Telemanagement Forum has set several objectives for a New Generation OSS (NGOSS). A framework is required to support the development of OSS solutions. The transition from a framework to a working system is achieved by specifying, acquiring

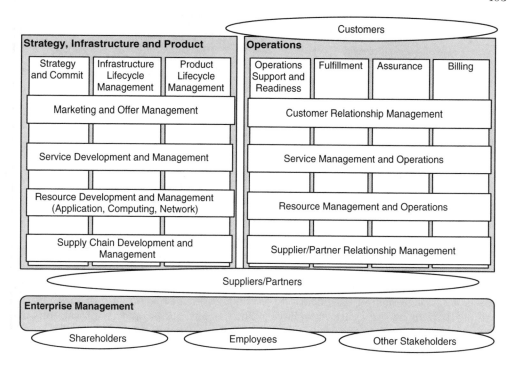

Figure 9.11. eTOM with level-1 processes.

and integrating a set of software components that work together to create end-to-end processes. These components must be implemented in existing software technologies. The components and method of integration must give cost-effective solutions. To this end, the NGOSS promotes the use of commercial off-the-shelf components.

The NGOSS approach to meeting these requirements has several threads. First, a set of views on the target system supports different phases in the lifecycle. Second, technology neutral methods are used initially before moving to technology-specific implementations. Third, a data model is used that supports all phases in the lifecycle.

The framework support the different views on the target system. Four views identified in Figure 9.12 cover the lifecycle of the OSS solution through its analysis, design, implementation and operation phases. Each view captures specific requirements.

The *business view* captures the requirements for the end-to-end processes that the service provider requires. The level 2 and lower processes in the eTOM are orchestrated to specify end-to-end processes. While the eTOM model assists the definition of the processes it is more than just an enterprise viewpoint.

The *system view* is concerned with the objects that make up their system, their behaviour and interaction. Here the functions and collaborations identified in the business view are defined in detail.

The *implementation view* is concerned with the components and their interactions. The way that components interact is defined using contracts. In the NGOSS context, a *contract* consists of three parts. First, a technology-neutral interface is defined,

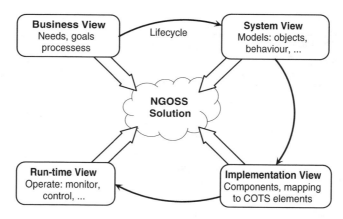

Figure 9.12. NGOSS views used to describe an OSS.

including the set of named operations and functionality that the interface supports. This is a service-oriented specification. Second, a definition is made of how entities that participate in the contract behave. Third, constraints on the interactions are defined. Component functionality and contracts are defined in technology-neutral form. A constraint could determine whether a particular operation may be used by a particular requesting party. The implementation view also deals with how the components are mapped to commercial-off-the-shelf components.

The *run-time view* is concerned with the proper operation of the system and mechanisms to monitor and adjust the system if required.

The four views of the NGOSS, while bearing strong similarities to the RM-ODP viewpoints and the modelling stages in MDA, have a more powerful mapping. Figure 9.13 shows the NGOSS specification, design and implementation process in relation to the RM-ODP and MDA methodology [198].

The eTOM defines enterprise and information viewpoint aspects but also defines requirements on processes and policies for the operation of the system. The information viewpoint or first stage of the MDA platform independent model is defined in the *Shared Information and Data Model* (SID). The SID model uses UML notations and is therefore a technology-neutral definition method. Extensive use is made of design patterns for various purposes: managed entities, management entities, products, services and resources, both physical and logical. Management entities can be of the FCAPS types as well as keeping state and statistical information about resources.

The shared information and data, expressed in SID, extends across the four NGOSS views. For each view, the process requirements specified in the eTOM call for information or data definitions in the SID. Types exist for defining business relationships. The information model defines managed objects, their behaviour and interactions. Managed objects are not confined to physical and logical entities as in SNMP or TMN. A service specification can be defined, including customer-facing attributes and resource-facing attributes. For the implementation view, the SID defines contracts and classes for specifying common services such as repository

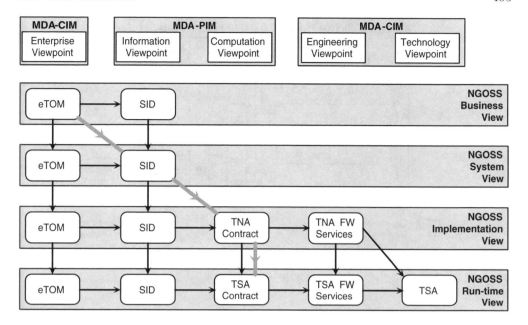

Figure 9.13. NGOSS process using NGOSS views and RM-ODP/MDA.

services. SID defines profiles for mapping technology-neutral components and service definitions.

The RM-ODP computational viewpoint maps into both implementation and run-time views. Technology-specific issues are restricted to the run-time viewpoint. The technology-neutral contract is defined in UML. This specification is mapped to technology specific forms to give the implementable components. Similarly, the technology-neutral engineering viewpoint defines the support for distributed computing, for example repository services. Mapping to the target implementation technology gives the technology-specific version of the engineering viewpoint.

Figure 9.13 shows that design and implementation of an OSS is not simply a step-by-step process through the RM-ODP viewpoints or MDA models, indicated by the shaded arrow.

Rather, NGOSS deals with the high complexity of an OSS within the entire enterprise context. The NGOSS analysis, design and implementation process for a particular operations support system is guided by the specific requirements of the services provider, within the framework provided by the eTOM, supported by the Shared Information and Data Model using the generic methodologies and tools provided within the Model Driven Architecture.

9.6 CONCLUSION

This review of the evolution and current state of OS/BS systems concentrates on the broad characteristics of the various architectures. Management is required at various scales, and systems based on SNMP, TMN and NGOSS systems will exist. For large-scale systems in all types of service providers, the NGOSS represents the convergence

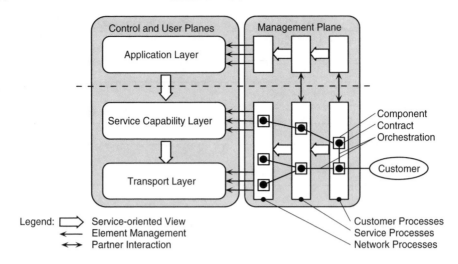

Figure 9.14. A unified view of a NGN with NGOSS-structured management plane.

of several streams of development to provide a unified approach to complex OS/BS systems. The communication-oriented TMN has matured into a distributed computing paradigm with full support for distributed processing defined as part of the information model. The SID model defines all information and data in a single UML form.

Figure 9.1 introduced OS/BS systems by enlarging on the concept of the management plane and its relationship to elements involved in delivering services, that is, those that inhabit the control and user planes. The NGOSS allows the relationship between management entities and managed entities to be summarised as in Figure 9.14.

Service orientation is taking root in the control plane. Services made available by one layer to its neighbour above are encapsulated in an open interface that is supported by discovery and access control methods. An instance is the Service Oriented Architecture with its WSDL descriptions, UDDI discovery and SOAP/HTTP transport. Other forms of service orientation such as the OSA/Parlay API are based on other mechanisms.

The OS/BS system designed following NGOSS, based on components and contracts is service oriented. A specific principle is that functionality must be made available as components, preferably available commercially off-the-shelf. The TMN layering principles are maintained in NGOSS. The components may therefore be classified as providing functionality particular to customer care, service management or network management. Communication between components is, however, governed by contracts rather than a hierarchy as in TMN. In general, service oriented views are presented between and within layers.

The eTOM embraces management processes between enterprises that partner to provide services. The OS/BS components may therefore be distributed across domains. In the case of an independent service provider using an OSA/Parlay platform, management information may have to be exchanged with the telco's OSS. The implementation of the Connection Management API may make use of management rather than control plane components.

Chapter 10

Migration from Legacy to Next Generation Networks

10.1 RETROSPECT

Chapter 1 identifies a number of present day networks that mark the starting point of the evolution toward next generation networks studied in this book. Convergence is identified as a process and a state of networks or services. This study is aided by the NGN Framework presented in Chapter 2 and by the specification and description methods for software processes reviewed in Chapter 3.

On the evolution path, we discovered a number of NGNs. Chapter 4 introduced the H.323 and canonical SIP IP multimedia call and session control protocols. Convergence enablers are the media and signalling gateways and their associated signalling protocols.

We examined enterprise networks, computer–telephony integration and trends in enterprise software in Chapter 5. Convergence exists between IT applications and the control of communications that originates or terminated in a private domain. Telephony APIs and network-level convergence are major influences.

Chapter 6 is an informative digression into network architectures that, despite their merits, were not adopted for development of products and deployment in the network: the Broadband ISDN, TINA and TIPHON.

Chapter 7 examined 3G mobile communication networks as an instance of a next generation network. This study reveals convergence in the form of adoption of an IP packet network in the core and service architectures that allows single and multimedia services to be offered.

Chapter 8 examines standards for supporting service creation and opening the capabilities of the service provider's network to third party service providers. We focussed principally on the OSA/Parlay and Parlay X interfaces, but also outlined JAIN, the OMA Service Environment and the general concept of a Service Delivery Platform. Here, convergence enablers are open interfaces that create a point of integration between diverse applications, inside and outside the telco domain, and abstracted network capabilities. Similar to the enterprise domain, convergence is enabled between IT application providers and public telecommunications infrastructure.

Chapter 9 provides a view of the long and difficult development of operations support systems and business support systems and locates these relative to the NGN Framework.

Chapters 4–9 provide a treatment of the chosen architectures grounded in the NGN Framework. We have touched on some issues that are the preoccupation of bodies involved in NGN standardisation, namely functional architectures, service capabilities, service control and network management.

Much detail, such as the three areas described next, could not be covered. First, several of the standards and their application to products and systems are still evolving. For example, the IMS described in Chapter 7 applied in fixed networks is such a case. Second, many topics could not be included, principally a detailed study of the evolution of Internet services, billing for complex services in NGNs, and a study of the multiservice packet network. Third, the lessons of Chapter 6 teach us that technological merit is often insufficient and factors outside the technology often determine its acceptance or rejection. Other factors such as timeliness, advocacy and marketing on the part of standards bodies and the ability to fulfill a perceived need, albeit imperfectly, can be equally important. These could not be treated in detail.

10.2   REFLECTING ON EVOLUTION AND CONVERGENCE

10.2.1   *THE EVOLUTION METAPHOR, WITH HINDSIGHT*

The evolution metaphor is used constantly in ICT to describe the progress of technology and the resulting services that end users enjoy and that generate revenue for service providers. In biology, evolution is the accumulation of traits by a process of inheritance over time [43]. Natural selection captures the observation that better-adapted living organisms are more likely to survive, resulting in a build up of favourable variation and a rejection of unfavourable characteristics. Humankind has exploited some of the diversity of lifeforms that have evolved. From early agrarian societies to modern agribusiness, man has helped the development of advantageous traits in farm plants and animals by breeding, hybridising and more recently by genetic engineering.

Evolution in ICT is in similar fashion driven by people and organisations seeking advantageous innovations. Different parts of the total infrastructure evolve in a loosely coupled way: core networks, access networks, service architecture and management systems. A modern ICT system relies on the co-ordinated use of different facilities to provide a service. Over time, facilities evolve and new services are developed from existing services, driven by changes in the environment. Developments range from supportive to disruptive. Some changes are successful while others are not: a process similar to natural selection operates. This observation applies to both architectures and services. Third generation mobile communication system architectures have gained acceptance while B-ISDN and TINA did not. A diversity of successful services, for example some of the many listed in Table 10.1, results from the evolutionary process.

While technological evolution results in a growing suite of viable services, the range of underlying facilities should, however, not proliferate. Rather, through a process of convergence, the combination of facilities required to support an expanding set of services, such as those listed in Table 10.1, should grow at a limited rate. This is a *scaling principle* for multiservice networks: only a limited change of facilities should be required for a larger expansion in services offered and users served.

Technologies may encounter evolutionary dead-ends at different stages of their lifecycle. A mature technology may not be able to take on more functions, as in the case of circuit-switched voice services. A new architecture may fail to gain acceptance

Table 10.1. Current or near future services and capabilities required by end users

PSTN and ISDN Services	Internet Services
Voice: analogue loop	Electronic Mail
Voice VPN	Web Browsing
Dialup-to PoP	Chat
ISDN Voice	Transactional Services
ISDN dialup	e-commerce
ISDN videoconference	Search
IP VPN	Voice-on-the-Net
ADSL Broadband Internet	Videoconferencing
	Peer-to-peer filesharing
2G and 3G Mobile	Music download
Voice	Instant Messaging
Data: CS	Video-on-demand
GPRS Data	Games
Short messaging	Weblogs (blogs)
Multimedia Messaging	Wiki
Broadband Internet Access	Voting
Push-to-talk	Advertising
Video on demand	Betting
Pay per view video clips	
t-commerce	**Work Support Tools**
Virtual [Mobile] PBX	Calendar
[Virtual] Call Centre	Wordprocessing
Multimedia VPN	Spreadsheet
	Presentation
Mandatory Services	Whiteboarding
Legal interception	
Emergency Service	**Entertainment**
	Music Player
Bundled Services	Personal video recorder
Triple play (Voice+Internet+Video)	Interactive TV
Fixed-mobile convergence	Enhanced HDTV
Quad Play (Triple play+mobile phone)	Home networking
Interactive news ticker	News
Rich Voice (Voice+image+ ..)	My content any time
Community Services	
Generating content	
Local events	
Community TV viewing	

for one or more reasons: it may not be sufficiently understood, it may be too complex or too simple, it may be too early or too late to find acceptance in the marketplace and achieve an adequate scale of operation.

Evolution and convergence are influenced by the interplay between the development of service enablers and the goals of the market. Market pull is the primary driver and has been since the inception of the Intelligent Network: the ISDN marked the end of a technology-push approach. A deregulated, competitive ICT market ensures that this situation will prevail well into the future. Developers of new architectures and standards must be market-oriented.

10.2.2 MARKET-ORIENTED OBJECTIVES

The AIN and the ITU-T Intelligent Network were the first service architectures to be proposed as telecommunications changed from a monopolistic to competitive environment. Several objectives were formulated for AIN and IN and are still valid today when new service architectures are being deployed. While the emphasis may have changed, the market-oriented objectives of service architectures remain as follows:

- *Rapid service creation* was an objective of the classical IN [112]. There it was expressed as reduced time-to-market: the time from conception of a new service to deployment. This goal persists in current service architectures, for example OSA/Parlay [150]. The envisaged services are potentially more complex.

- *Software reuse* is related to rapid service creation. While the approach to reuse has changed, for example from SIBs to components, the goal remains important.

- *Network independence* was, interestingly, an objective of the classical IN. The same service provisioning methods would be applied when new bearer networks take over from the circuit switched type. This actually occurs in 3G voice networks with CAMEL-based services using packet bearer networks. With multiservice packet networks and softswitch control, network independence is attained.

- Allowing *subscriber control* of subscriber-specific attributes of a service and *end user control* of user-specific attributes were objectives of the IN. This concept has broadened into user-centricity in next generation networks and extends to both service management and real-time service control.

- *Vendor independence* in the sense of interoperability of equipment from different vendors was an objective of the original Advanced Intelligent Network (AIN) and remains an objective of new service architectures such as OSA/Parlay and Parlay X.

- Support for *independent software vendors* to develop applications that enhance telecommunications or that are enhanced by network capabilities is an objective of current software architectures. In the IN, this objective was limited to reducing the dependency of telcos on equipment vendors for developing service logic and avoiding the inflexibility and long development cycle of switch-based services.

- Related to the two previous objectives, *broadening the pool of applications developers* is a new and ambitious goal today. The strategy is to use widely understood software technologies to access a large pool of programmers.

To achieve these objectives, the IN located value-adding service logic on a separate computing platform, the Service Control Point. This separation has consolidated with the general application of the softswitch and open service access principles, the latter based on service-independent interfaces.

10.2.3 PROMOTING CONVERGENCE

A number of characteristics of convergence are identified in Section 1.3.2. The object of working toward or discovering convergence is the integration of facilities, services or business models to provide new, improved or more economical services. Integration takes several forms: multiservice networks, multifunction terminals, digital media encoding and interfaces that support integration or interworking. The treatment of next generation networks in Chapters 4–8 reveals three strategic approaches to promote convergence.

First, and also paradoxically, integration is promoted by seeking meaningful *separations*. Useful points of integration occur between carefully separated functional groupings. We therefore promote the use of layering and the definition of open interfaces.

A second form of integration is interworking, for example between different technology domains or different types of terminals accessing a multifunction core network. Interdomain interworking is well established, for example between circuit and packet-switched telephony. Interlayer interworking is also important, for example allowing different types of service platforms to access network capabilities. We therefore also seek *points of integration* between domains and between layers.

Third, integration is promoted by a common approach to the development of standards through *technology-independent definitions*, allowing implementation in the most appropriate technology.

10.3 TECHNOLOGY MIGRATION

The evolution of architectures, technologies, services and applications is the focus of Chapters 4–9. Most service providers must take legacy facilities and services into account in assessing technologies for the future. New technology may take over the functions of legacy technology to sustain existing services, for example telephony, while expanding the range of services offered, for example unified messaging. Candidate technologies in different areas are listed in Table 10.2. The process of introducing new technologies to take over functions of legacy infrastructure is termed *migration*. For decision makers in service providers, an understanding of the migration process must complement an understanding of the technology. This chapter therefore examines salient features the technology migration process. We revisit the question of managing complexity: we have described means for controlling architectural complexity; we now need to examine the complexity of the migration process.

Migration generally involves substitution of infrastructure or service implementations to deliver an existing service, for example voice, and to open opportunities for

Table 10.2. Current candidate technologies for various aspects of the NGN

Area	Candidate technology
Access	Numerous wireline and wireless technologies
Core network	MPLS, Ethernet and Next Generation SDH
Call/session control	IMS profile of SIP
Service capabilities	OSA/Parlay SCFs, JAIN and OSE
Application servers	OSA/Parlay, J2EE, SIP Servlets
Network Management	NGOSS

new service offerings on the new infrastructure. Migration can be driven by different economic factors: preservation or increase of revenue or containment of operating costs. In particular, an operator may migrate technology that supports existing and new services to avoid having to operate two technologies simultaneously.

Migration is also referred to as *network transformation*. Four categories of transformation are identified. First, the operator's business model may change, for example a voice service provider expanding into consumer information and entertainment. Second, at the network layer, transformation generally takes the form of conversion to a packet network with the view to this being the sole transport network in the longer term. Third, service platforms and the service creation methodology change. Finally, operations support, business support and customer care systems also change, as outlined in Chapter 9.

We first revisit the technology evolution process to gain insight into migration. New technologies variously replace existing technologies to provide a given service, interwork with existing technologies or provide completely new services. The last case is rare and operators are faced mostly with the migration of one technology to another.

10.3.1   THE MIGRATION PROCESS

A service provider wishing to offer a service must have the necessary resources to support the service. Two main processes are required to offer a service. The *resource management process* contains the selection of the resource, both hardware and software, the integration of the resource into the providers infrastructure, provisioning of resources for the service and ultimate withdrawal of the resource from service. Figure 10.1 shows the resource management process. The *provisioning lifecycle* incorporates the planning, building and operation of the resources required to support the service as well as the recovery process when unwanted resources are returned to the pool of resources.

The *service management lifecycle* starts with the selection and integration of the service, supported by the provisioned resources. The service is then available for subscription by customers. The *customer care lifecycle* covers ordering, fulfilment assurance and billing with deactivation when the customer no longer uses the service. the service is withdrawn at the end of its lifecycle.

Figure 10.2 shows the main evolution paths for mobile communication systems showing the resources required at each generation. Evolution paths are identified for the access network, the core network, call control, call/session signalling and

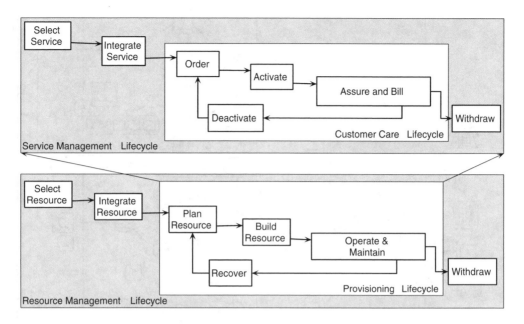

Figure 10.1. Resource and Service Management Lifecycles.

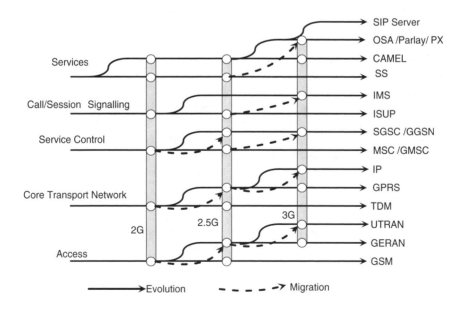

Figure 10.2. Evolution paths for different network constituents and examples of migration processes.

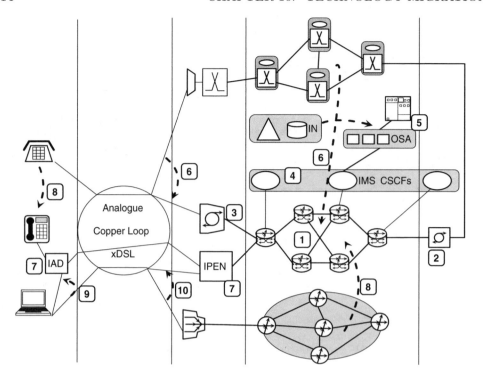

Figure 10.3. Migration of PSTN and Internet to converged multiservice network.

value-added services. Other areas such as OSS are not shown. For each generation of communication system, one or more technology in each of these areas is used to make up the system, as shown by the circles. The original 2G system uses a GSM (TDM) access network, a time division multiplex circuit-switched core network controlled by Mobile-system Switching Centres, ISUP signalling, supplementary services and CAMEL-based services. Selection of technologies from the individual evolution paths to make a system is an aspect of the integration process.

A number of migration processes are superimposed on Figure 10.2. The 2.5G system adds GPRS core transport and the access network migrates to GERAN. Major migrations are required to provide facilities for the 3G system. The UTRAN access network is added and must coexist with the GERAN network to accommodate users with 2G handsets. The core network, if not already one, must become an IP network. ISUP call/session signalling is still used in the CS mode core network but IMS signalling must be added to offer multimedia services. While a CAMEL Service Environment may be used, telcos are likely to introduce an OSA/Parlay service platform or, when standards mature, a SIP server with a view to migrating all services.

An example of a migration scenario for a fixed network operator is shown in Figure 10.3. The starting point is the existence of a legacy PSTN with a large number of users served by copper loops, connected to subscriber line concentrators that may be distributed or co-located with the end exchanges. In addition, the telco has an

Internet backbone network and supports ADSL subscribers connected to DSLAMs. An important requirement is that the existing telephone subscribers must continue to receive service over copper loops after migration to a packet core network. Telephone subscribers must continue to enjoy value-added IN services. A possible sequence of migration steps is as follows:

1. Introduce a core network capable of supporting quality of services for each future class of traffic it will carry. Packet voice is one such class.

2. Introduce media gateways to support interworking with legacy switched circuit networks (including the operator's own network until migration is complete).

3. Introduce access gateways to receive the analogue copper loops. Voice digitisation, encoding and packetisation take place in the access gateway.

4. Introduce IMS call/session signalling. The access gateways must generate IMS messages in response to loop signalling on the subscriber line side.

5. Introduce a new OSA/Parlay service platform. Develop services identical to those already supported on the IN platform.

6. Subscribers on analogue loops can then be transferred to the NGN. IMS call/session control takes over and bearer traffic moves to the multiservice network.

7. Intelligent phones can be offered via an integrated access device (IAD) at the customer premises and multiservice IP edge node (IPEN). IMS signalling now involves the terminal.

Several migration processes are not shown in Figure 10.3. Three example follow. First, Signalling System No. 7 is replaced by signalling carried in the IP network. Signalling paths with adequate performance and reliability must be set up in the QoS packet network. Second, subscriber profile data requires enhancement for use with IMS call/session control. Third, billing information must be collected at appropriate points and transferred to the off-line billing system after mediation.

The Internet backbone network function may be integrated into a single IP core as a best effort traffic class (8). The IAD and IPEN allow access to data services (9, 10).

10.4 IS THERE A TARGET NGN?

The evolution metaphor is indicative of a continuing process. However, the history of telecommunications reveals long though decreasing periods of comparative stability in the infrastructure required to support the main services. For example, electromechanical telephone switching was little changed for about five decades. In the 1980s, fully electronic switching was introduced and was supported by digital transmission. For three decades, this technology will have supported voice communication. The question must be asked whether a particular architecture is emerging that may provide comparative stability in network infrastructure for a period, probably shorter than in the past, due to the increased pace of innovation.

Put differently, is there a migration target that vendors and service providers should focus on?

Our first recourse in attempting to answer the question is the standards bodies. The evolution of mobile communication systems up to the third generation is described in Chapter 7. The fourth generation is frequently mentioned but it remains a vision rather than an definitive architecture. For fixed networks, an authoritative source is the ETSI Telecommunications and Internet converged Services and Protocols for Advanced Networking (TISPAN) initiative [88]. Similarly, the ITU-T presents a more abstract view of next generation networks. The NGN overview [138], the general principles and reference model build on the concept of a Global Information Infrastructure (GII) and other key ITU-T standards.

10.4.1 *WHITHER FIXED NETWORKS?*

First a definition of a fixed network is needed. Fixed networks tether the terminal by means of a wireline connection (metallic or fibre) or provide a wireless connection that constrains the area where the terminal's incoming and outgoing service requests can be processed. Fixed networks support user mobility, for example universal personal telecommunications in PSTN/IN networks and IP Mobility in IP networks. These forms of mobility exhibit *nomadicity*: a user who is known to the network is able to move from terminal to terminal and to access services and be reached. Fixed networks may support *roaming*: a user accesses services provided in the home network from a terminal located in a visited network. A traditional fixed network does not support terminal mobility in real time across wide geographic areas and across operator domains. Thus, the least problematic definition of a fixed network is one that does not support real-time terminal mobility; wireless access is not a defining attribute. Otherwise a fixed network has most of the features of a mobile network.

The current circuit-switched network is faced with an evolutionary dead-end due to the inflexibility and inefficiency of 64 kbit/s TDM channels. The TIPHON initiative described in Section 6.8 sought a framework for applying Internet protocols to telephony within an architectural framework. TIPHON has been subsumed into the ETSI initiative TISPAN [88]. TISPAN also draws on 3GPP concepts and standards. The approach to standardisation is based on service and transport layer functions.

Within the service layer, TISPAN defines several core network subsystems. The IP Multimedia Subsystem is adapted from the 3GPP standards. A PSTN/ISDN Emulation Subsystem provides support for interworking between packet-based core networks and legacy terminals and networks [90]. The PSTN Emulation Subsystem is similar to the 3GPP Circuit-Switched core network. TISPAN recognises that future services involve media streaming to single terminals and broadcasting to multiple terminals. Streaming and Broadcasting Subsystems are therefore identified but not defined in the TISPAN standards. Common functions located in the service layer are user profile management, charging, security and routing.

TISPAN identifies an Application Server functional entity that would be located in the user's home network or in a third party domain. OSA/Parlay and SIP application servers are possible implementations [89].

The transport layer distinguishes between transport control and transfer functions. Two transport control functional entities are identified. The Network Attachment

Subsystem deals with network level authentication and addressing issues. The Resource and Admission Control Subsystem supports admission control for network level connection requests on the basis of subscriber profiles, policies and resource availability. Resource and admission control therefore play a pivotal role in delivering connections with quality of service. Transfer functions within TISPAN are concerned with interworking already familiar from the 3GPP standards: Media Gateway Function, Border Gateway Function (access-core border and inter-core network border), Signalling Gateway Function and the Media Resource Function Processor.

The various IP multimedia standards do not provide architectures for fixed access networks analogous to GERAN and UTRAN in 3G mobile communication systems. A number of types of access network exist for use with fixed NGNs: the xDSL family, cable systems, wireless systems such as WiFi and WiMax as well as interactive broadcast. TISPAN therefore defines a number of functional entities for controlling access that allow specific access network architectures to be defined.

TISPAN thus provides a framework for a near-term fixed NGN, for example using IMS and OSA/Parlay that would interwork with 3G mobile communication systems as well as legacy networks. TISPAN also looks forward to more general multimedia services.

10.4.2 ARE FIXED AND MOBILE NETWORKS DISTINCT?

Roaming is no longer a concept restricted to mobile communication systems. Future fixed networks support nomadicity. The distinguishing feature of a mobile network is the ability to provide connections to a given terminal, identified by a SIM card, with real time changes of location, involving handovers at various levels. A topic of current interest is handovers involving user equipment based on different technologies, for example handover between WiFi and 3G mobile station or a 3G mobile station and a wired LAN connection. The ITU-T NGN Framework and Functional Architecture [138] introduces the concept of generalised mobility involving handover due to location change and between terminals of different technologies while ensuring that services are delivered consistently.

Fixed and mobile networks have been regarded as providing different services. For example, voice services in the two networks generally involve separate subscriptions to different providers. No technological constraints prevent services being offered by a single provider via fixed and mobile access. The example of 3G multimedia services supported by IMS and a TISPAN-based NGN using an IMS core network is one enabling environment. Increasingly, we will talk only of differences between fixed and mobile access and not of distinctive types networks.

10.4.3 FOURTH GENERATION MOBILE COMMUNICATION SYSTEMS

The evolution of mobile communication systems is conveniently described as progressing through three generations: first generation analogue systems – now a historical artifact; a second generation circuit switched digital system; and the third generation progressing to a packet-switched multiservice network. Intermediate stages are labelled: 2.5G for the introduction of GPRS; C3G flagging the introduction of IMS; and B3G indicating developments such as higher data rate access networks.

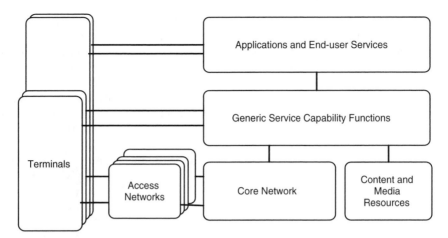

Figure 10.4. Elements of the Fourth Generation Network Concept.

At the time of writing, the fourth generation (4G) mobile communication system is more a vision than a definitive set of standards. What prompts the definition of a new generation of mobile communication system?

- *New access networks* offering increased data rates, seamless handover between different types of access network (session mobility) and the use of personal area networks (PAN) or *ad hoc* networks for access to the communication system.

- *An all-IP core network* as the designated multiservice core network. Interworking with circuit-switched networks which looms large in 3G systems, recedes in emphasis.

- *User-centricity* through personalisation of services and user-defined services.

- *Application-oriented QoS* through mechanisms that link the application layer to the transport layer.

- *New handover mechanisms* both vertical, involving interaction between functional layers in the access network, and horizontal, between access networks that use different technologies.

- *A variety of terminals.*

As 3G communications systems develop, several of the 4G objectives are being met, at least partially.

10.5   MANAGING COMPLEXITY: AVOIDING PITFALLS

Technology development and application in ICT systems is a complex process and the resulting systems are also complex. The NGN Framework and abstraction are general methods of controlling complexity. For example, understanding of the 3GPP and 3GPP2 architectures, depicted in the standards in unlayered form, benefitted from mapping to the NGN Framework in Chapter 7. The NGN Framework and abstraction are tools that the analyst, designer, researcher or student can use. In addition, a number of points of best practice described in the following sections may prove useful in complex problem solving.

10.5.1 DO NOT FORGET THE PRINCIPLES

A number of principles can be distilled from the NGN Framework and its application to describe various architectures. These principles endure while detailed technologies change.

- The *softswitch principle* conveys the decoupling between call and media session control and the underlying transport network mechanism. Packet telephony networks as described in Chapter 4 and 3G mobile networks from Release 5 are based on the softswitch principle. Details of softswitch control vary. In the 3G CS domain, call control involves the legacy call state models while in IMS the emphasis is on media session control, with the end-to-end association between call parties as a by-product.

- The *open service access principle* decouples programmable logic that enhances communications or that is enhanced by communications capabilities from the generic services and network capabilities.

- The principle of *access independence* means that users should be able to access services using any terminal with the required capability.

- A *service oriented* approach to service exposure provides good abstraction of underlying resources. Applications are network agnostic. Service orientation is more general that Web services: encapsulation of reusable functionality behind an open, described and discoverable interface with access control. The interface should preferably be defined in a technology-neutral form.

- *Content is treated as a resource*, that can be accessed via suitable services, for example OSA or SOA interfaces. This principle applies to all kinds of data from video libraries to user profiles.

10.5.2 BE AWARE OF LEAKY ABSTRACTIONS

In Chapter 2, we cautioned that the essential property of a complex system is that, for proper analysis, prediction and design, it cannot be simplified [179]. Complexity arises from the interactions of different parts of a complex system. Layering provides a way of limiting the coupling between parts of the communication system to inter-layer interfaces and providing abstractions. Defining reference points limits the interactions and the data that can flow between layers.

The degree of abstraction depends on the application layer protocol or API used as well as the distribution mechanism. A CORBA-based system in which the API is defined in the technology-independent IDL and where serving objects are identified by network-independent names provides complete hiding of the underlying communication mechanism. Some abstractions provide imperfect hiding and are

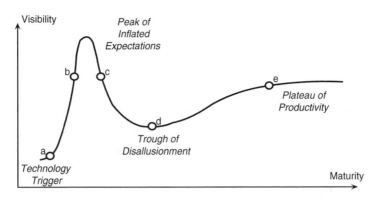

Figure 10.5. The Gartner Hype Cycle.

described as *leaky*. A SIP message, by contrast to a CORBA invocation, contains information on network addresses and the transport protocol for signalling messages. The route followed by the signalling messages must be recorded. The application must configure and process information that spans several conventional abstraction layers in addition to the application layer information (request or response) and data (headers). SIP therefore provides a leaky abstraction of the transport and network layers.

10.5.3   SILOS MAY STILL BE CREATED

Convergence generally tries to avoid the vertically integrated or silo architectures of legacy systems. However, in an NGN context it is still possible to construct silo or stovepipe systems that may hinder convergence. A possible situation is in an IMS-based system with a SIP server hosting applications. The SIP-based ISC interface allows call/session signalling to interact with the service platform. A SIP-oriented technology, namely SIP servlets, is advocated. Two effects are apparent. The application and service layers are coupled by the SIP paradigm. The application receives the SIP messages. Programming may take into account the rich, low-level detail of the protocol. While the detailed formatting of the messages may be hidden by an API, the low-level parameters are accessible. Second, the SIP server is incapable of receiving requests and controlling calls with other signalling protocols, for example INAP. A SIP-only silo can be created.

10.5.4   BE AWARE OF HYPE

Development and standardisation of complex ICT technologies is a protracted process. Frequently, the promise of the technology attracts attention well before it is mature. Technologies are subject to *hype:* an exaggerated view, often based on limited information, of the potential benefits of a development rather than one formed by sober, informed appraisal.

The inevitable hype surrounding a new development changes in intensity and focus as the technology goes through its lifecycle. Figure 10.5 shows a construction developed by the Gartner Group [98] to position new technologies with respect to their potential adoption dates and to indicate the stages that the technology must still go through.

As a technology matures, its visibility changes. Visibility is an expression of the attention the technology attracts. The maturity cycle has a number of phases. Initially, the start or *trigger* of the technology's development attracts little attention, as at point (a). Over time the visibility increases sharply, for example to point (b), with excitement about its potential benefits that have not yet been demonstrated. Visibility rises to a peak before waning. The turning point is called the *peak of inflated expectations*. Inevitably, attention to the technology wanes as time passes and the development proceeds, a shown at point (c). Visibility continues to decline while the long, hard process of maturing the technology continues. Visibility passes through the *trough of disillusionment* at point (d) before once again attracting attention, albeit at a more sober level, as a mature technology. A mature technology attains the *plateau of productivity* (e).

Development of ICT technologies, systems and services is a long complex process and developers and potential users must be aware that they are subject to the hype cycle.

10.5.5 *HANDLING THE GREAT DEBATES*

Convergence brings players from different roles and backgrounds down from their traditional towers and makes them engage with paradigms different to their own. Great debates ensue that are often pursued from a particular point of view. One such debate is whether intelligence should be in the network, IN-style or at the edge of the network, Internet style. For example, OSA/Parlay seeks to open telecommunication network capabilities to information technology applications, especially via Web service interfaces. Is the intelligence 'in the network' in OSA/Parlay and therefore anathema from the Internet perspective? This and other debates illustrate that convergence brings new perspectives to old arguments. In the OSA/Parlay case, a new type of network edge has been created. The application service provider can be in a domain outside the telco yet interact with the network capabilities in a secure and abstracted way. A new type of network edge has been created but one that relies on abstractions of the considerable capability in the network.

10.5.6 *EVOLUTION IS AN ONGOING PROCESS*

Voice on the Net is a disruptive technology and is often seen as a threat to telephony services offered by telcos. The telco-based form of telephony seeks to achieve carrier gradeness: availability, reliability, billability, scalability and meeting legal requirements such as legal intercept and emergency services. The Internet-based form of telephony seeks cheapness and convenience. Many issues must play out before some equilibrium appears between the telco-based form and Internet-based form of services such as telephony.

Along the way, a number of ironies become apparent, as the history of SIP and IMS reveals. SIP was conceived in the Internet world as a general session initiation protocol. Internet telephony – allowing players in the Internet community to do what historically was the preserve of telcos – was only one of its potential applications. SIP was developed into IMS to allow mobile and fixed networks to provide multimedia services; it supported their ability to provide highly available, assured quality and billable services. SIP, developed in the Internet and developed into IMS, provides telcos with a defensive strategy in the form of multimedia services.

10.5.7   REVISIT THE TREASURE HOUSE

Architectures are sufficiently complex for standards writers to leave particular detail undefined, as an implementation detail, as the responsibility of another unidentified or net yet realised standard or 'for further study'. When faced with filling such a gap, a treasure house of solutions exists in standards such as TINA that were never implemented. The problem of assigning more intelligence to the terminal provides such an example.

The plethora of terminals already found in ICT systems is identified in Section 2.1.1. Layering of functions in the core network is well accepted. Figure 10.4 shows the application, service capability and transport network layers of the NGN Framework. Terminals have progressed from a telephone capable of loop signalling only to devices with significant computing power. Layering could be extended to the terminal domain. Call session signalling supports SCF-layer functions. The terminal may host application-layer logic.

The TINA architecture introduces the idea of some of the service logic, the User Application Part (UAP) in Figure 6.14, being located in the consumer domain. For example, a multiparty conference could be controlled from the user domain by adding, suspending, resuming and dropping parties by method calls from the user via the UAP to the conference control application. Current architectures such as the IMS rely on call/session signalling that is logically placed at the service capability layer. Invocation of applications takes place via a triggering mechanism: parameters of the signalling messages having values in a defined range cause a call/session signalling message to be intercepted and a notification directed to the application. Interactions with the user are controlled by the application, not by the user. The TINA approach removes this limitation.

10.6   A LAST WORD

Technology *evolution* in ICT is a process of incremental improvement of facilities and their capability to support services for end users – punctuated by occasional disruptive episodes. *Convergence* is a strategy for steering evolution to maximise benefits by bringing services or facilities together. A beneficial result of such a coming together is also called convergence. Three principles that help attaining convergence are *open service access, softswitch separation* and *access independence. Migration* is the process of implementing evolved, preferably converged, technologies to better offer existing services and add new services. The core competence of a engineer who works with evolving, converging ICT and the migration of solutions is the ability to handle complexity. This competence is supported by tools such as the NGN Framework and abstraction and modelling. In particular, a disciplined approach to analysis and synthesis that follows a process such as RM-ODP or MDA is commended. *Best practice* dictates that one proceed from system requirements, to a technology-neutral model before plunging into specific technologies.

It is hoped that this book has provided a way of seeing next generation networks that supports successful research, standardisation, technology evaluation, migration and operations of ICT systems.

Glossary

Abstraction: a process of hiding detail of a system or component that is not required for the purpose at hand.

Access transparency: a distribution transparency which masks differences in data representation and invocation mechanisms to enable interworking between objects.

Accounting: the process of collecting information on the usage of services and network resources to enable billing.

Address of record (AOR): in the SIP context, an address in the form of a SIP URI that identifies a user that would be published and may require resolution to a direct route to the user.

Application: a service that is hosted on a computing node separate from the call or session control that is invoked to enhances basic telecommunication services or is enhanced by network services.

Architecture: identifies the constituents of a system, network, network element or software and the way that they relate to each other; a set of concepts and principles for the constituents of a network or software system and the way they are brought together and interact.

Asynchronous: events, frames, bits or bytes that do not occur according to a time frame; in computing, operations of communicating processes that may be initiated without waiting for previous operations to be completed.

Asynchronous Transfer Mode (ATM): a packet switching method based on fixed length packets called cells having a 5 byte header and a 48 byte payload, giving high bandwidths and low delays required for a multiservice network.

Automatic call distribution (ACD): a process in a private network, PBX or call centre for directing incoming calls to an appropriate agent who is logged into the system.

Availability: the fraction of the total time that a given resource is available for use averaged over a suitably long interval.

Basic Call Process (BCP): the process in the PSTN switch concerned with controlling and routing calls but not with *supplementary services*.

Basic Call State Model (BCSM): is a high level representation of the PSTN switch call control process required to establish and maintain connections in a switched circuit network and consists of *points in call* and *detection points*.

Basic telecommunications service: the capability offered by a network to transfer data streams or messages between end users in fixed and mobile networks.

Bearer service: a type of telecommunications service that provides the capability for the transmission of signals between user-network interfaces.

Behaviour: a constrained set of actions that an object completes; behaviour may be internal or external.

Best effort: a term describing a packet transport network that offers no Quality of Service guarantees.

BORSCHT: an acronym for the constituents of the analogue subscriber loop interface to a digital telephone exchange: battery power supply, overvoltage protection, ringing current supply, signalling across the interface, codec function, hybrid function, and testing functions.

Broadband: an access connection to the Internet that is significantly faster than a dial-up connection.

Broadband Integrated Services Digital Network (B-ISDN): an Integrated Services Digital Network intended to provide multimedia services at high bit rates, involving broadband access and switching based on Asynchronous Transfer Mode.

Canonical: the most general or usual form of an entity, not specialised for a particular context.

Carrier-grade: a collective description of network performance metrics such as availability, post dial delay and speech quality, that are generally accepted as appropriate to a public telecommunications operator.

Chat: a form of call in which text-based messages are exchanged between participants.

Circuit: in a switched circuit network context, a 64 kbit/s transmission path that is available for the entire duration of a call formed by allocating a time slot on each TDM link between source and destination and transferring the bit stream from one link to another by each switch.

Circuit switching: a means of routing signals from a source to a destination by allocating dedicated resources for the duration of the connection.

Closed network: one in which service providers outside the network operator cannot control network resources to provide services with telecommunications features.

Common channel signalling: a method of signalling relating to numerous circuits, calls or services is conveyed over a single dedicated channel or network by means of messages.

Computer Supported Telephony Applications (CSTA): is specification for an application programming interface for monitoring and controlling calls, devices and media processing in a (private) telecommunications network.

Computer–telephony integration (CTI): originally meaning the use of computer applications for routing telephone calls in a call centre or PBX but now applied to all forms of computer-based telephony applications in the enterprise environment.

Contact address: in the SIP context, an address that provides a direct route to a user, consisting of a username and a fully qualified domain name.

Content: describes all forms of information and the way that it is organised and delivered. Content includes text, hypertext, data of various forms, audio signals, images, video, possibly organised as Web pages, databases, libraries, news, programmes.

Customer premises equipment (CPE): facilities in the telco customer domain consisting of terminals and possibly a private network, connecting to the access network at the user-to-network interface.

De-marshalling: the process of restoring parameters received in a language-independent flat format across a network to a language-specific typed format. Synonymous with un-marshalling.

Dependability: the ability of a network to preform transport and service functions reliably.

Detection Point (DP): a point in basic call processing at which an event may be reported to the service control function and transfer of processing may take place.

Digital Subscriber Line Access Module (DSLAM): a unit for terminating digital subscriber lines, for example using ADSL technology, at the edge of the core network.

Disruptive technology: describes a technology that, while not necessarily offering better performance, challenges existing markets and regulation, by virtue of some or all of the following attributes: dynamic, varied, personalised, adaptable, novel, offering new opportunities, or being radically different.

Distribution transparency: the property of a distributed system in which particular aspects of the behaviour of the system are hidden from a specified user.

Domain Name Service (DNS): is a network application that enables nodes to query the IP address corresponding to a given application or host name.

Dynamic Host Configuration Protocol (DHCP): an IETF protocol (RFC2131) providing a framework for passing configuration information to hosts on a TCP/IP network

E.164 Number: a telephone number conforming to ITU-T Recommendation E.164, having Country Code, National Destination Code and Subscriber Number fields.

Encapsulation: in the object orientation context, the process of hiding information in an object and providing access through interfaces.

End-to-end principle: the traditional principle of the Internet that all intelligence should be in the client and server terminals, while the network simply provides transport of packets.

Engineering: consideration in the distributed system context are concerned with the form interfaces and their patterns of use to support the interaction of computing entities.

Enterprise service bus (ESB): a standards-based, implementation-independent shared messaging mechanism for integrating enterprise applications by supporting invocation of coarse-grained, loosely coupled services using XML-based messaging.

Extranet: an intranet-like network that is accessible, with possible restrictions, by authorised users outside the enterprise.

Federation: a process in which service providers operate jointly for defined purposes and under agreed procedures.

Formal method: a means of describing a system or process that is based on a restricted syntax with defined semantics, where necessary based on mathematical concepts.

Functional entity (FE): a unique group of functions in a single location and a subset of the total functions required to provide a service in the context of the IN Conceptual Model of the ITU-T Q.1200 series.

Gateway: in general, an element which allowing communication between two systems operating according to different standards.

Global System for Mobile Communications (GSM): a second generation digital mobile telephony standard originally developed in Europe but now deployed in many countries.

Handoff: a process in mobile network in which some control function is passed from one entity to another without affecting the service as seen by the users.

Hyperlink: a reference (URL) to a file on a computer anywhere on the Internet, coupled with the ability of a browser and other programmes to download and display the contents of the file.

Hypertext: pages of information connected by hyperlinks in an organisation that is in general nonlinear.

Hypertext Markup Language (HTML): a computer (script) language using standardised tags which allows a user to write the content and define the format of a Web page, insert graphics and link with other pages.

Hypertext Transfer Protocol (HTTP): a communications protocol used to access and download Web Pages from a server to a client's browser.

The Internet Engineering Task Force (IETF): A body of the Internet Society responsible for developing standards for all aspects of the Internet.

Information Flow (IF): an interaction between a pair of functional entities in the context of the IN Conceptual Model of the ITU-T Q.1200 series.

Information technology (IT): is concerned with the processes of creating, acquiring, processing, storing and disseminating information using computing and communications systems, using underlying engineering and computer science concepts.

Information and communications technologies (ICT): encompasses the scope of Information Technology but gives greater emphasis to the communications aspects.

Instant messaging (IM): a text based, conference-like service in which participants exchange messages that are delivered to the user screen on arrival.

Intelligent Network Application Protocol (INAP): an Application Layer protocol supporting operations and responses involving physical entities in the Intelligent Network defined in ITU-T Recommendations Q1218 and Q.1228.

Interface: the boundary between two systems, subsystems, devices, functional units, processes or persons that is defined by specific functional and physical attributes; the physical realisation of a reference point or part of its specification.

Internet service provider (ISP): is a company that provides individuals and enterprises with access to the Internet and, possibly, services such as Web site hosting.

Interoperability: allowing objects that are implemented on different platforms to interact to meet specified objectives

Intranet: a private network that serves an enterprise that is logically restricted to users within an enterprise but may extend physically across several networks.

IP Multimedia Subsystem (IMS): in the 3G network context, a set of core network elements that support multimedia services, that is services involving audio, video, text and other forms of information in combination.

Least cost routing (LCR): in the telephony context, a process that determines the route or routes toward the destination that minimises the call cost; in the IP network context, selecting a next hop that meets an optimality criterion such as giving the smallest number of hops toward the destination.

Legacy system: telco facilities of an obsolescent technology which still serves a purpose in the network.

Local area network (LAN): a shared network medium using a defined communications protocol at layer 2 usually confined to a geographic area.

Managed network: a network operated by an identified authority that offers stated levels of service to customers though planning and operation that includes usage accounting, performance monitoring, fault reporting and proactive response to resolving problems.

Marshalling: the process of mapping typed parameters onto a flat, language-independent format for transport across a network in a CORBA or Java RPC system.

Model Driven Architecture (MDA): an OMG standard underpinning all OMG specifications that separates the functional specification of software from the specification of how the software is to be implemented using a specific technology platform.

Message: a self-contained unit of data normally carrying control or management information or limited end-user data, usually transported in a single packet.

Middleware: software that is used between two computing entities in order to provide a standard way for disparate entities to communicate.

Multicast Backbone on the Internet (MBONE): an experimental extension of the Internet that supports efficient multicasting of packets.

Multicasting: a process for sending packets from a host to multiple destinations using only one address, involving only one connection and sending each packet once with networking resources delivering the packet to all recipients.

Multiservice network: is a single switching and transmission infrastructure network which supports a range of real time and information services.

Name: a linguistic identifier for an entity that is unique in a stated naming context, including the global context.

Northbound interface: an interface in a layered architecture such as Parlay toward the served layer, usually shown above the serving layer, hence northbound.

Packet switching: a means of routing signals from a source to a destination by segmenting into units called packets, transmitting the packets through shared links, switching a packet at a time at each node, and reconstructing the original information at the destination node.

Peer-to-peer computing: a mode of structuring applications in which all parties have equal rank.

Platform: is a set of defined interfaces to the infrastructure underlying a computing application and their patterns of use and may be generic, specific to technologies or specific to individual vendors implementations.

Portability: a property of an object that, by virtue of its reference points, allows it to be applied in different configurations or on different platforms.

Private branch exchange (PBX): a digital TDM or IP switching unit in a private domain, connecting multiple telephone users, and in some cases data users, within the domain and to the public network via trunks, supports Direct Inward Dialling and services such as voice mail, call queueing and conferencing.

Profile: a definition that adds specialised function to or specialises a canonical entity or standard.

Protocol: a set of messages by which two entities communicate together with a specification of the expected behaviour of an entity as a result of receiving a message.

Real-time Transport Protocol (RTP): provides end-to-end network transport functions suitable for real-time data such as IP telephony and video conferencing over unicast and multicast network.

Reference point: a logical or conceptual point between two, nonoverlapping functional groups having defined characteristics; a point of interaction between two entities at which conformance with a specification can be tested.

Remote access server (RAS): a network element, usually located close to a PSTN end-exchange and connected by a TDM trunk which terminates dial-up modem circuits and ISDN B-Channels and provides data access to the Internet.

Reusable: the property of an already implemented software object that allows it to used as part of a new object or system without redesign.

Screen pop: information presented to a PBX attendant or call centre agent relating to the caller or service called.

Script: a programme or sequence of instructions that is compiled or interpreted or carried out by other programmes.

Service architecture: a consistent set of software concepts, principles, structures and standards that supports efficient creation, deployment and maintenance of services.

Service delivery platform (SDP): a part of a fixed or mobile network operator's information technology infrastructure that supports Web services style provision of value added services, that has suitably abstracted access to but does not include the core network and interworks with the operator's OSS/BSS and user access control (AAA) systems.

Service mediation: providing access for service developers to use core network capabilities in a unified, open standards based, secure and controlled manner.

Service Oriented Architecture (SOA): an architecture for a software system conforming to the Web Service model.

SIP Application Server: a server with an interface responding to SIP methods or responses that performs normal SIP server roles (Proxy, Redirect, Registrar and Back-to-back User Agent) as well as other functions.

Southbound interface: an interface in a layered architecture such as Parlay toward the next lower layer, usually shown below the serving layer, hence southbound.

Store and forward messaging: a method of message transfer in which the sender connects to a server and transmits the message, the server stores the message and at a time determined by operating configuration forwards the message to the recipient. Examples are SMS and MMS.

Store and retrieve messaging: a method of message transfer in which the sender connects to a server and transmits the message and the server stores the message until the recipient downloads the message from the server. An example is e-mail.

Stub: A local procedure that a client calls to invoke a remote procedure that handles the transmission of the request parameters to the remote procedure and receives the response, both without the knowledge of the mechanisms involved.

Supplementary services: standardised services, usually implemented as switch-based services relying on the call signalling protocol, offered in the ISDN and GSM contexts in addition to basic bearer services.

Sustaining technology: describes a technology with several of the following characteristics: its development is evolutionary, incremental, it is standardised, it interoperates with existing technologies, it consolidates or grows existing systems or technologies, and complies with conventional regulation.

Switched circuit network (SCN): a public or private switched telecommunications network based on circuit switching of bearer channels, for example the PSTN, N-ISDN, or 2G GSM networks.

Synchronous: events, frames, bits or bytes that occur according to a predictable time frame; in computing, the operation of two or more communicating processes is related to specific events or timing signals.

Synchronous Digital Hierarchy (SDH): a time division multiplexing method specified in ITU-T recommendation G.707 as a hierarchical set of digital transport structures, standardised for the transport of suitably adapted payloads over physical transport networks.

Synchronous Optical Network (SONET): a set of synchronous digital transport structures or optical carriers conforming with minor variations to the SDH standards.

Technology: in the distributed system context is the specific hardware and software that implements and supports the distribution of the various computing entities.

Telecom Web service: a Web service supported by a platform that allows an IT-type application to invoke telecommunications services such as bearer connections and messaging.

Telephone URI (tel URI): in the SIP context, a telephone number that identifies a resource and may be a global number or a local number within an identified context.

Teleservice: a type of telecommunication service that uses basic telecommunications services but relies on the interaction of intelligent terminals.

Third party application: is hosted on a computing node in a separate domain from the call or session control and uses basic telecommunications services and network data via an application protocol or API.

Time-based charging: a principle for charging for services according to the duration of usage and possibly weighted by the time of day/day of week.

Time-division multiplexing (TDM): a method of multiplexing dividing the available time for transmission of a signal sample into time slots which are allocated to individual connections.

Transaction: an interaction between two entities, for example a client and a database, involving request and response, carried out in a consistent manner independent of other transactions.

Transfer mode: the method of information transfer across a network including the structuring of the information (frame or packet) and the method of switching at nodes.

Transmission Control Protocol (TCP): an Internet Layer 4 protocol that operates over IP using datagrams supporting multiple upper layer entities (ports) designed to ensure reliable transfer of stream data over an unreliable transport network.

Transparency: a condition in a system where specified aspects of the system behaviour are hidden.

Triple-play: a bundled set of voice telephony, broadband Internet and television services offered to telco customers.

Unified messaging: a service that manages existing messaging methods, including e-mail, fax and voice messaging, as a single service adapting messages to formats suited to the user terminal and preferences.

Uniform Resource Identifier (URI): an extensible compact string of characters with a restricted syntax for identifying abstract or physical resources by location, name or both.

User Datagram Protocol (UDP): is a connectionless Transport Layer protocol that operates over IP using datagrams supporting multiple upper layer entities (ports) but provides no transport guarantees or flow control.

Value-added service: a service involving processing information which enhances the value of the service beyond that of the basic telecommunications bearer function.

View: a representation of a system from a defined viewpoint.

Viewpoint: an abstraction of a system based on selected concepts and principles; a way of describing or modelling a system.

Virtual: an attribute of a network or other entity that indicates that it is constructed logically, usually by a management operation.

Virtual private network (VPN): a data network service created using public operator facilities that has the characteristics of a private network and provides private and secure communication for a defined community of users; in the telephony environment, a PBX-like network spanning multiple geographic sites implemented using telco infrastructure.

Volume-based charging: A principle for charging for services according to the amount of data transferred.

Web service: a software system that is identified by a URI and has a public interface defined using XML that is published and can be discovered and invoked by other software systems using XML-based messages.

Wide area network: the physical and logical interconnection of multiple geographically separate LANs by means of dedicated lines or a networking technology such as Frame Relay.

X.25: An ITU-T standard defining at Layer 2 and Layer 3 the interconnection of a terminal called Data Terminal Equipment (DTE) with packet network interface called Data Communication Equipment (DCE) for end to end connection over an undefined packet network using virtual circuits.

References

[1] 3GPP. *TR 21.902: Evolution of 3GPP System.* 3rd Generation Partnership Project, 2003.

[2] 3GPP. *TR 21.905: Vocabulary for 3GPP Specifications.* 3rd Generation Partnership Project, 2003.

[3] 3GPP. *TR 29.998-01: Open Service Access (OSA) Application Programming Interface (API) Mapping for Open Service Access; Part 1: General Issues on Application Programme Interface (API) Mapping.* 3rd Generation Partnership Project, 2003.

[4] 3GPP. *TR 29.998-04: Open Service Access (OSA) Application Programming Interface (API) Mapping for Open Service Access; Part 4: Call Control Service Mapping; Subpart 4: Multiparty Call Control ISC.* 3rd Generation Partnership Project, 2003.

[5] 3GPP. *TR 29.998-08: Open Service Access (OSA) Application Programming Interface (API) Mapping for Open Service Access; Part 8: Data Session Control Service Mapping to CAP.* 3rd Generation Partnership Project, 2003.

[6] 3GPP. *TS 22.060: General Packet Radio Service (GPRS); Service Description; Stage 1.* 3rd Generation Partnership Project, 2003.

[7] 3GPP. *TS 22.078: Customised Applications for Mobile Network Enhanced Logic (CAMEL); Service Description, Stage 1.* 3rd Generation Partnership Project, 2003.

[8] 3GPP. *TS 22.228: Service Requirements for the IP Multimedia Core Network Subsystem.* 3rd Generation Partnership Project, 2003.

[9] 3GPP. *TS 23.002: Network Architecture (Release 6).* 3rd Generation Partnership Project, 2003.

[10] 3GPP. *TS 23.060: General Packet Radio Service (GPRS); Service Description; Stage 2.* 3rd Generation Partnership Project, 2003.

[11] 3GPP. *TS 23.078: Customised Applications for Mobile Network Enhanced Logic (CAMEL) Phase 3 – Stage 2.* 3rd Generation Partnership Project, 2003.

[12] 3GPP. *TS 23.218: IP Multimedia (IM) Session Handling; IP Multimedia (IM) call model.* 3rd Generation Partnership Project, 2003.

[13] 3GPP. *TS 23.228: IP Multimedia Subsystem (IMS); Stage 2.* 3rd Generation Partnership Project, 2003.

[14] 3GPP. *TS 23.234: 3GPP System to Wireless Local Area Network (WLAN) Interworking.* 3rd Generation Partnership Project, 2003.

[15] 3GPP. *TS 24.228: Signalling Flows for the IP Multimedia Call Control Based on SIP and SDP.* 3rd Generation Partnership Project, 2003.

[16] 3GPP. *TS 29.229: Cx and Dx Interfaces Based on Diameter – Protocol Details.* 3rd Generation Partnership Project, 2003.

[17] 3GPP. *TS 23.198: Open Service Access (OSA); Stage 2.* 3rd Generation Partnership Project, 2004.

[18] 3GPP. *TS 29.078: Customised Applications for Mobile Network Enhanced Logic (CAMEL) Phase 4; CAMEL Application Part (CAP) Specification.* 3rd Generation Partnership Project, 2004.

[19] 3GPP. *TS 23.018: Basic Call Handling; Technical Realization.* 3rd Generation Partnership Project, 2005.

[20] 3GPP. *TS 25.323: Packet Data Convergence Protocol (PDCP) Specification.* 3rd Generation Partnership Project, 2005.

[21] 3GPP. *TS 29.414: Core Network Nb Data Transport and Transport Signalling.* 3rd Generation Partnership Project, 2005.

[22] C. Abarca, P. Farley, J. Forslow, et al. *Service Architecture, version 5.0.* The TINA Consortium, 1997.

[23] F. Andreasen and B. Foster. *Media Gateway Control Protocol (MGCP) Version 1.0.* Internet Engineering Task Force, RFC3435, 2003.

[24] M. Appeldorn, R. Kung, and R. Saracco. TMN + IN = TINA. *IEEE Communications Magazine*, 31(3):78–85, March 1993.

[25] J.C. Arnbak. Technology trends and their implications for telecom regulation. In W.H. Melody, editor, *Telecom Reform: Principles, Policies and Regulatory Practices*, chapter 6, pages 66–80. Den Private Ingeniorfond, Technical University of Denmark, Lyngby, Denmark, 1997.

[26] J.-L. Bakker and F. Anjum. Service Capability APIs. In K.J. Turner, E.H. Magill, and D.J. Marples, editors, *Service Provision: Technologies for Next Generation Communications*, chapter 11, pages 189–216. Wiley, Chichester, 2004.

[27] J.-L. Bakker, D. Tweedie, and M.R. Unmehopa. Evolving service creation: new developments in network intelligence. *Telektronikk*, pages 58–68, 2002.

[28] K. Ballinger et al. (Eds.). *Basic Profile, Version 1.0.* Web Services Interoperability Organisation, http://www.ws-i.org/Profiles/BasicProfile-1.0-2004-04-16.html, 2004.

[29] D.J. Berg and J.S Fritzinger. *Advanced Techniques for Java Developers.* Wiley, New York, revised edition, 1999.

[30] R. Bhat and R.Gupta. JAIN Protocol APIs. *IEEE Communications Magazine*, pages 100–107, 2000.

[31] P. Calhoun, J. Loughney, E. Guttman, G. Zorn, and J. Arkko. *Diameter Base Protocol*. Internet Engineering Task Force, RFC3588, 2003.

[32] B. Campbell (Ed.). *Session Initiation Protocol (SIP) Extension for Instant Messaging*. Internet Engineering Task Force, RFC3428, 2002.

[33] J. Case, M. Fedor, M. Schoffstall, and J. Davin. *A Simple Network Management Protocol*. Internet Engineering Task Force, RFC1157, 1990.

[34] V. Cerf. *IAB Recommendations for the Development of Internet Network Management Standards*. Internet Engineering Task Force, RFC1052, 1988.

[35] M. Chapman and S. Montesi. *Overall concepts and principles of TINA, version 1.0*. The TINA Consortium, 1995.

[36] R. Chinnichi et al. (Eds.). *Web Services Description Language (WSDL) Version 2.0 Part 1: Core Language*. W3C Recommendation, www.w3c.org, 2006.

[37] Y.-H. Choe. Next Generation Network activities in ITU-T. In *Proceedings of the 8th International Conference in Intelligence in Next Generation Networks*, pages 46–51, Bordeaux, April 2003.

[38] Cisco. *Gigabit campus network–design principles and architecture*. White Paper, Cisco Systems, http://www.cisco.com, 1999.

[39] D. Collins. *Carrier Grade Voice Over IP*. McGraw Hill, New York, 2nd edition, 2003.

[40] S. Craggs. Best-of-Breed ESBs: Identifying best-of-breed characteristics in Enterprise Services Buses (ESBs). Saint Consulting, http://http://www.integrationconsortium.org/, 2003.

[41] J. Craveur, F. Dupuy, J-M. Pageot, and S. Pensivy. TINA: Concepts that are actually turning into reality with IP and NGN. In *Proceedings of TINA 2000 Conference*, pages 88–93. TINA Consortium, 2000.

[42] D.H. Crocker. *Standard for the Format of ARPA Internet Text Messages*. Internet Engineering Task Force, RFC822, 1982.

[43] R. Dawkins. *The Ancestor's Tale*. Orion, London, 2004.

[44] M. Day, J. Rosenberg, and H. Sugano. *A Model for Presence and Instant Messaging*. Internet Engineering Task Force, RFC2778, 2000.

[45] S. Determan and C. Arthmann. Carrier class availability for enterprises. *Alcatel Telecommunications Review*, pages 292–297, 4th Quarter 2002.

[46] S. Donovan. *The SIP INFO Method*. Internet Engineering Task Force, RFC2976, 2000.

[47] D. Durham (Ed.). *The COPS (Common Open Policy Service) Protocol.* Internet Engineering Task Force, RFC2748, 2000.

[48] ECMA. *Standard ECMA-285: Protocol for Computer Supported Telecommunications Applications (CSTA) Phase III.* ECMA International, Geneva, 3rd edition, 2002.

[49] ECMA. *Standard ECMA-269: Services for Computer Supported Telecommunications Applications (CSTA) Phase III.* ECMA International, Geneva, 6th edition, 2004.

[50] F. Steegmans (Ed.). *The TCon Reference Point, version 1.* The TINA Consortium, 1996.

[51] F. Steegmans (Ed.). *Network Resource Architecture, version 3.* The TINA Consortium, 1997.

[52] H. Mulder (Ed.). *TINA Business Model and Reference Points, version 4.* The TINA Consortium, 1997.

[53] J. Ellsberger, D. Hogrefe, and A. Sarma. *SDL: Formal Object-Oriented Language for Communicating Systems.* Prentice Hall, Hemel-Hempsted, 1997.

[54] Enterasys. Best-in-class solutions for mySAP.com implementation. Enterasys Networks Solution Overview, http://www.enterasys.com, 2005.

[55] D. Estrin, D. Farinacci, A. Helmy, et al. *Protocol Independent Multicast-Sparse Mode (PIM-SM): Protocol Specification.* Internet Engineering Task Force, RFC2362, 1998.

[56] ETSI. *Intelligent Network (IN); Intelligent Network Capability Set 1 (CS-1) Core Intelligent Network Applications Protocol (INAP); Part 1: Protocol Specification, ETS 300 374-1.* ETSI, Sophia Antipolis, France, 1994.

[57] ETSI. *Digital Cellular Telecommunications System (Phase 2+): Mobile Application Part (MAP) Specification, ETS 300 974.* ETSI, Sophia Antipolis, France, 1998.

[58] ETSI. *Telecommunications and Internet Protocol Harmonisation over Networks (TIPHON) Release 3; Network architecture and reference configurations; Phase II: Scenario 1 + Scenario 2. TS 101 313.* ETSI, Sophia Antipolis, France, February 1999.

[59] ETSI. *Digital Cellular Telecommunications System (Phase 2+) (GSM); Customised Applications for Mobile Network Enhanced Logic (CAMEL) Phase 2; Stage 2 (GSM 03.78), TS 101 441.* ETSI, Sophia Antipolis, France, July 2000.

[60] ETSI. *Telecommunications and Internet Protocol Harmonisation over Networks (TIPHON) Release 3; Abstract Architecture and reference Points Definition; Network Architecture and Reference Points. TS 101 314.* ETSI, Sophia Antipolis, France, June 2001.

[61] ETSI. *Telecommunications and Internet Protocol Harmonisation over Networks (TIPHON) Release 3; Functional Entities, Information Flow and Reference Point Definitions; Guidelines for Application of TIPHON Functional Architecture to Interdomain Services. TS 101 315*. ETSI, Sophia Antipolis, France, June 2001.

[62] ETSI. *Telecommunications and Internet Protocol Harmonisation over Networks (TIPHON) Release 3; Requirements Definition Study; Scope and Requirements for a Simple Call. TR 101 877*. ETSI, Sophia Antipolis, France, June 2001.

[63] ETSI. *Telecommunications and Internet Protocol Harmonisation over Networks (TIPHON) Release 3; Protocol Framework Definition; General (Metaprotocol). TS 101 882*. ETSI, Sophia Antipolis, France, June 2002.

[64] ETSI. *Telecommunications and Internet Protocol Harmonisation over Networks (TIPHON) Release 3; Technology Mapping; Implementation of TIPHON Architecture using SIP. TS 101 884*. ETSI, Sophia Antipolis, France, September 2002.

[65] ETSI. *Telecommunications and Internet Protocol Harmonisation over Networks (TIPHON) Release 3; Service Capability Definition; Service Capabilities for a Multimedia Call. TR 101 878*. ETSI, Sophia Antipolis, France, February 2004.

[66] ETSI. *Open Service Access (OSA); Applications Programming Interface (API); Part 1: Overview (Parlay 5), ES 203 915-1*. ETSI, Sophia Antipolis, France, April 2005.

[67] ETSI. *Open Service Access (OSA); Applications Programming Interface (API); Part 10: Connectivity Manager SCF (Parlay 5), ES 203 915-10*. ETSI, Sophia Antipolis, France, April 2005.

[68] ETSI. *Open Service Access (OSA); Applications Programming Interface (API); Part 11: Account Management SCF (Parlay 5), ES 203 915-11*. ETSI, Sophia Antipolis, France, April 2005.

[69] ETSI. *Open Service Access (OSA); Applications Programming Interface (API); Part 12: Charging SCF (Parlay 5), ES 203 915-12*. ETSI, Sophia Antipolis, France, April 2005.

[70] ETSI. *Open Service Access (OSA); Applications Programming Interface (API); Part 13: Policy Management SCF (Parlay 5), ES 203 915-13*. ETSI, Sophia Antipolis, France, April 2005.

[71] ETSI. *Open Service Access (OSA); Applications Programming Interface (API); Part 14: Presence and Availability Management SCF (Parlay 5), ES 203 915-14*. ETSI, Sophia Antipolis, France, April 2005.

[72] ETSI. *Open Service Access (OSA); Applications Programming Interface (API); Part 15: Multi-Media Messaging SCF (Parlay 5), ES 203 915-15*. ETSI, Sophia Antipolis, France, April 2005.

[73] ETSI. *Open Service Access (OSA); Applications Programming Interface (API); Part 2: (Parlay 5), ETSI ES 203 915-2.* ETSI, Sophia Antipolis, France, April 2005.

[74] ETSI. *Open Service Access (OSA); Applications Programming Interface (API); Part 3: Framework (Parlay 5), ES 203 915-3.* ETSI, Sophia Antipolis, France, April 2005.

[75] ETSI. *Open Service Access (OSA); Applications Programming Interface (API); Part 4: Call Control; Sub-part 1: Call Control Common Definitions (Parlay 5), ES 203 915-4-1.* ETSI, Sophia Antipolis, France, April 2005.

[76] ETSI. *Open Service Access (OSA); Applications Programming Interface (API); Part 4: Call Control; Sub-part 2: Generic Call Control SCF (Parlay 5), ES 203 915-4-2.* ETSI, Sophia Antipolis, France, April 2005.

[77] ETSI. *Open Service Access (OSA); Applications Programming Interface (API); Part 4: Call Control; Sub-part 3: Multi-Party Call Control SCF (Parlay 5), ES 203 915-4-3.* ETSI, Sophia Antipolis, France, April 2005.

[78] ETSI. *Open Service Access (OSA); Applications Programming Interface (API); Part 4: Call Control; Sub-part 4: Multi-Media Call Control SCF (Parlay 5), ES 203 915-4-4.* ETSI, Sophia Antipolis, France, April 2005.

[79] ETSI. *Open Service Access (OSA); Applications Programming Interface (API); Part 4: Call Control; Sub-part 5: Conference Call Control SCF (Parlay 5), ES 203 915-4-5.* ETSI, Sophia Antipolis, France, April 2005.

[80] ETSI. *Open Service Access (OSA); Applications Programming Interface (API); Part 5: User Interaction SCF (Parlay 5), ES 203 915-5.* ETSI, Sophia Antipolis, France, April 2005.

[81] ETSI. *Open Service Access (OSA); Applications Programming Interface (API); Part 6: Mobility SCF (Parlay 5), ES 203 915-6.* ETSI, Sophia Antipolis, France, April 2005.

[82] ETSI. *Open Service Access (OSA); Applications Programming Interface (API); Part 7: Terminal Capabilities SCF (Parlay 5), ES 203 915-7.* ETSI, Sophia Antipolis, France, April 2005.

[83] ETSI. *Open Service Access (OSA); Applications Programming Interface (API); Part 8: Data Session Control SCF (Parlay 5), ES 203 915-8.* ETSI, Sophia Antipolis, France, April 2005.

[84] ETSI. *Open Service Access (OSA); Applications Programming Interface (API); Part 9: Generic Messaging SCF (Parlay 5), ES 203 915-9.* ETSI, Sophia Antipolis, France, April 2005.

[85] ETSI. *Open Service Access (OSA); Parlay X Web Services; Part 1: Common. ES 202 391-1.* ETSI, Sophia Antipolis, France, March 2005.

[86] ETSI. *Open Service Access (OSA); Parlay X Web Services; Part 11: Audio Call. ES 202 391-11.* ETSI, Sophia Antipolis, France, March 2005.

[87] ETSI. *Open Service Access (OSA); Parlay X Web Services; Part 2: Third Party Call. ES 202 391-2.* ETSI, Sophia Antipolis, France, March 2005.

[88] ETSI. *Telecommunications and Internet Converged Services and Protocols for Advanced Networking (TISPAN); NGN Functional Architecture Release 1, ES 282 001.* ETSI, Sophia Antipolis, France, August 2005.

[89] ETSI. *Telecommunications and Internet Converged Services and Protocols for Advanced Networking (TISPAN); IP Multimedia Subsystem (IMS); Functional Architecture, ES 282 007.* ETSI, Sophia Antipolis, France, March 2006.

[90] ETSI. *Telecommunications and Internet Converged Services and Protocols for Advanced Networking (TISPAN); PSTN/ISDN Emulation Sub-system (PES); Functional Architecture, ES 282 002.* ETSI, Sophia Antipolis, France, March 2006.

[91] G. Fantuzzi. *Digital Switching Control Architectures.* Artech House, Norwood MA, 1990.

[92] P. Farley, S. Hogg, L. Christiansen, et al. *Ret Reference Point Specifications, version 1.1.* The TINA Consortium, 1999.

[93] I. Faynberg, L.R. Gabuzda, M.R. Kaplan, and N.J. Shah. *Intelligent Network Standards: Their Application to Services.* McGraw-Hill, New York, 1997.

[94] S.P. Ferguson. Implications of SONET and SDH. *Electronics and Communications Engineering Journal*, 6(6):133–142, September 1994.

[95] R. Fielding, J. Gettys, J. Mogul, H. Frystyk, L. Masinter, P. Leach, and T. Berners-Lee. *Hypertext Transfer Protocol – HTTP/1.1.* Internet Engineering Task Force, RFC2616, 1999.

[96] ATM Forum. *Private Network–Network Interface Specification Version 1.1 (PNNI 1.1).* Number af-pnni-0055.002. The ATM Forum, 2002.

[97] C. Gallon. *Quality of Service for Next Generation Voice over IP Networks.* Number MSF-TR-QoS-OO1-Final. Multiservice Switching Forum, Geneva, February 2003.

[98] Gartner. Hype Cycle of Emerging Trends and Technologies. Research Note T-13-8665, Gartner Group, July 2001.

[99] A. Gillwald. *National Convergence Policy in a Globalised World: Preparing South Africa for Next Generation Networks, Services and Regulation.* LINK Centre, University of the Witwatersrand, Johannesburg, http://link.wits.ac.za, 2004.

[100] J.M. Griffiths. *ISDN Explained: Worldwide Network and Applications Technology.* Wiley, Chichester, 2 edition, 1992.

[101] Parlay Group. *Parlay X Interface.* White Paper, The Parlay Group, http://www.parlay.org, 2002.

[102] Parlay Group. *Parlay/OSA and the Intelligent Network.* While Paper, The Parlay Group, http://www.parlay.org, 2005.

[103] V. Gurbani (Ed.). *The SPIRITS (Services in PSTN requesting Internet Services) Protocol.* Internet Engineering Task Force, RFC3910, 2004.

[104] M. Handley, J. Crowcroft, C. Bormann, and J. Ott. Very large congerences on the Internet: The Internet Multimedia Conferencing Architecture. *The International Journal of Computer and Telecommunications Networking,* 31(3):191–204, 1999.

[105] M. Handley and V. Jacobson. *SDP: Session Description Protocol.* Internet Engineering Task Force RFC 2327, 1998.

[106] P.F. Hansen and C.A. Licciardi (Eds.). *Service Component Specifications: Computational Model and Dynamics, version 1.0b.* The TINA Consortium, 1998.

[107] D. Harrington, R. Preshun, and B. Wijnen. *An Architecture for Describing Simple Network Management Protocol (SNMP) Management Frameworks.* Internet Engineering Task Force, RFC3411, 2002.

[108] IPCC. *Packet Communications Reference Architecture v2.* International Packet Communications Consortium, http://www.packetcomm.org, 2003.

[109] IPDR. *IPDR Business Solution Requirements – Network Data Management-Usage (NDM-U), Version 3.5.0.1.* IPDR.Inc, http://www.ipdr.org, 2004.

[110] ITU-T. *Recommendation Q.931: ISDN User-Network Interface Layer 3 Specification for Basic Call Control.* International Telecommunications Union, Geneva, 1988.

[111] ITU-T. *Recommendation M.3200: TMN Management Services: Overview.* International Telecommunication Union, Geneva, 1992.

[112] ITU-T. *Recommendation Q.1201: Principles of Intelligent Network Architecture.* International Telecommunications Union, Geneva, October 1992. Also published as Recommendation I.312.

[113] ITU-T. *Recommendation Q.1202: Intelligent Network – Service Plane Architecture.* International Telecommunications Union, Geneva, October 1992. Also published as Recommendation I.328.

[114] ITU-T. *Recommendation I.326: Integrated Services Digital Network (ISDN): Overall Network Aspects And Functions: B-ISDN Functional Architecture.* International Telecommunications Union, Geneva, 1993.

[115] ITU-T. *Recommendation Q.1211: Introduction to Intelligent Network Capability Set 1.* International Telecommunications Union, Geneva, March 1993. 30pp.

[116] ITU-T. *ITU-T Recommendation X.680 & ISO IEC 8824-1, Abstract Syntax Notation ONE (ASN.1).* International Telecommunications Union, Geneva, 1994.

[117] ITU-T. *ITU-T Recommendation X.902 ISO/IEC: Information technology – Open Distributed Processing – Reference Model: Foundations.* International Telecommunication Union, Geneva, 1995.

[118] ITU-T. *ITU-T Recommendation X.903 ISO/IEC: Information technology – Open Distributed Processing – Reference Model: Architecture.* International Telecommunication Union, Geneva, 1995.

[119] ITU-T. *Recommendation M.3100: Generic Network Information Model.* International Telecommunication Union, Geneva, 1995.

[120] ITU-T. *Recommendation Q.1214: Distributed Functional Plane for Intelligent Network CS-1.* International Telecommunications Union, Geneva, October 1995.

[121] ITU-T. *ITU-T Recommendation Z.106: Common Interchange Format for SDL.* International Telecommunication Union, Geneva, 1996.

[122] ITU-T. *Recommendation M.3010: Principles for a Telecommunications Management Network.* International Telecommunication Union, Geneva, 1996.

[123] ITU-T. *ITU-T Recommendation X.901/ISO IEC: Information Technology – Open Distributed Processing – Reference Model: Overview.* International Telecommunication Union, Geneva, 1997.

[124] ITU-T. *Recommendation Q.1221: Introduction to Intelligent Network Capability Set 2.* International Telecommunications Union, Geneva, September 1997. 39pp.

[125] ITU-T. *Recommendation Q.1224: Distributed Functional Plane for Intelligent Network Capability Set 2.* International Telecommunications Union, Geneva, September 1997.

[126] ITU-T. *Recommendation Q.65: The Unified Functional Methodology for the Characterisation of Services and Network Capabilities.* International Telecommunications Union, Geneva, September 1997.

[127] ITU-T. *Recommendation H.225.0: Call Signalling Protocols and Media Stream Packetization for Packet-based Multimedia Communication Systems.* International Telecommunications Union, Geneva, 1998.

[128] ITU-T. *Recommendation H450.1: Generic Functional Protocol for the Support of Supplementary Services in H.323.* International Telecommunications Union, Geneva, 1998.

[129] ITU-T. *Recommendation Y.110: Global Information Infrastructure Principles and Framework Architecture.* International Telecommunications Union, Geneva, June 1998.

[130] ITU-T. *Recommendation H.450.6: Call Waiting Supplementary Service for H.323.* International Telecommunications Union, Geneva, 1999.

[131] ITU-T. *Recommendation I.150: Integrated Services Digital Network (ISDN): General Structure – General Description of Asynchronous Transfer Mode: B-ISDN Asynchronous Transfer Mode Functional Characteristics.* International Telecommunications Union, Geneva, 1999.

[132] ITU-T. *Recommendation Q.1231: Introduction to Intelligent Network Capability Set 3.* International Telecommunications Union, Geneva, December 1999. 27pp.

[133] ITU-T. *Recommendation G.805, Generic Functional Architectures of Transport Networks.* International Telecommunicatons Union, Geneva, 2000.

[134] ITU-T. *Recommendation Q.1237: Extensions to Intelligent Network Capability Set 3 in Support of B-ISDN.* International Telecommunications Union, Geneva, June 2000. 88pp.

[135] ITU-T. *Recommendation Q.1901: Bearer Independent Call Control Protocol.* International Telecommunications Union, Geneva, 2000.

[136] ITU-T. *Recommendation H.248.1: Gateway Control Protocol: version 2.* International Telecommunications Union, Geneva, 2002.

[137] ITU-T. *Recommendation H.323: Packet-based Multimedia Communications systems.* International Telecommunications Union, Geneva, 2003.

[138] ITU-T. *Recommendation Y.2001: Next Generation Networks – Frameworks and Functional Architecture Models: General Overview of NGN.* International Telecommunications Union, Geneva, December 2004.

[139] R. Jain, F.M. Anjum, P. Missier, and S. Shastry. Java Call Control, Coordination and Transactions. *IEEE Communications Magazine*, 19(1):108–114, 2000.

[140] R. Jain, J.-L. Bakker, and F. Anjum. Java Call Control (JCC) and Session Initiation Protocol (SIP). *IEICI Transactions on Communications*, E84-B(12):3096–3103, December 2001.

[141] R. Jain, J.-L. Bakker, and F. Anjum. *Programming Converged Networks: Call Control in Java, XML, and Parlay/OSA.* Wiley, New York, 2004.

[142] N.O. Johannesson. The ETSI computational model: A tool for transmission planning of telephone networks. *IEEE Communications Magazine*, 35(1):70–79, January 1997.

[143] A. Johnston, S. Dononan, R. Sparks, C. Cunningham, and K. Summers. *Session Initiation Protocol (SIP) Basic Call Flow Examples.* Internet Engineering Task Force, RFC3665, 2003.

[144] A. Johnston, S. Dononan, R. Sparks, C. Cunningham, and K. Summers. *Session Initiation Protocol (SIP) Public Switched telephone Network (PSTN) Call Flows.* Internet Engineering Task Force, RFC3666, 2003.

[145] J.Rosenberg. *The Session Initiation Protocol (SIP) UPDATE Method.* Internet Engineering Task Force, RFC3311, 2002.

[146] J. Klensin. *Simple Mail Transfer Protocol.* Internet Engineering Task Force, RFC2821, 2001.

[147] M Laitinen and J Rantala. Integration of Intelligent Network services into future GSM networks. *IEEE Communications Magazine*, 33(6):76–86, June 1995.

[148] J. Lennox, X. Wu, and H. Schulzrinne. *Call Processing Language (CPL): A Language for User Control of Internet Telephony Services.* Internet Engineering Task Force, RFC3880, 2004.

[149] C.A. Licciardi, R. Minerva, and A. Cuda. TINA is dead, long live TINA: toward programmable solutions for Next Generation Services. In *Proceedings of TINA 2000 Conference, Paris*, pages 16–20. TINA Consortium, TINA Consortium, 2000.

[150] Z. Lozinski. *Parlay/OSA – a New Way to Create Wireless Services.* While Paper, The Parlay Group, http://www.parlay.org, 2005.

[151] T. Magedanz and R. Popescu-Zeletin. *Intelligent Networks: Basic Technology, Standards and Evolution.* International Thompson Computer Press, London, 1996.

[152] A. Mankin, S. Bradner, R. Mahy, D. Willis, J. Ott, and B. Rosen. *Change Process for the Session Initiation Protocol (SIP).* Internet Engineering Task Force, RFC3427, 2002.

[153] K. McCloghrie and M. Rose. *Management Information Base for Network Management of TCP-based Internets.* Internet Engineering Task Force, RFC1213, 1991.

[154] S. McGlashan et al. (Eds.). *Voice Extensible Markup Language (VoiceXML) Version 2.0.* W3C Recommendation, www.w3c.org, 2004.

[155] D. McIntosh. Building a Packet Cable network: A Comprehensive Design for the Delivery of VoIP Services. Cable Television Laboratories, http://www.cablelabs.com, 2002.

[156] Microsoft. Telephony Application Programming Interfaces (TAPI, TSPI, MSPI). Microsoft Corporation, http://msdn.microsoft.com/library/en-us/tapi/, 2002.

[157] J. Miller and J. Mukerji (Eds.). *MDA Guide version 1.0.1*. Object Management Group, 2003.

[158] A.R. Modarressi and R.A. Skoog. Signaling System No. 7: A tutorial. *IEEE Communications Magazine*, 28(7):19–35, July 1990.

[159] A.-J. Moerdijk and L. Klostermann. Opening the networks with Parlay/OSA: Standards and aspects behind the APIs. *IEEE Network*, pages 58–64, May/June 2003.

[160] Moriana. *Service Delivery Platforms and Telecom Web Services*. The Moriana Group, http://www.morianagroup.com, 2004.

[161] A. Nadalin et al. (Eds.). *Web Services Security: SOAP Message Security 1.0*. OASIS, Standard 200401, http://docs.oasis-open.org/wss/2004/01/oasis-200401-wss-soap-message-security-1.0.pdf, 2004.

[162] N. Natarajan and L. Demounem (Eds.). *The ConS Reference Point, version 1*. The TINA Consortium, 1997.

[163] A. Niemi. *Session Initiation Protocol (SIP) Extension for Event State Publication*. Internet Engineering Task Force, RFC3903, 2004.

[164] P. O'Doherty. JAIN SLEE principles. http://java.sun.com/products/jain/, 2003.

[165] OMA. *OMA Service Environment Version 1.0.2*. Open Mobile Alliance, http://www.openmobilealliance.com/, August 2005.

[166] OMA. *Utilisation of IMS capabilities architecture Version 1.0*. Open Mobile Alliance, http://www.openmobilealliance.com/, August 2005.

[167] OMG. *A Discussion of the Object Management Architecture, version 3*. Object Management Group, www.omg.org, 1997.

[168] L. Ong and J. Yoakum. *An Introduction to the Stream Control Transmission Protocol*. Internet Engineering Task Force, RFC3286, 2002.

[169] C. Perkins (Ed.). *IP Mobility Support for IPv4*. Internet Engineering Task Force, RFC3344, 2002.

[170] S. Petrack and L. Conroy. *The PINT Service Protocol: Extensions to SIP and SDP for IP Access to Telephone Call Services*. Internet Engineering Task Force, RFC2848, 2000.

[171] R. Pinheiro and S.Tsang. Advanced Intelligent Networks. In K.J. Turner, E.H. Magill, and D.J. Marples, editors, *Service Provision: Technologies for Next Generation Communications*, chapter 4, pages 53–71. Wiley, Chichester, 2004.

[172] M. Poiksela, G. Mayer, H. Khartabil, and A. Niemi. *The IMS: IP Multimedia Concepts and Services in the Mobile Domain*. Wiley, Chichester, 2004.

[173] J. Postel (Ed.). *Transmission Control Protocol*. Internet Engineering Task Force, RFC793, 1981.

[174] R. Preshun, J. Case, K. McCloghrie, M. Rose, and S. Waldbusser. *Management Information Base (MIB) for the Simple Network Management Protocol (SNMP)*. Internet Engineering Task Force, RFC3418, 2002.

[175] R. Preshun, J. Case, K. McCloghrie, M. Rose, and S. Waldbusser. *Protocol Operations for the Simple Network Management Protocol (SNMP)*. Internet Engineering Task Force, RFC3416, 2002.

[176] P. Redman, K. Dulaney, and W. Clark. Hype Cycle for mobile and wireless networking, 2003. Strategic Analysis Report R-20-0115, Gartner Inc., 2003.

[177] J.-F. Rey and C. Thyrland. SIP technology in the enterprise. *Alcatel Telecommunications Review*, pages 255–260, 4th Quarter 2002.

[178] K. Richardson, P. Cilliers, and M. Lissack. Complexity science: A 'gray' science for the 'stuff in between'. *Emergence*, 3(2):6–18, 2001.

[179] K. Richardson, G. Mathieson, and P. Cilliers. The theory and practice of complexity science: Epistemological considerations for military operational analysis. http://kurtrichardson.com.

[180] U. Riemers. *DVB: the Family of International Standards for Digital Video Broadcasting*. Springer, Berlin, 2nd edition, 2005.

[181] C. Rigney, S. Willens, A. Rubens, and W. Simpson. *Remote Authentication Dial In User Service (RADIUS)*. Internet Engineering Task Force, RFC2865, 2000.

[182] A.B. Roach. *Session Initiation Protocol (SIP)-Specific Event Notification*. Internet Engineering Task Force, RFC3265, 2002.

[183] S. Rogers and S.D. Hendrick. *Oracle Builds Comprehensive SOA*. White Paper, IDC, http://www.idc.com, 2005.

[184] M. Rose and K. McCloghrie. *Concise MIB Definitions*. Internet Engineering Task Force, RFC1212, 1991.

[185] J. Rosenberg, J. Peterson, H. Schulzrinne, and G. Camarillo. *Best Current Practices for Third Party Call Control (3pcc) in the Session Initiation Protocol*. Internet Engineering Task Force, RFC3725, 2004.

[186] J. Rosenberg and H. Schultzrinne. *An Offer/answer Model with SDP*. Internet Engineering Task Force, RFC3264, 2002.

[187] J. Rosenberg and H. Schultzrinne. *Session Initiation Protocol (SIP): Locating SIP Servers*. Internet Engineering Task Force, RFC3263, 2002.

[188] J. Rosenberg, H. Schultzrinne, G. Camarillo, et al. *SIP: Session Initiation Protocol*. Internet Engineering Task Force, RFC3261, 2002.

[189] T. Russell. *Signalling System No 7*. McGraw Hill, New York, 1995.

[190] H. Schultzrinne, S. Casner, R. Fredrick, and V. Jacobson. *RTP: A Transport Protocol for Real-time Applications*. Internet Engineering Task Force, RFC3550, 1996.

[191] H. Schulzrinne and J.D. Rosenberg. The Session Initiation Protocol: providing advanced telephony services across the Internet. *Bell Labs Technical Journal*, 19(3):144–160, 1998.

[192] M. Sexton and A. Reid. *Broadband Networking: ATM, SDH and SONET*. Artech House, Norwood MA, 1997.

[193] G. Sidebottom, K. Morneault, and J. Pastor-Balbas (Eds.). *Signaling System 7 (SS7) Message Transfer Part 3 (MTP3) – User Adaptation Layer (M3UA)*. Internet Engineering Task Force, RFC3332, 2002.

[194] R. Sparks. *The Session Initiation Protocol (SIP) Refer Method*. Internet Engineering Task Force, RFC3515, 2003.

[195] T. Speakman, J. Crowcroft, and J. Gemmell et al. *PGM Reliable Transport Protocol Specification*. Internet Engineering Task Force, RFC3208, 2001.

[196] M. Spencer. Introduction to the Asterisk Open Source PBX. Linux Support Services Inc, available http://www.asterisk.org, 2002.

[197] R. Stewart, Q. Xie, and K. Morneault et al. *Stream Control Transmission Protocol*. Internet Engineering Task Force, RFC2960, 2002.

[198] J. Strassner, J. Fleck, J. Huang, C. Faurer, and T. Richardson. *TMF White Paper on NGOSS and MDA version 1.0*. TeleManagement Forum, 2003.

[199] Sun. JTAPI 1.4 Specification (Final Release), JSR-000043. Sun Microsystems, http://java.sun.com/products/jtapi/, 2002.

[200] Sun. *JAIN and Open Networks*. White Paper, Sun Microsystems, http://java.sun.com/products/jain/, August 2003.

[201] Sun. *JAIN and Java in Communications*. White Paper, Sun Microsystems, http://java.sun.com/products/jain/, March 2004.

[202] TMF. *Telecom Operations Map, GB910*. Telemanagement Forum, http://www.tmforum.org, 1999.

[203] TMF. *Enhanced Telecom Operations Map (eTOM): The Business process Framework for the Information and Communications Service Industry, GB929, v6.1*. Telemanagement Forum, http://www.tmforum.org, 2005.

[204] D.K. Udupa. *TMN: Telecommunications Management Network*. McGraw-Hill, New York, 1999.

[205] M. Unmehopa, K. Vemuri, and A. Bennett. *Parlay/OSA: From Standards to Reality*. Wiley, 2006.

[206] A. Vemuri and J. Peterson. *Session Initiation Protocol for Telephones (SIP-T): Context and Architectures.* Internet Engineering Task Force, RFC3372, 2002.

[207] VON Coalition. *Unleashing the promise of Internet voice communication.* White Paper, Voice on the Net Coalition, www.von.org, 2004.

[208] R. Yavatkar, D. Pendarakis, and R. Guerin. *A Framework for Policy-based Admission Control.* Internet Engineering Task Force, RFC2753, 2000.

[209] Y. Yemini. The OSI network management model. *IEEE Communications Magazine*, 31:20–29, September 1993.

[210] O. Zimmerman, M. Tomlinson, and S. Peuser. *Perspectives on Web Services: Applying SOAP, WSDL and UDDI to Real World Projects.* Springer, Berlin, 2003.

[211] J. Zuidweg. *Next Generation Intelligent Networks.* Artech House, Norwood, MA, 2002.

Index

3rd Generation Partnership Project, 273

3rd Generation Partnership Project 2, 273, 324
- all-Ip network, 324
- architecture, 324
- in NGN Framework, 324
- Mobile IP, 324

A-interface, 281
Abstract Syntax Notation One, 102
Abstraction, **423**
- at reference point, 83
- leaky, 83, 420
- network, 50
- packet switch, 26
- perpetuating silos, 420
- PSTN switch, 51
- router, 26
- SIP as leaky abstraction, 420
Access gateway, 140
Access network
- as technical functional domain, 62
- EDGE, 281
- enterprise network, 216
- GERAN, 281
- second generation GSM, 275
- third generation mobile, 275
Access transparency, 86, **423**
Accounting, **423**
Activity diagram, 114
Address of record, 172, 177, **423**
Address Resolution Protocol, 26
All-IP network, 324
Always-on, 134
Analogue-to-digital conversion, 135
Application, **423**
- communications-enhanced, 328
Application programming interface, 124

Application server, 60
- SIP, 310
Application service provider, 60
Architecture, **423**
- 3G CCS-domain, 284
- 3G PS-domain, 297
- 3GPP2, 324
- Asterix, 212
- B-ISDN, 226
- GPRS, 297
- H.323 multimedia communications, 157
- IMS, 308
- JTAPI, 210
- OSA/Parlay, 332
- Parlay X, 369
- service delivery platform, 384
- SIP multimedia system, 172
- success factors, 408
- TINA, 232
ASN.1, 102
- managed object definition, 105
- protocol specification, 103
Asterisk Open Source PBX, 212
- application programming interfaces, 213
- architecture, 212
Asynchronous, **423**
Asynchronous Transfer Mode, 227, **423**
- Adaptation layer, 228
- connection control, 229
- PNNI, 230
- routing, 230
- switch model, 229
ATM adaptation layer, 228
Audio bridge, 137
Authentication, Authorization and Accounting protocol, 25